Study Guide
Student Solutions Manual
to accompany

Principles of Physics

second edition

volume 2

Serway

Saunders College Publishing
Harcourt Brace College Publishers
Fort Worth Philadelphia San Diego New York Orlando Austin
San Antonio Toronto Montreal London Sydney Tokyo

Study Guide
Student Solutions Manual
to accompany

Principles of Physics

second edition

volume 2

Serway

John R. Gordon
James Madison University

Ralph V. McGrew
Broome Community College

Steve Van Wyk
Olympic College

Raymond A. Serway
James Madison University

Saunders College Publishing
Harcourt Brace College Publishers
Fort Worth Philadelphia San Diego New York Orlando Austin
San Antonio Toronto Montreal London Sydney Tokyo

Printed in the United States of America

Serway, Gordon, McGrew, and Vanwyk: Study Guide and Student Solutions Manual, Volume Two, to accompany *Principles of Physics*.

0-03-024628-8

567 202 7654321

Preface

This Student Solutions Manual and Study Guide has been written to accompany the textbook **Principles of Physics**, Second Edition, Volume II, by Raymond A. Serway. The purpose of this Student Solutions Manual and Study Guide is to provide students with a convenient review of the basic concepts and applications presented in the textbook, together with solutions to selected end-of-chapter problems. This is not an attempt to rewrite the textbook in a condensed fashion. Rather, emphasis is placed upon clarifying typical troublesome points, and providing further practice in methods of problem solving.

Each chapter is divided into several parts, and every textbook chapter has a matching chapter in this book. Very often, reference is made to specific equations or figures in the textbook. Each feature of this Study Guide has been included to insure that it serves as a useful supplement to the textbook. Most chapters contain the following components:

- **Notes From Selected Chapter Sections:** This is a summary of important concepts, newly defined physical quantities, and rules governing their behavior.

- **Equations and Concepts:** This is a review of the chapter, with emphasis on highlighting important concepts and describing important equations and formalisms.

- **Suggestions, Skills, and Strategies:** This offers hints and strategies for solving typical problems that the student will often encounter in the course. In some sections, suggestions are made concerning mathematical skills that are necessary in the analysis of problems.

- **Review Checklist:** This is a list of topics and techniques the student should master after reading the chapter and working the assigned problems.

- **Answers to Selected Conceptual Questions:** Suggested responses are provided for approximately half of the Conceptual Questions.

- **Solutions to Selected End-of-Chapter Problems:** Solutions are given for approximately half of the odd-numbered problems from the text. Problems were selected to illustrate important concepts in each chapter.

- **Tables:** A list of selected Physical Constants is printed on the inside front cover; and a table of some Conversion Factors is provided on the inside back cover.

An important note concerning significant figures: The answers to all end-of-chapter problems are stated to three significant figures, though calculations are carried out with as many digits as possible. For uniformity of style, this practice has been followed even in those cases where the data in the corresponding problem statement is given to only two significant figures. In the latter case, your instructor may request that you "round off" your calculations to two significant figures.

We sincerely hope that this Student Solutions Manual and Study Guide will be useful to you in reviewing the material presented in the text, and in improving your ability to solve problems and score well on exams. We welcome any comments or suggestions which could help improve the content of this study guide in future editions; and we wish you success in your study.

John R. Gordon
Raymond A. Serway
Dept. of Physics
James Madison University
Harrisonburg, VA 22807

Ralph McGrew
Dept. of Engineering Science, AT-101
Broome Community College
Binghamton, NY 13902-1017

Steve Van Wyk
Olympic College
Bremerton, WA 98310

Acknowledgments

It is a pleasure to acknowledge the excellent work of Michael Rudmin whose attention to detail in the preparation of the camera-ready copy did much to enhance the quality of this edition of the Student Solutions Manual and Study Guide. His graphics skills and technical expertise combined to produce illustrations for the first edition which continue to add much to the appearance and usefulness of this volume.

Special thanks go to Senior Developmental Editor, Susan Dust Pashos and Ancillary Editor, Alexandra Buczek for managing all phases of this project. Finally, we express our appreciation to our families for their inspiration, patience, and encouragement.

Suggestions for Study

Very often we are asked "How should I study this subject, and prepare for examinations?" There is no simple answer to this question, however, we would like to offer some suggestions which may be useful to you.

1. It is essential that you understand the basic concepts and principles before attempting to solve assigned problems. This is best accomplished through a careful reading of the textbook before attending your lecture on that material, jotting down certain points which are not clear to you, taking careful notes in class, and asking questions. You should reduce memorization of material to a minimum. Memorizing sections of a text, equations, and derivations does not necessarily mean you understand the material. Perhaps the best test of your understanding of the material will be your ability to solve the problems in the text, or those given on exams.

2. Try to solve as many problems at the end of the chapter as possible. You will be able to check the accuracy of your calculations to the odd-numbered problems, since the answers to these are given at the back of the text. Furthermore, detailed solutions to approximately half of the odd-numbered problems are provided in this study guide. Many of the worked examples in the text will serve as a basis for your study.

3. The method of solving problems should be carefully planned. First, read the problem several times until you are confident you understand what is being asked. Look for key words which will help simplify the problem, and perhaps allow you to make certain assumptions. You should also pay special attention to the information provided in the problem. In many cases a simple diagram is a good starting point; and it is always a good idea to write down the given information before trying to solve the problem. After you have decided on the method you feel is appropriate for the problem, proceed with your solution. If you are having difficulty in working problems, we suggest that you again read the text and your lecture notes. It may take several readings before you are ready to solve certain problems, though the solved problems in this Study Guide should be of value to you in this regard. However, your solution to a problem does not have to look just like the one presented here. A problem can sometimes be solved in different ways, starting from different principles. If you wonder about the validity of an alternative approach, ask your instructor.

4. After reading a chapter, you should be able to define any new quantities that were introduced, and discuss the first principles that were used to derive fundamental formulas. A review is provided in each chapter of the Study Guide for this purpose, and the marginal notes in the textbook (or the index) will help you locate these topics. You should be able to correctly associate with each physical quantity the symbol used to represent that quantity (including vector notation, if appropriate) and the SI unit in which the quantity is specified. Furthermore, you should be able to express each important formula or equation in a concise and accurate prose statement.

5. We suggest that you use this Study Guide to review the material covered in the text, and as a guide in preparing for exams. You should also use the Chapter Review, Notes From Selected Chapter Sections, and Equations and Concepts to focus in on any points which require further study. Remember that the main purpose of this Study Guide is to improve upon the efficiency and effectiveness of your study hours and your overall understanding of physical concepts. However, it should not be regarded as a substitute for your textbook or individual study and practice in problem solving.

Contents

Contents

Chapter

16

The glass vessel contains dry ice (solid carbon dioxide). The white cloud is carbon dioxide vapor, which is denser than air and hence falls from the vessel. *(R. Falwell / Science Photo Library / Photo Researchers)*

TEMPERATURE AND THE KINETIC THEORY OF GASES

Chapter 16

TEMPERATURE AND THE KINETIC THEORY OF GASES

INTRODUCTION

A quantitative description of thermal phenomena requires a careful definition of the concepts of temperature, heat, and internal energy. The laws of thermodynamics provide us with relationships among heat flow, work, and the internal energy of a system.

The composition and structure of a body are important factors when dealing with thermal phenomena. For example, liquids and solids expand only slightly when heated, whereas gases expand appreciably when heated. If the gas is not free to expand, its pressure rises when heated. Certain substances may melt, boil, burn, or explode.

This chapter concludes with a study of ideal gases. We shall approach this study on two levels. The first will examine ideal gases on the macroscopic scale. Here we shall be concerned with the relationships among such quantities as pressure, volume, and temperature. On the second level, we shall examine gases on a microscopic scale, using a model that pictures the components of a gas as small particles. The latter approach, called the kinetic theory of gases, will help us to understand what happens on the atomic level to affect such macroscopic properties as pressure and temperature.

In the current view of gas behavior, called the **kinetic theory**, gas molecules move about in a random fashion, colliding with the walls of their container and with each other. Perhaps the most important consequence of this theory is that it shows the equivalence between the kinetic energy of molecular motion and the internal energy of the system. Furthermore, the kinetic theory provides us with a physical basis upon which the concept of temperature can be understood.

NOTES FROM SELECTED CHAPTER SECTIONS

16.1 Temperature and the Zeroth Law of Thermodynamics

Thermal physics is the study of the behavior of solids, liquids and gases, using the concepts of heat and temperature. Two approaches are commonly used in this area of science. The first is a **macroscopic** approach, called **thermodynamics,** in which one explains the bulk thermal properties of matter. The second is a **microscopic** approach, called **statistical mechanics,** in which properties of matter are explained on an atomic scale. Both approaches require that you understand some basic concepts, such as the concepts of temperature and heat. As we will see, **all** thermal phenomena are manifestations of the laws of mechanics as we have learned them. For example, thermal energy (or heat energy) is actually a consequence of the random motions of a large number of particles making up the system.

The concept of the **temperature** of a system can be understood in connection with a measurement, such as the reading of a thermometer. Temperature, a scalar quantity, is a property which can only be defined when the system is in thermal equilibrium with another system. Thermal equilibrium implies that two (or more) systems are at the same temperature.

The **zeroth law of thermodynamics** states that if two systems are in thermal equilibrium with a third system, they must be in thermal equilibrium with each other. The third system is usually a calibrated thermometer whose reading determines whether or not the systems are in thermal equilibrium. There are several types of thermometers which can be used.

2 Thermometers and Temperature Scales

Thermometers are devices used to define and measure the temperature of a system. All thermometers make use of a change in some physical property with temperature. Some of the physical properties used are (1) the change in volume of a liquid, (2) the change in length of a solid, (3) the change in pressure of a gas held at constant volume, (4) the change in volume of a gas held at constant pressure, (5) the change in electric resistance of a conductor, and (6) the change in color of a very hot object. For a given substance, a temperature scale can be established based on any one of these physical quantities.

The **gas thermometer** is a standard device for defining temperature. In the constant-volume gas thermometer, a low density gas is placed in a flask, and its volume is kept constant while it is heated. The pressure is measured as the gas is heated or cooled. Experimentally, one finds that the temperature is proportional to the absolute pressure.

The **thermodynamic temperature scale** is based on a scale for which the reference temperature is taken to be the **triple point of water**; that is, the temperature and pressure at which water, water vapor, and ice coexist in equilibrium. On this scale, the SI unit of temperature is the **kelvin**, defined as the fraction 1 / 273.16 of the temperature of the triple point of water.

16.5 The Kinetic Theory of Gases

A microscopic **model of an ideal gas** is based on the following assumptions:

- **The number of molecules is large, and the average separation between them is large** compared with their dimensions. Therefore, the molecules occupy a negligible volume compared with the volume of the container.

- **The molecules obey Newton's laws of motion, but the individual molecules move in a random fashion.** By random fashion, we mean that the molecules move in all directions with equal probability and with various speeds. This distribution of velocities does not change in time, despite the collisions between molecules.

- **The molecules undergo elastic collisions with each other.** Thus, the molecules are considered to be structureless, and in the collisions both kinetic energy and momentum are conserved.

- **The forces between molecules are negligible except during a collision.** The forces between molecules are short-range, so that the only time the molecules interact with each other is during a collision.

- **The gas under consideration is a pure gas.** That is, all molecules are identical.

- **The gas is in thermal equilibrium with the walls of the container.** The collisions of molecules with the walls are perfectly elastic.

EQUATIONS AND CONCEPTS

Pressure is force per unit area and has units of N/m^2. The SI unit of pressure is the pascal (Pa).

$$1\ Pa \equiv 1\ N/m^2$$

One atmosphere of pressure is atmospheric pressure at sea level.

$$1\ atm = 1.013 \times 10^5\ Pa$$

The **Celsius temperature** T_C is related to the absolute temperature T (in kelvins) according to Equation 16.1, where 0°C corresponds to 273.15 K.

$$T_C = T - 273.15 \qquad (16.1)$$

The **Fahrenheit temperature** T_F can be converted to degrees Celsius using Equation 16.2. Note that 0°C = 32°F and 100°C = 212°F.

$$T_F = \frac{9}{5} T_C + 32°F \qquad (16.2)$$

If a body has a length L_o, the **change in its length** ΔL due to a change in temperature is proportional to the change in temperature and the length. The proportionality constant α is called the **average coefficient of linear expansion.**

$$\Delta L = \alpha L_o \Delta T \qquad (16.4)$$

or

$$\alpha = \frac{1}{L_o}\frac{\Delta L}{\Delta T}$$

If the temperature of a body of volume V changes by an amount ΔT at constant pressure, **the change in its volume** is proportional to ΔT and the original volume. The constant of proportionality β is the **average coefficient of volume expansion**. For an isotropic solid, $\beta \cong 3\alpha$.

$$\Delta V = \beta V \Delta T \qquad (16.6)$$

A sample of any substance contains a number of moles which equals the ratio of the mass of the sample to the molar mass characteristic of that particular substance.

$$n = \frac{m}{M} \tag{16.8}$$

If n moles of a dilute gas occupy a volume V, the **equation of state** which relates the variables P, V, and T at equilibrium is that of an **ideal gas,** Equation 16.9, where R is the **universal gas constant.**

$$PV = nRT \tag{16.9}$$

$$R = 8.315\ \text{J/mol·K} \tag{16.10}$$

or

$$R = 0.0821\ \text{L·atm / mol·K}$$

The **equation of state of an ideal gas** can also be expressed in the form of Equation 16.11, where N is the total number of gas molecules and k_B is **Boltzman's constant.**

$$PV = Nk_BT \tag{16.11}$$

$$k_B = 1.38 \times 10^{-23}\ \text{J/K} \tag{16.12}$$

In the kinetic theory of an ideal gas, one finds that the pressure of the gas is proportional to the number of molecules per unit volume and the average translational kinetic energy per molecule. Note that m here represents the mass of a single molecule.

$$P = \frac{2}{3}\left(\frac{N}{V}\right)\left(\frac{1}{2}m\overline{v^2}\right) \tag{16.13}$$

From Equation 16.13, and the equation of state for an ideal gas, $PV = Nk_BT$, we find that **the absolute temperature of an ideal gas is a direct measure of the average molecular kinetic energy.**

$$\frac{1}{2}m\overline{v^2} = \frac{3}{2}k_BT \tag{16.15}$$

The **total translational kinetic energy** E of N molecules (or n moles) of a monatomic ideal gas is proportional to the absolute temperature.

$$E = \frac{3}{2}Nk_BT = \frac{3}{2}nRT \tag{16.17}$$

The expression for the root-mean-square speed of molecules shows that at a given temperature, lighter molecules move faster on the average than heavier molecules.

$$v_{rms} = \sqrt{\frac{3k_BT}{m}} = \sqrt{\frac{3RT}{M}} \qquad (16.18)$$

REVIEW CHECKLIST

- ▷ Understand the concepts of thermal equilibrium and thermal contact between two bodies, and state the zeroth law of thermodynamics.

- ▷ Discuss some physical properties of substances which change with temperature, and the manner in which these properties are used to construct thermometers.

- ▷ Describe the operation of the constant-volume gas thermometer and how it is used to define the ideal gas temperature scale.

- ▷ Convert between the various temperature scales, especially the conversion from degrees Celsius into kelvins, degrees Fahrenheit into kelvins, and degrees Celsius into degrees Fahrenheit.

- ▷ Provide a qualitative description of the origin of thermal expansion of solids and liquids; define the linear expansion coefficient and volume expansion coefficient for an isotropic solid, and learn how to deal with these coefficients in practical situations involving expansion or contraction.

- ▷ State and understand the assumptions made in developing the molecular model of an ideal gas.

- ▷ Recognize that the temperature of an ideal gas is proportional to the average molecular kinetic energy. State the theorem of equipartition of energy, noting that each degree of freedom of a molecule contributes an equal amount of energy, of magnitude $\frac{1}{2}k_BT$.

ANSWERS TO SELECTED CONCEPTUAL QUESTIONS

2. In the movie **Apollo 13** (Universal, 1995), the character who depicts astronaut Jim Lovell says, in describing his upcoming trip to the Moon, "I'll be walking in a place where there's 400° difference between sunlight and shadow." What exactly is it that is hot in sunlight and cold in shadow? Suppose an astronaut standing on the Moon holds a thermometer in his or her gloved hand. Is it reading the temperature of the vacuum at the Moon's surface? Does it read a temperature? If so, what object or substance has that temperature?

Answer The temperature difference mentioned refers to the material on the lunar surface. It does not refer to atmospheric temperatures, as we interpret such temperatures on the Earth, since to a good approximation no atmosphere is present on the Moon. The thermometer held by the astronaut does not measure the temperature of the vacuum—temperature is a concept related to the average energy of molecules. Since there are no molecules in a vacuum, the concept of temperature has no meaning there. The thermometer does indeed read a temperature. As with any thermometer, regardless of its location, it is reading the temperature of **itself**. The reading on the thermometer will correspond to an equilibrium between energy emitted by the thermometer by radiation and that absorbed by radiation from the Sun, the lunar surface, space, stars, planets, etc.

□ □ □ □

3. When the metal ring and metal sphere in Figure Q16.3 are both at room temperature, the sphere does not fit through the ring. After the ring is heated, the sphere can be passed through the ring. Explain.

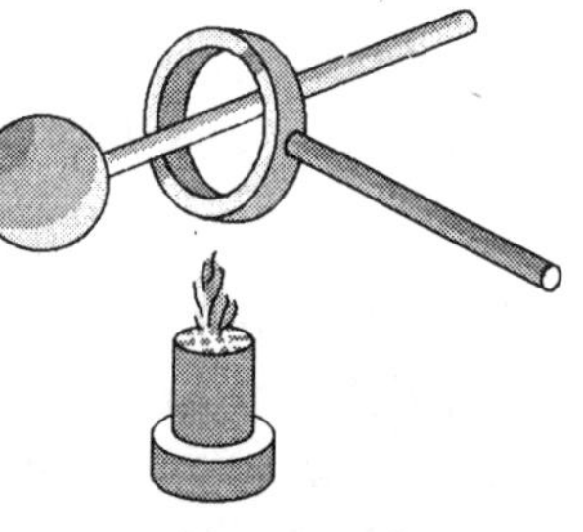

Figure Q16.3

Answer The thermal expansion of the ring is not like the inflation of a blood-pressure cuff. Rather, it is like a photographic enlargement; every linear dimension, including the hole diameter, increases by the same factor. The reason for this is that the atoms everywhere, including those around the inner circumference, push away from each other. The only way that the atoms can accommodate the greater distances is for the circumference—and corresponding diameter— to grow. This property was once used to fit metal rims to wooden wagon and horse-buggy wheels.

□ □ □ □

6. If a helium-filled balloon initially at room temperature is placed in a freezer, will its volume increase, decrease, or remain the same?

Answer Helium is a fairly ideal gas. Therefore, we think in terms of the ideal gas law.

$$PV = nRT$$

In this case, the pressure stays nearly constant, being equal to 1 atm. Since the temperature may decrease by 10% (from 293 K to 263 K), the volume should also **decrease** by 10%. This process is called "isobaric cooling," or "isobaric contraction."

□ □ □ □

8. What happens to a helium-filled balloon released into the air? Will it expand or contract? Will it stop rising at some height?

Answer Imagine the balloon rising into the air. The air cannot be uniform in pressure, because the lower layers support the weight of all the air above. Therefore, as the balloon rises, the pressure will decrease. At the same time, the temperature decreases slightly, but not enough to overcome the effects of the pressure.

By the ideal gas law, $PV = nRT$. Since the decrease in temperature has little effect, the decreasing pressure results in an increasing volume; thus the balloon expands quite dramatically. By the time the balloon reaches an altitude of 10 miles, its volume will be 90 times larger.

Long before that happens, one of two events will occur. The first possibility, of course, is that the balloon will break, and the pieces will fall to the earth. The other possibility is that the balloon will come to a spot where the average density of the balloon and its payload is equal to the average density of the air. At that point, buoyancy will cause the balloon to "float" at that altitude, until it loses helium and descends.

People who remember releasing balloons with pen-pal notes will perhaps also remember that replies most often came back when the balloons were low on helium, and were barely able to take off. Such balloons found a fairly low floatation altitude, and were the most likely to be found intact at the end of their journey.

□ □ □ □

12. The shore at a particular point of the ocean is very rocky. At one location, the rock formation is such that a cave is located with an opening that is underwater, as shown in figure Q16.12a. (a) Inside the cave, there is a pocket of trapped air. As the level of the ocean rises and falls due to the tides, will the level of the water in the cave rise and fall? If so, will it have the same amplitude as that of the ocean? (b) Suppose the cave is deeper in the water, so that it is completely submerged and filled with water at high tide, as shown in Figure Q16.12b. At low tide, will the level of the water in the cave be the same as that of the ocean?

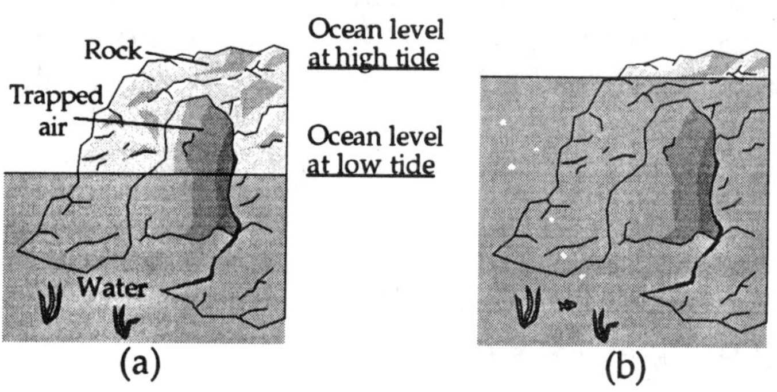

Figure Q16.12

Answer (a) As the water level rises due to the tides, the trapped air in the cave will be compressed into a smaller volume. As a result, its pressure will increase. Thus, the water surface in the cave will experience a larger air pressure than that of the atmosphere on the water surface external to the cave. The water surface in the cave will not rise as high as the water surface outside. As the tide falls, the air in the cave will expand and experience a lower pressure. This will result in the water level in the cave not falling as low as the surface outside the cave. Thus, the amplitude of variation will be smaller in the cave than for the external surface.

(b) In this case, there is no air in the cave. Thus, the air pressure on the water surface in the cave is zero. As the external water level falls due to the tides, the water in the cave will not fall, and will continue to fill the cave. Atmospheric pressure on the external water surface is sufficient to keep the water level in the cave at the top of the cave. Atmospheric pressure can support a column of water of height about 10 meters. Thus, if the water level were to fall more than this, the water surface in the cave would begin to drop, leaving a vacuum at the top of the cave. This is much larger, however, than the normal variation of water level associated with the tides.

□ □ □ □

13. An alcohol rub is sometimes used to help lower the temperature of a sick patient. Why does this help?

Answer As the alcohol evaporates, high speed molecules leave the liquid. This reduces the average speed of the remaining molecules. Since the average speed is lowered, the temperature of the alcohol is reduced. This process helps to carry energy away from the skin of the patient, resulting in cooling of the skin. The alcohol plays the same role of evaporative cooling as does perspiration, but alcohol evaporates much more quickly than perspiration.

□ □ □ □

SOLUTIONS TO SELECTED END-OF-CHAPTER PROBLEMS

3. A constant-volume gas thermometer is calibrated in dry ice (which is carbon dioxide in the solid state and has a temperature of –80.0 °C) and in boiling ethyl alcohol (78.0°C). The two pressures are 0.900 atm and 1.635 atm. (a) What value of absolute zero does the calibration yield? What is the pressure at (b) the freezing point of water and (c) the boiling point of water?

Solution Since we have a linear graph, the pressure is related to the temperature as $P = A + BT$, where A and B are constants. To find A and B, we use the given data:

$$0.900 \text{ atm} = A + (-80.0°\text{C})B \quad \text{and} \quad 1.635 \text{ atm} = A + (78.0°\text{C})B$$

Solving these simultaneously, we find

$$A = 1.272 \text{ atm} \quad \text{and} \quad B = 4.652 \times 10^{-3} \text{ atm/°C}$$

Therefore, $P = 1.272 \text{ atm} + (4.652 \times 10^{-3} \text{ atm/°C})T$

(a) At absolute zero, $P = 0 = 1.272 \text{ atm} + (4.652 \times 10^{-3} \text{ atm/°C})T$

which gives $T = -273.5 \text{ °C}$ ◊

(b) At the freezing point, $P = 1.272 \text{ atm} + 0 = 1.27 \text{ atm}$ ◊

(c) and at the boiling point, $P = 1.272 \text{ atm} + (4.652 \times 10^{-3} \text{ atm/°C})(100°\text{C}) = 1.74 \text{ atm}$ ◊

5. Liquid nitrogen has a boiling point of –195.81 °C at atmospheric pressure. Express this temperature in (a) degrees Fahrenheit, and (b) kelvin.

Solution

(a) By Equation 16.2, $T_F = \frac{9}{5}T_C + 32.0°\text{F} = \frac{9}{5}(-195.81) + 32.0 = -320°\text{F}$ ◊

(b) Applying Equation 16.1, $273.15\text{ K} - 195.81\text{ K} = 77.3\text{ K}$ ◊

Related Comment A convenient way to remember Equations 16.1 and 16.2 is to remember the freezing and boiling points of water, in each form:

$T_{\text{Freeze}} = 32.0°\text{ F} = 0°\text{ C} = 273.15\text{ K}$ $\qquad$ $T_{\text{boil}} = 212.0°\text{ F} = 100°\text{ C}$

To convert from Fahrenheit to Celsius, subtract 32 (the freezing point), and then adjust the scale by the liquid range of the water.

$$\text{Scale} = \frac{100 - 0\text{ °C}}{212 - 32\text{ °F}} = \frac{5\text{ °C}}{9\text{ °F}}$$

A kelvin is identical to a degree Celsius, except that it takes its zero point at **absolute zero**, instead of the freezing point of water. Therefore, to convert from kelvin to Celsius, subtract 273.15 K.

---

11. A copper telephone wire has essentially no sag between poles 35.0 m apart on a winter day when the temperature is -20.0 °C. How much longer is the wire on a summer day with $T_C = 35.0$ °C?

Solution The change in length between cold and hot conditions is

$$\Delta L = \alpha L_o \Delta T = \left(17 \times 10^{-6}\,(°\text{C})^{-1}\right)(35.0\text{ m})(35.0°\text{C} - (-20.0°\text{C}))$$

$$\Delta L = 3.27 \times 10^{-2}\text{ m} \qquad \text{or} \qquad \Delta L = 3.27\text{ cm} \quad ◊$$

---

17. The active element of a certain laser is made of a glass rod 30.0 cm long by 1.50 cm in diameter. If the temperature of the rod increases by 65.0 °C, find the increase in (a) its length, (b) its diameter, and (c) its volume. (Take $\alpha = 9.00\times10^{-6}(°\text{C})^{-1}$)

Solution

(a) $$\Delta L = \alpha L_o\,\Delta T = \left(9.00\times10^{-6}(°\text{C})^{-1}\right)(0.300\text{ m})(65.0\text{ °C}) = 1.76\times10^{-4}\text{ m} \quad \lozenge$$

(b) The diameter is a linear dimension, so the same equation applies:

$$\Delta D = \alpha D_o \Delta T = \left(9.00\times10^{-6}(°\text{C})^{-1}\right)(0.0150\text{ m})(65.0\text{ °C}) = 8.78\times10^{-6}\text{ m} \quad \lozenge$$

(c) The original volume $V = \pi r^2 L = \frac{\pi}{4}(0.0150\text{ m})^2(0.300\text{ m}) = 5.30\times10^{-5}\text{ m}^3$

Using the volumetric coefficient of expansion, β,

$$\Delta V = \beta V\,\Delta T \cong 3\alpha V\,\Delta T$$

$$\Delta V \cong 3\left(9.00\times10^{-6}(°\text{C})^{-1}\right)\left(5.30\times10^{-5}\text{ m}^3\right)(65.0\text{ °C}) = 93.1\times10^{-9}\text{ m}^3 \quad \lozenge$$

Related Calculation The above calculation ignores ΔL^2 and ΔL^3 terms. Calculate the change in volume exactly, and compare your answer with the approximate solution above.

The volume will increase by a factor of

$$\frac{\Delta V}{V} = \left(1+\frac{\Delta D}{D}\right)^2\left(1+\frac{\Delta L}{L}\right) - 1 = \left(1+\frac{8.78\times10^{-6}\text{ m}}{0.015\text{ m}}\right)^2\left(1+\frac{1.76\times10^{-4}\text{ m}}{0.300\text{ m}}\right) - 1 = 1.76\times10^{-3}$$

So, $$\Delta V = \frac{\Delta V}{V}V = \left(1.76\times10^{-3}\right)\left(5.30\times10^{-5}\text{ m}^3\right) = 93.1\times10^{-9}\text{ m}^3 = 93.1\text{ mm}^3 \quad \lozenge$$

The answer is virtually identical, and the approximation $\beta \cong 3\alpha$ is a good one.

21. An auditorium has dimensions 10.0 m × 20.0 m × 30.0 m. How many molecules of air fill the auditorium at 20.0 °C and 101 kPa pressure?

Solution The air is far from liquifaction, so we can model it as an ideal gas: $PV = nRT$

$$n = \frac{PV}{RT} = \frac{(1.01\times10^5\ \text{N/m}^2)[(10.0\ \text{m})(20.0\ \text{m})(30.0\ \text{m})]}{(8.315\ \text{J/mol}\cdot\text{K})(293\ \text{K})} = 2.49\times10^5\ \text{mol}$$

$$N = n(N_A) = (2.49\times10^5\ \text{mol})\left(6.022\times10^{23}\ \frac{\text{molecules}}{\text{mol}}\right) = 1.50\times10^{29}\ \text{molecules} \quad \Diamond$$

---

23. The mass of a hot-air balloon and its cargo (not including the air inside) is 200 kg. The air outside is at 10.0 °C and 101 kPa. The volume of the balloon is 400 m³. To what temperature must the air in the balloon be heated before the balloon will lift off? (Air density at 10.0 °C is 1.25 kg/m³.)

Solution The air displaced by the balloon has a volume of 400 m³ at 10.0 °C, and a mass of

$$m_d = \rho V = (1.25\ \text{kg/m}^3)(400\ \text{m}^3) = 500\ \text{kg}$$

The weight of the displaced air is

$$w_d = m_d g = (500\ \text{kg})(9.80\ \text{m/s}^2) = 4900\ \text{N}$$

The weight of the balloon and cargo is

$$w_b = m_b g = (200\ \text{kg})(9.80\ \text{m/s}^2) = 1960\ \text{N}$$

The weight of the displaced air (the buoyancy force) must equal the weight of the balloon and the air inside the balloon, w_a. Therefore, $w_d = w_b + m_a g$:

$$m_a = \frac{w_d - w_b}{g} = \frac{4900\ \text{N} - 1960\ \text{N}}{9.80\ \text{m/s}^2} = 300\ \text{kg}$$

We estimate dry air to be 20% O_2 and 80% N_2, with a little Argon. We calculate the air's average molar mass based on the O_2 and N_2, and round it up to 29.0 g/mol to account for the Argon; we then calculate the number of moles of air that are present.

$$n = \frac{m_a}{M_a} = \frac{m_a}{0.80\left(28.0\ \frac{\text{g}}{\text{mol}}\right) + 0.20\left(32.0\ \frac{\text{g}}{\text{mol}}\right)}$$

$$n = \frac{(300\ \text{kg})(10^3\ \text{g/kg})}{(29.0\ \text{g/mol})} = 1.03\times10^4\ \text{mol}$$

Applying the ideal gas law, and solving for the temperature, $T = PV/nR$:

$$T = \frac{(1.01\times10^5\ \text{N/m}^2)(400\ \text{m}^3)}{(1.03\times10^4\ \text{mol})(8.31\ \text{J/mol}\cdot\text{K})} = 471\ \text{K} \ \Diamond$$

---

25. An automobile tire is inflated using air originally at 10.0 °C and normal atmospheric pressure. During the process, the air is compressed to 28.0 percent of its original volume and the temperature is increased to 40.0 °C. (a) What is the tire pressure? (b) After the car is driven at high speed, the tire air temperature rises to 85.0 °C and the interior volume of the tire increases by 2.00%. What is the new tire pressure in pascals (absolute)?

Solution

(a) Taking $PV = nRT$ in the initial (*i*) and final (*f*) states, and dividing the two equations, we have

$$P_iV_i = nRT_i \quad \text{and} \quad P_fV_f = nRT_f \quad \text{yielding} \quad \frac{P_fV_f}{P_iV_i} = \frac{T_f}{T_i}$$

So

$$P_f = P_i\frac{V_iT_f}{V_fT_i} = (1.013\times10^5 \text{ Pa})\left(\frac{V_i}{0.28V_i}\right)\left(\frac{273\text{ K}+40\text{ K}}{273\text{ K}+10\text{ K}}\right)$$

$$P_f = 4.00\times10^5 \text{ Pa} \quad \lozenge$$

(b) Introducing the hot (*h*) state, $\dfrac{P_hV_h}{P_fV_f} = \dfrac{T_h}{T_f}$

So

$$P_h = P_f\left(\frac{V_f}{V_h}\right)\left(\frac{T_h}{T_f}\right) = (4.00\times10^5 \text{ Pa})\left(\frac{V_f}{1.02V_f}\right)\left(\frac{358\text{ K}}{313\text{ K}}\right)$$

and

$$P_h = 4.49\times10^5 \text{ Pa} \quad \lozenge$$

33. A cylinder contains a mixture of helium and argon gas in equilibrium at 150 °C. (a) What is the average kinetic energy of each gas molecule? (b) What is the root-mean-square speed of each type of molecule?

Solution

(a) Both kinds of molecules have the same average kinetic energy, sharing it in collisions:

$$\tfrac{1}{2}m\overline{v^2} = \tfrac{3}{2}k_B T = \tfrac{3}{2}(1.38\times10^{-23}\ \mathrm{J/K})(273+150)\mathrm{K}$$

$$\overline{K} = 8.76\times10^{-21}\ \mathrm{J}\quad\Diamond$$

(b) The root-mean square velocity can be calculated from the kinetic energy:

$$v_{rms} = \sqrt{\overline{v^2}} = \sqrt{\frac{2\overline{K}}{m}}$$

These two gases are noble, and therefore monatomic. The mass of each molecule, in kilograms, is:

$$m_{\mathrm{He}} = \frac{(4.00\ \mathrm{g/mol})(10^{-3}\ \mathrm{kg/g})}{6.02\times10^{23}\ \mathrm{atoms/mol}} = 6.64\times10^{-27}\ \mathrm{kg}$$

$$m_{\mathrm{Ar}} = \frac{(39.9\ \mathrm{g/mol})(10^{-3}\ \mathrm{kg/g})}{6.02\times10^{23}\ \mathrm{atoms/mol}} = 6.63\times10^{-26}\ \mathrm{kg}$$

Substituting these values,

$$v_{rms,\ \mathrm{He}} = \sqrt{\frac{2(8.76\times10^{-21}\ \mathrm{J})}{6.64\times10^{-27}\ \mathrm{kg}}} = 1.62\times10^{3}\ \mathrm{m/s}$$

$$v_{rms,\ \mathrm{Ar}} = \sqrt{\frac{2(8.76\times10^{-21}\ \mathrm{J})}{6.63\times10^{-26}\ \mathrm{kg}}} = 514\ \mathrm{m/s}\quad\Diamond$$

43. A mercury thermometer is constructed as in Figure P16.43. The capillary tube has a diameter of 0.00400 cm, and the bulb has a diameter of 0.250 cm. Neglecting the expansion of the glass, find the change in height of the mercury column for a temperature change of 30.0 °C.

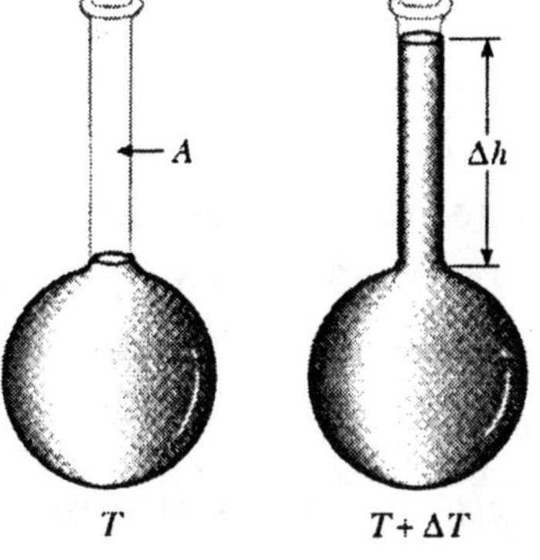

Figure P16.43

Solution Neglecting the expansion of the glass, the volume of liquid in the capillary tube will be $\Delta V = A(\Delta h)$ where A is the cross-sectional area of the capillary tube and V is the volume of the bulb.

$$\Delta V = V\beta\,\Delta T$$

$$\Delta h = \left(\frac{V}{A}\right)\beta\,\Delta T = \left(\frac{\frac{4}{3}\pi R_{\text{bulb}}^3}{\pi R_{\text{cap}}^2}\right)\beta\,\Delta T$$

$$\Delta h = \frac{4}{3}\frac{(0.125\text{ cm})^3}{(0.0020\text{ cm})^2}\left(1.82\times10^{-4}\ (°\text{C})^{-1}\right)(30\ °\text{C}) = 3.55\text{ cm} \quad \Diamond$$

---

45. The rectangular plate shown in Figure P16.45 has an area A equal to lw. If the temperature increases by ΔT, show that the increase in area is $\Delta A = 2\alpha A\,\Delta T$, where α is the average coefficient of linear expansion. What approximation does this expression assume? (**Hint:** Note that each dimension increases according to $\Delta l = \alpha l\,\Delta T$.)

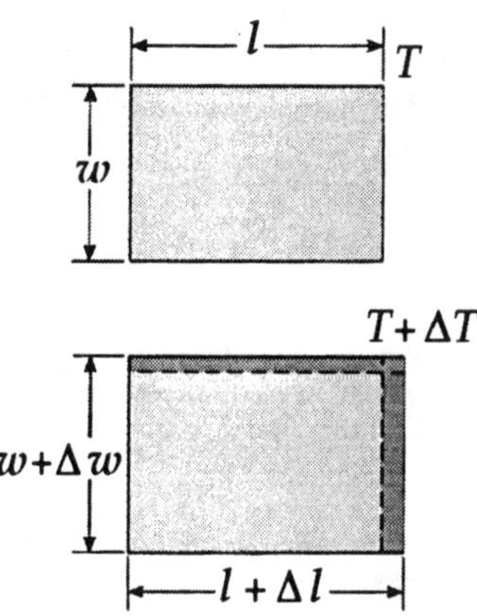

Figure P16.45

Solution From the diagram in Figure P16.45, we see that the **change** in area is

$$\Delta A = l\,\Delta w + w\,\Delta l + \Delta w\,\Delta l$$

Since Δl and Δw are each small quantities, the product $\Delta w\,\Delta l$ will be very small. Therefore, we assume $\Delta w\,\Delta l \approx 0$.

Since $\Delta w = w\alpha\,\Delta T$ and $\Delta l = l\alpha\,\Delta T$, we then have $\Delta A = lw\alpha\,\Delta T + wl\alpha\,\Delta T$

Finally, since $A = lw$, we find $\Delta A = 2\alpha A\,\Delta T \quad \Diamond$

---

47. A liquid has a density ρ. (a) Show that the fractional change in density for a change in temperature ΔT is $\Delta\rho/\rho = -\beta\Delta T$. What does the negative sign signify? (b) Fresh water has a maximum density of 1.000 g/cm³ at 4.0 °C. At 10.0 °C, its density is 0.9997 g/cm³. What is β for water over this temperature interval?

Solution We start with the two equations:

$$\rho = \frac{m}{V} \quad \text{and} \quad \frac{\Delta V}{V} = \beta\,\Delta T.$$

(a) Differentiating the first equation, $$d\rho = -\frac{m}{V^2}\,dV$$

For very small changes in V and ρ, this can be written as $$\Delta\rho = -\frac{m}{V}\frac{\Delta V}{V} = -\left(\frac{m}{V}\right)\frac{\Delta V}{V}$$

Substituting both of our initial equations, we find that $$\Delta\rho = -\rho\beta\,\Delta T$$

The negative sign means that if β is positive, any increase in temperature causes the density to decrease and vice versa. ◊

(b) We apply the equation $\beta = -\frac{\Delta\rho}{\rho\,\Delta T}$ for the specific case of water:

$$\beta = -\frac{(1.0000\ \text{g/cm}^3 - 0.9997\ \text{g/cm}^3)}{(1.0000\ \text{g/cm}^3)(4.00\ °\text{C} - 10.0\ °\text{C})}$$

Solving, we find that $\beta = 5.00\times10^{-5}\ (°\text{C})^{-1}$ ◊

---

49. A vertical cylinder of cross-sectional area A is fitted with a tight-fitting, frictionless piston of mass m (Figure P16.49). (a) If there are n mol of an ideal gas in the cylinder at a temperature of T, determine the height h at which the piston is in equilibrium under its own weight. (b) What is the value for h if n = 0.200 mol, T = 400 K, A = 0.00800 m^2 and m = 20.0 kg?

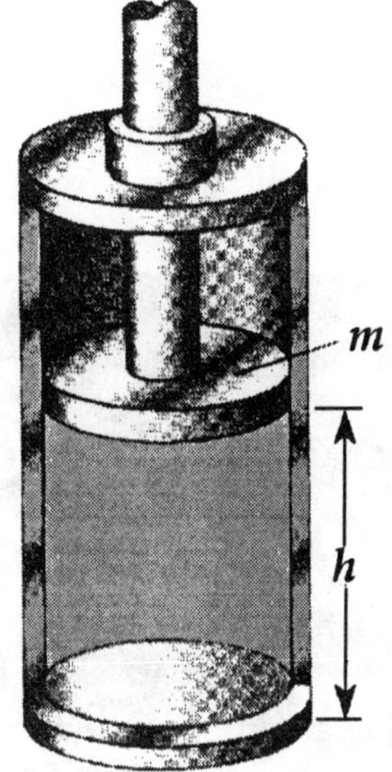

Figure P16.49

Solution

(a) We suppose that the piston does not fit tightly into its guide, and thus the air above the piston remains at a pressure of P_{atm}. For the piston's equilibrium, $\Sigma F_y = ma_y$ yields

$$-P_{atm}A - mg + PA = 0$$

where P is the pressure exerted by the gas contained.

Noting that $V = Ah$, and that n, T, m, g, A, and P_{atm} are given,

$PV = nRT$ becomes $\quad P = \dfrac{nRT}{Ah} \quad$ and $\quad -P_{atm}A - mg + \dfrac{nRT}{Ah}A = 0$

$$h = \frac{nRT}{P_{atm}A + mg} \quad \lozenge$$

(b) $$h = \frac{(0.200 \text{ mol})(8.315 \text{ J/mol}\cdot\text{K})(400 \text{ K})}{\left(1.013\times10^5 \text{ N/m}^2\right)\left(0.00800 \text{ m}^2\right) + (20.0 \text{ kg})\left(9.80 \text{ m/s}^2\right)}$$

$$h = \frac{665 \text{ N}\cdot\text{m}}{810 \text{ N} + 196 \text{ N}} = 0.661 \text{ m} \quad \lozenge$$

Chapter
17

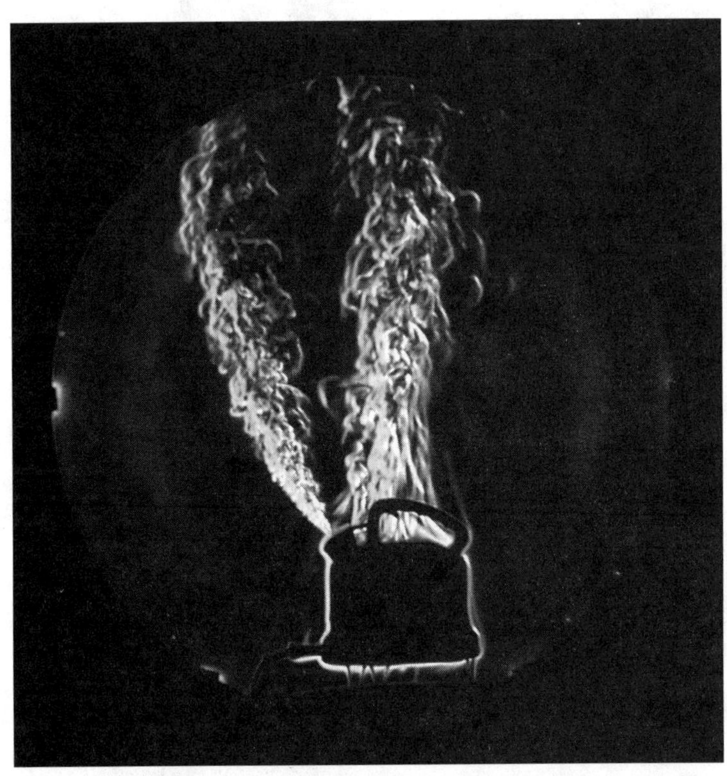

This is a thermogram of a teakettle showing hot areas in white and cooler areas in violet and black. (Gary Settles / Science Source / Photo Researchers)

HEAT AND THE FIRST LAW OF THERMODYNAMICS

Chapter 17

HEAT AND THE FIRST LAW OF THERMODYNAMICS

INTRODUCTION

This chapter focuses on the concept of heat, the first law of thermodynamics, processes by which thermal energy is transferred, and some important applications. The first law of thermodynamics is essentially the law of conservation of energy. It tells us only that an increase in one form of energy must be accompanied by an equal decrease in some other form of energy. The first law places no restrictions on the types of energy conversions that can occur. Furthermore, it makes no distinction between the effects of heat and work. According to the first law, a system's internal energy can be increased either by transfer of heat to the system or by work done on the system. An important difference between thermal energy and mechanical energy is not evident from the first law; it is possible to convert work completely to thermal energy but impossible to convert thermal energy completely to mechanical energy.

NOTES FROM SELECTED CHAPTER SECTIONS

17.1 Heat, Thermal Energy, and Internal Energy

When two systems at different temperatures are in contact with each other, energy will transfer between them until they reach the same temperature (that is, until they are in thermal equilibrium with each other). This energy is called heat, and the term "heat flow" refers to an energy transfer as a consequence of a temperature difference.

The unit of heat is the **calorie** (cal), defined as the amount of heat necessary to increase the temperature of 1 g of water from 14.5°C to 15.5°C. The **mechanical equivalent of heat,** first measured by Joule, is given by 1 cal = 4.186 J.

17.2 Specific Heat

The **specific heat,** c, of any substance is defined as the amount of heat required to increase the temperature of 1 kg of that substance by one Celsius degree. Its units are J/kg·°C.

17.3 Latent Heat and Phase Changes

The **heat of fusion** is the amount of heat required to melt (or freeze) 1 kg of a specific substance, with no temperature change in the substance. The **heat of vaporization** characterizes the liquid-to-gas change in a similar manner. Both of these parameters have units of J / kg.

17.4 Work and Thermal Energy in Thermodynamic Processes

The work done in the expansion from the initial state to the final state is the area under the curve in a PV diagram, as shown in Figure 17.1.

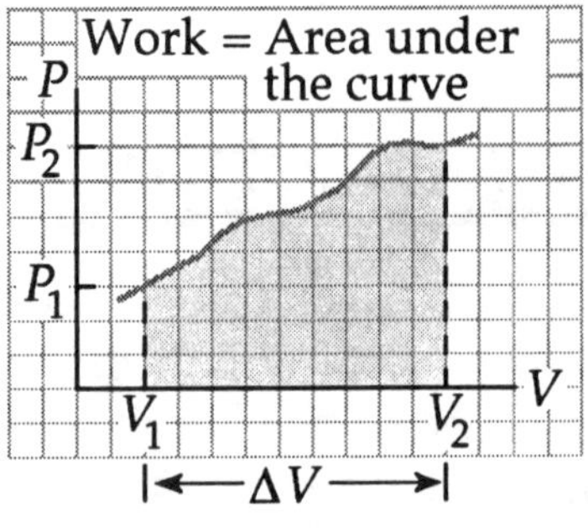

Figure 17.1

If the gas is compressed, $V_f < V_i$, and the work is negative. That is, work is done **on** the gas. If the gas expands, $V_f > V_i$, the work is positive, and the gas does work on the piston. If the gas expands at **constant pressure**, called an **isobaric process**, then $W = P(V_f - V_i)$.

The work done by a system depends on the process by which the system goes from the initial to the final state. In other words, the work done depends on the initial, final, and intermediate states of the system.

17.5 The First Law of Thermodynamics

In the first law of thermodynamics, $\Delta U = Q - W$, Q is the heat added to the system and W is the work done by the system. Note that by convention, Q is **positive** when heat enters the system and **negative** when heat is removed from the system. Likewise, W can be positive or negative as mentioned earlier. The initial and final states must be **equilibrium** states; however, the intermediate states often are not equilibrium states since the thermodynamic coordinates may be impossible to determine during the thermodynamic process. For an **infinitesimal change of the system,** the **first law of thermodynamics** becomes $dU = dQ - dW$. It is important to note that dQ and dW **are not exact differentials,** since both Q and W are not functions of the system's coordinates. That is, both Q and W depend on the **path** taken between the initial and final equilibrium states, during which time the system interacts with its environment. On the other hand, dU is an **exact differential** and the internal energy U is a **state function**. The function U is analogous to the potential energy function used in mechanics when dealing with conservative forces.

17.6 Some Applications of the First Law of Thermodynamics

An **isolated system** is one which does not interact with its surroundings. In such a system, $Q = W = 0$, so it follows from the first law that $\Delta U = 0$. That is, the internal energy of an isolated system cannot change.

A **cyclic process** is one that originates and ends up at the same state. In this situation, $\Delta U = 0$, so from the first law we see that $Q = W$. That is, the work done per cycle equals the heat added to the system per cycle. This is important to remember when dealing with heat engines in the next chapter.

An **adiabatic process** is a process in which no heat enters or leaves the system; that is, $Q = 0$. The first law applied to this process gives $\Delta U = -W$. A system may undergo an adiabatic process if it is thermally insulated from its surroundings, or if the process is so rapid that negligible heat has time to flow.

An **isobaric process** is a process which occurs at constant pressure.

An **isovolumetric process** is one which occurs at constant volume. By definition, $W = 0$ for such a process (since $dV = 0$), so from the first law it follows that $\Delta U = Q$. That is, all of the heat added to the system kept at constant volume goes into increasing the internal energy of the system.

A process that occurs at constant temperature is called an **isothermal process**, and a plot of P versus V at constant temperature for an ideal gas yields a hyperbolic curve called an **isotherm.** The internal energy of an ideal gas is a function of temperature only. Hence, in an isothermal process of an ideal gas, $\Delta U = 0$.

17.7 Heat Transfer

There are three basic processes of heat transfer. These are (1) conductio
(2) convection, and (3) radiation.

Conduction is a heat transfer process which occurs when there is a **te**
gradient across the body. That is, conduction of heat occurs only wh
temperature is **not** uniform. For example, if you heat a metal rod
flame, heat will flow from the hot end to the colder end. If the
(that is, along the rod), and we define the **temperature gra**
quantity of heat dQ will flow in a time dt along the rod
along the rod is proportional to the cross-sectional ar
gradient, and k, the **thermal conductivity** of the m

When heat transfer occurs as the result o
mixing of hot and cold fluids, the proces
heating is used in conventional hot-ai
currents produce changes in weat
mix in the atmosphere.

Heat transfer by **radiation** is
electromagnetic radiation by all bodies.

EQUATIONS AND CONCEPTS

The **specific heat** c of a substance is a measure of the heat energy required to change its temperature by 1°C.

$$c \equiv \frac{Q}{m\,\Delta T} \qquad (17.2)$$

The **heat energy** Q transferred between a system of mass m and its surroundings, for a temperature change ΔT, varies with the substance.

$$Q = mc\,\Delta T \qquad (17.3)$$

A substance may undergo a phase change when heat is transferred between the substance and its surroundings. The heat Q required to change the phase of a unit of mass of the substance is called the **latent heat**, L. The process occurs at constant temperature. The value of L depends on the nature of the phase change and the properties of the substance.

$$Q = mL \qquad (17.5)$$

The **work done by a gas** which undergoes an expansion or compression from initial volume V_i to final volume V_f depends on the path taken between the initial and final states. The pressure is generally not
[illegible]onstant, so exercise care in evaluating W
[illegible]om this equation. In general, the work
[illegible]e equals the area under the PV curve
[illegible] 17.2, on the following page) bounded
[illegible]and V_f, and the pressure function.

$$W = \int_{V_i}^{V_f} P\,dV \qquad (17.7)$$

[illegible]t **law of thermodynamics** is a
[illegible]tion of the law of conservation
[illegible]hat includes possible changes in
[illegible]gy. It states that the **change** in
[illegible]y of a system, ΔU, equals the
[illegible], where Q is the heat added
[illegible] The quantity $(Q - W)$ is
[illegible]e path taken between the
[illegible]tes.

$$\Delta U = Q - W \qquad (17.8)$$

In an **adiabatic process,** $Q = 0$, and the change in internal energy equals the negative of the work done by the gas.

$$\Delta U = -W \tag{17.10}$$

In a constant volume (isovolumetric) process, the work done is zero and all heat added to the system goes into increasing the internal energy.

$$\Delta U = Q \tag{17.11}$$

An **isothermal process** is one which occurs at constant temperature. During such a process, an ideal gas absorbs heat and puts out an equal amount of work.

$$W = nRT \ln\left(\frac{V_f}{V_i}\right) \tag{17.12}$$

SUGGESTIONS, SKILLS, AND STRATEGIES

Many applications of the first law of thermodynamics deal with the work done by (or on) a system which undergoes a change in state. For example, as a gas is taken from a state whose initial pressure and volume are P_i, V_i, and to a state whose final pressure and volume are P_f, V_f, the work can be calculated if the process can be drawn on a PV diagram as in Figure 17.2. The work done during the expansion is given by the integral expression

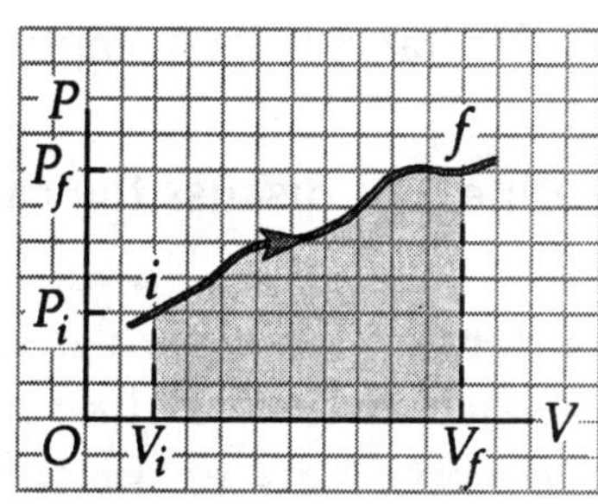

Figure 17.2

$$W = \int_{V_i}^{V_f} P\,dV$$

which numerically represents the **area** under the PV curve (the shaded region) shown in Figure 17.2. It is important to recognize that the work **depends on the path taken as the gas goes from** i **to** f. That is, W depends on the specific manner in which the pressure P changes during the process.

Problem-Solving Strategy: Calorimetry Problems

If you are having difficulty with calorimetry problems, consider the following factors:

- Be sure your units are consistent throughout. For instance, if you are using specific heats in cal/g·°C, be sure that masses are in grams and temperatures are Celsius throughout.
- Heat loss and gains are found by using $Q = mc\Delta T$ only for those intervals in which no phase changes are occurring. The equations $Q = mL_f$ and $Q = mL_v$ are to be used only when phase changes **are** taking place.
- Often sign errors occur in heat loss = heat gain equations. One way to check your equation is to examine the signs of all ΔT's that appear in it.

REVIEW CHECKLIST

▷ Understand the concepts of heat, internal energy, and thermodynamic processes.

▷ Define and discuss the calorie, specific heat, and latent heat.

▷ Understand how work is defined when a system undergoes a change in state, and the fact that work (like heat) depends on the path taken by the system. You should also know how to sketch processes on a *PV* diagram, and calculate work using these diagrams.

▷ State the first law of thermodynamics ($\Delta U = Q - W$), and explain the meaning of the three forms of energy related by this statement.

▷ Discuss the implications of the first law of thermodynamics as applied to (a) an isolated system, (b) a cyclic process, (c) an adiabatic process, and (d) an isothermal process.

ANSWERS TO SELECTED CONCEPTUAL QUESTIONS

3. Using the first law of thermodynamics, explain why the **total** energy of an isolated system is always constant.

Answer The first law of thermodynamics says that the net change in energy of a system is equal to the heat added, minus the work done by the system.

$$\Delta U = Q - W$$

However, an isolated system is defined as an object or set of objects that exchanges neither work nor heat with its surroundings. Specifically, it says that $Q = W = 0$, so the change in total energy of the system at all times must be zero.

It should be noted that although $\Delta U = 0$, the energy may change from one form to another or move from one object to another within the system. For example, "bomb calorimeters" are closed systems that consist of a sturdy steel container, a water bath, an item of food, and oxygen. The food is burned in the presence of an excess of oxygen, and chemical energy is converted to heat energy. In the process of oxidation, heat energy is absorbed by the water bath, raising its temperature. The change in temperature of the water bath is used to determine the caloric content of the food. The process works specifically **because** the total energy remains unchanged in a closed system.

□ □ □ □

7. What is wrong with the statement: "given any two bodies, the one with the higher temperature contains more heat"?

Answer The statement shows a misunderstanding of the concept of heat. Heat is a form of energy that is being transferred, not a form of energy that is held or contained. If you wish to speak of energy that is "contained," you speak of **internal energy**, not **heat**.

Further, even if the statement used the term "internal energy," it would still be incorrect, since the effects of specific heat and mass are both ignored. A 1-kg mass of water at 20 °C has more internal energy than a 1-kg mass of air at 30° C. Similarly, the earth has far more internal energy than a drop of molten titanium metal.

Correct statements would be: (1) "given any two bodies in thermal contact, the one with the higher temperature will transfer heat to the other." (2) "given any two bodies of equal mass, the one with the higher product of absolute temperature and specific heat contains more internal energy."

□ □ □ □

9. The air temperature above coastal areas is profoundly influenced by the large specific heat of water. One reason is that the heat released when 1 cubic meter of water cools by 1.0 °C will raise the temperature of an enormously larger volume of air by 1.0 °C. Estimate this volume of air. The specific heat of air is approximately 1.0 kJ/kg · °C. Take the density of air to be 1.3 kg/m³.

Answer:

The thermal energy released by one cubic meter of water, cooling by one Celsius degree, is $Q = m_w c_w \Delta T$. The thermal energy absorbed by one cubic meter of air, warming by one Celsius degree, is $Q = m_a c_a \Delta T = \rho_a V_a c_a \Delta T$. Setting these two equations equal, $m_w c_w \Delta T = \rho_a V_a c_a \Delta T$:

$$V_a = \frac{m_w c_w \Delta T}{\rho_a c_a \Delta T} = \frac{(1000\text{ kg})\left(4186\ \frac{\text{J}}{\text{kg}\cdot{}^\circ\text{C}}\right)(1\ {}^\circ\text{C})}{\left(1.3\ \frac{\text{kg}}{\text{m}^3}\right)\left(1000\ \frac{\text{J}}{\text{kg}\cdot{}^\circ\text{C}}\right)(1\ {}^\circ\text{C})} = 3220\text{ m}^3$$

□ □ □ □

SOLUTIONS TO SELECTED END-OF-CHAPTER PROBLEMS

3. The temperature of a silver bar rises by 10.0 °C when it absorbs 1.23 kJ of heat. The mass of the bar is 525 g. Determine the specific heat of silver.

Solution $\Delta Q = m c_{\text{silver}} \Delta T$ or $c_{\text{silver}} = \dfrac{Q}{m\,\Delta T}$

$$c_{\text{silver}} = \frac{1.23 \times 10^3\text{ J}}{(0.525\text{ kg})(10.0{}^\circ\text{C})} = 234\text{ J/kg}\cdot{}^\circ\text{C} \quad \lozenge$$

5. A 1.50-kg iron horseshoe initially at 600°C is dropped into a bucket containing 20.0 kg of water at 25.0 °C. What is the final temperature of the water and horseshoe? (Neglect the heat capacity of the container and assume that a negligible amount of water boils away.)

Solution The iron loses as much heat as the water gains. Therefore,

$$\Delta Q_{\text{iron}} = -\Delta Q_{\text{water}} \quad \text{or} \quad (mc\,\Delta T)_{\text{iron}} = -(mc\,\Delta T)_{\text{water}}$$

$$(1.50\text{ kg})(448\text{ J/kg}\cdot{}^\circ\text{C})(T - 600\ {}^\circ\text{C}) = -(20.0\text{ kg})(4186\text{ J/kg}\cdot{}^\circ\text{C})(T - 25.0\ {}^\circ\text{C})$$

$T = 29.6\ {}^\circ\text{C}$ ◊

---

9. A 3.00-g copper penny at 25.0 °C drops 50.0 m to the ground. (a) If 60.0% of its initial potential energy goes into increasing the internal energy, determine its final temperature. (b) Does the result depend on the mass of the penny? Explain.

Solution Take f to be the fraction of energy that goes into raising the temperature of the penny. The energy that goes into the copper penny can then be calculated from the work equation:

$$\Delta U = f(mgh)$$

This raises the temperature of the penny according to the equation $Q = mc\,\Delta T$, just as if it were heat $Q = \Delta U$ put into the penny from a stove.

$$\Delta T = \frac{Q}{mc} = \frac{\Delta U}{mc} = \frac{f(mgh)}{mc} = f\left(\frac{gh}{c}\right)$$

Noting that $f = 0.600$ and $c = 387$ J/kg · °C,

$$\Delta T = 0.600\left(\frac{(9.80\text{ m/s}^2)(50.0\text{ m})}{387\text{ J/kg}\cdot{}^\circ\text{C}}\right)$$

$$\Delta T = 0.760\ {}^\circ\text{C}$$

(a) Since the initial temperature was 25.0 °C, the final temperature will be 25.8 °C. ◊

(b) From our second equation, we see that the result is independent of the mass. This is because every gram of mass falls the same distance, and in effect must speed up its own molecules to the final temperature. ◊

---

15. In an insulated vessel, 250 g of ice at 0 °C is added to 600 g of water at 18.0 °C. (a) What is the final temperature of the system? (b) How much ice remains when the system reaches equilibrium?

Solution When 250 g of ice is melted,

$$\Delta Q_f = mL_f = (0.250 \text{ kg})(3.33\times 10^5 \text{ J/kg}) = 83.3 \text{ kJ}$$

The heat energy released when 600 g of water cools from 18.0 °C to 0 °C is

$$\Delta Q = mc\,\Delta T = (0.600 \text{ kg})(4186 \text{ J/kg}\cdot°\text{C})(18.0\ °\text{C}) = 45.2 \text{ kJ}$$

(a) Since the heat required to melt 250 g of ice at 0°C **exceeds** the heat released by cooling 600 g of water from 18.0 °C to 0 °C, the final temperature of the system (water + ice) must be 0 °C. ◊

(b) The 45.2 kJ released by the water will melt m grams of ice, where $Q = mL_f$

Solving for m, $$m = \frac{Q}{L_f} = \frac{45.2\times 10^3 \text{ J}}{3.33\times 10^5 \text{ J/kg}} = 0.136 \text{ kg}$$

Therefore, the ice remaining is equal to $m' = 0.250 \text{ kg} - 0.136 \text{ kg} = 0.114 \text{ kg}$ ◊

19. A sample of ideal gas is expanded to twice its original volume of 1.00 m³ in a quasi-static process for which $P = \alpha V^2$, with α = 5.00 atm/m⁶, as shown in Fig. P17.19. How much work was done by the expanding gas?

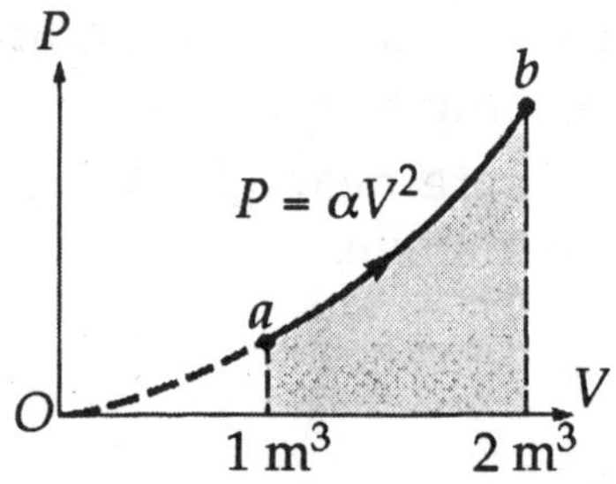

Figure P17.19

Solution $$W_{ab} = \int_a^b P\,dV$$

$$V_b = 2V_a = 2\left(1.00 \text{ m}^3\right) = 2.00 \text{ m}^3$$

The work done by the gas is the area under the curve $P = \alpha V^2$, from V_a to V_b.

$$W_{ab} = \int_{V_a}^{V_b} \alpha V^2\,dV = \tfrac{1}{3}\alpha\left(V_b^{\,3} - V_a^{\,3}\right)$$

$$W_{ab} = \tfrac{1}{3}\left(5.00\,\frac{\text{atm}}{\text{m}^6}\right)\left(1.013\times 10^5\,\frac{\text{Pa}}{\text{atm}}\right)\left(\left(2.00 \text{ m}^3\right)^3 - \left(1.00 \text{ m}^3\right)^3\right) = 1.18\times 10^6 \text{ J}$$ ◊

27. An ideal gas initially at 300 K undergoes an isobaric expansion at 2.50 kPa. If the volume increases from 1.00 m^3 to 3.00 m^3 and 12.5 kJ of thermal energy is transferred to the gas, find (a) the change in its internal energy and (b) its final temperature.

Solution

(a) $\Delta U = Q - W$ where $W = P\Delta V$ so that

$$\Delta U = Q - P\,\Delta V = 1.25\times10^4\ \text{J} - \left(2.50\times10^3\ \text{N/m}^2\right)\left(3.00\ \text{m}^3 - 1.00\ \text{m}^3\right) = 7500\ \text{J} \quad \lozenge$$

(b) Since $\dfrac{V_1}{T_1} = \dfrac{V_2}{T_2}$, $T_2 = \left(\dfrac{V_2}{V_1}\right)T_1 = \left(\dfrac{3.00\ \text{m}^3}{1.00\ \text{m}^3}\right)(300\ \text{K}) = 900\ \text{K} \quad \lozenge$

---

29. How much work is done by the steam when 1.00 mol of water at 100 °C boils and becomes 1.00 mol of steam at 100°C at 1.00 atm pressure? Determine the change in internal energy of the steam as it vaporizes. Consider the steam to be an ideal gas.

Solution $W = P\,\Delta V = P(V_s - V_w)$

Substituting, $V_s = \dfrac{nRT}{P}$ and $V_w = n\left(\dfrac{M}{\rho}\right)$

We have $W = n\left(RT - \dfrac{PM}{\rho}\right)$

$$W = (1.00\ \text{mol})\left((8.315\ \text{J/K}\cdot\text{mol})(373\ \text{K}) - \frac{\left(1.013\times10^5\ \text{N/m}^2\right)(18.0\ \text{g/mol})}{1.00\times10^6\ \text{g/m}^3}\right)$$

$W = 3.10\ \text{kJ} \quad \lozenge$

$Q = mL_V = (0.0180\ \text{kg})\left(2.26\times10^6\ \text{J/kg}\right) = 40.7\ \text{kJ}$

$\Delta U = Q - W = 37.6\ \text{kJ} \quad \lozenge$

---

31. Two moles of helium gas initially at 300 K and 0.400 atm are compressed isothermally to 1.20 atm. Find (a) the final volume of the gas, (b) the work done by the gas, and (c) the thermal energy transferred. Consider the helium to behave as an ideal gas.

Solution $PV = nRT$, so $V_i = \frac{nRT}{P_i} = \frac{(2.00 \text{ mol})(8.31 \text{ J/mol}\cdot\text{K})(300 \text{ K})}{(0.400 \text{ atm})(1.01\times10^5 \text{ N/m}^2/\text{atm})} = 0.123 \text{ m}^3$

(a) For isothermal compression (or expansion), PV = constant, so $P_iV_i = P_fV_f$:

$$V_f = V_i\left(\frac{P_i}{P_f}\right) = (0.123 \text{ m}^3)\left(\frac{0.400 \text{ atm}}{1.20 \text{ atm}}\right) = 0.0410 \text{ m}^3 \quad \Diamond$$

(b) $W = \int P\,dV = \int \frac{nRT}{V}dV = nRT\ln\left(\frac{V_f}{V_i}\right) = (4986 \text{ J})\ln\left(\frac{0.0410 \text{ m}^3}{0.123 \text{ m}^3}\right) = -5.48 \text{ kJ} \quad \Diamond$

(c) $\Delta U = 0 = Q - W$ and $Q = -5.48 \text{ kJ} \quad \Diamond$

---

37. A bar of gold is in thermal contact with a bar of silver of the same length and area (Fig. P17.37). One end of the compound bar is maintained at 80.0 °C, and the opposite end is at 30.0 °C. When the heat flow reaches steady state, find the temperature at the junction.

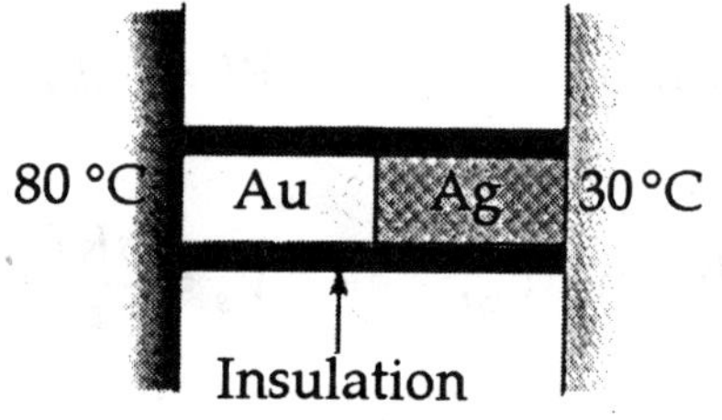

Figure P17.37

Solution Call the gold bar Object 1 and the silver bar Object 2. When heat flow reaches a steady state, the flow rate through each will be the same: $H_1 = H_2$ or

$$\frac{k_1A_1\,\Delta T_1}{L_1} = \frac{k_2A_2\,\Delta T_2}{L_2}$$

In this case, $L_1 = L_2$ and $A_1 = A_2$, so $k_1\,\Delta T_1 = k_2\,\Delta T_2$

Let T_3 = temperature at the junction; then $k_1(80.0 \text{ °C} - T_3) = k_2(T_3 - 30.0 \text{ °C})$

Solving, $T_3 = \frac{(80.0 \text{ °C})k_1 + (30.0 \text{ °C})k_2}{k_1 + k_2}$

$$T_3 = \frac{(80.0 \text{ °C})(314 \text{ W/m}\cdot\text{°C}) + (30.0 \text{ °C})(427 \text{ W/m}\cdot\text{°C})}{(314 \text{ W/m}\cdot\text{°C} + 427 \text{ W/m}\cdot\text{°C})} = 51.2 \text{ °C} \quad \Diamond$$

---

43. An aluminum rod, 0.500 m in length and of cross-sectional area 2.50 cm^2, is inserted into a thermally insulated vessel containing liquid helium at 4.22 K. The rod is initially at 300 K. (a) If one half of the rod is inserted into the helium, how many liters of helium boil off by the time the inserted half cools to 4.22 K? (Assume the upper half does not yet cool.) (b) If the upper end of the rod is maintained at 300 K, what is the approximate boil-off rate of liquid helium after the lower half has reached 4.22 K? (Note that aluminum has a thermal conductivity of 31.0 J/s·cm·K at 4.2 K, a specific heat of 0.210 cal/g·°C, and a density of 2.70 g/cm^3. The density of liquid helium is 0.125 g/cm^3.)

Solution As you solve this problem, be careful not to confuse L (the **conduction length** of the rod) with L_v (the **heat of vaporization** of the helium).

(a) Before heat conduction has time to become important, we suppose the heat energy lost by half the rod equals the heat energy gained by the helium. Therefore,

$$(mL_v)_{\text{He}} = (mc\,\Delta T)_{\text{Al}} \quad \text{or} \quad (\rho V L_v)_{\text{He}} = (\rho V c \Delta T)_{\text{Al}}$$

so that
$$V_{\text{He}} = \frac{(\rho Vc\,\Delta T)_{\text{Al}}}{(\rho L_v)_{\text{He}}} = \frac{(2.70\text{ g/cm}^3)(62.5\text{ cm}^3)(0.210\text{ cal/g}\cdot{}^\circ\text{C})(295.8\ {}^\circ\text{C})}{(0.125\text{ g/cm}^3)(4.99\text{ cal/g})}$$

and
$$V_{\text{He}} = 1.68\times 10^4\text{ cm}^3 = 16.8\text{ liters} \quad \Diamond$$

(b) Heat energy will be conducted along the rod at a rate of $\dfrac{dQ}{dt} = H = \dfrac{kA\,\Delta T}{L}$.

During any time interval, this will boil a mass of helium according to

$$Q = mL_v \quad \text{or} \quad \frac{dQ}{dt} = \left(\frac{dm}{dt}\right)L_v.$$

Combining these two equations gives us the "boil-off" rate: $\dfrac{dm}{dt} = \dfrac{kA\,\Delta T}{L\cdot L_v}$

The conduction length is $L = 25.0$ cm; use $k = 31.0\text{ J/s}\cdot\text{cm}\cdot\text{K} = 7.41\text{ cal/s}\cdot\text{cm}\cdot\text{K}$:

$$\frac{dm}{dt} = \frac{(7.41\text{ cal/s}\cdot\text{cm}\cdot\text{K})(2.50\text{ cm}^2)(295.8\text{ K})}{(25.0\text{ cm})(4.99\text{ cal/g})} = 43.9\text{ g/s}$$

or
$$\frac{dV}{dt} = \frac{dm/dt}{\rho_{\text{He}}} = \frac{43.9\text{ g/s}}{0.125\text{ g/cm}^3} = 351\text{ cm}^3\text{/s} = 0.351\text{ liters/s} \quad \Diamond$$

49. A "solar cooker" consists of a curved reflecting mirror that focuses sunlight onto the object to be heated (Fig. P17.49). The solar power-per-unit area reaching the Earth at the location is 600 W/m^2, and the cooker has a diameter of 0.600 m. Assuming that 40.0% of the incident energy is transferred to the water, how long would it take to completely boil off 0.500 liters of water initially at 20.0° C? (Neglect the heat capacity of the container.)

Figure P17.49

Solution The power incident on the solar collector is

$$P_i = IA = \left(600 \text{ W/m}^2\right)\pi(0.300 \text{ m})^2 = 170 \text{ W}$$

For a reflector which is 40.0% efficient, the collected power is $P_c = 67.9 \text{ J/s}$

The total energy required to increase the temperature of the water to the boiling point and to evaporate it is

$$Q = mc\Delta T + mL_v$$

$$= (0.500 \text{ kg})(4186 \text{ J/kg}\cdot°\text{C})(80.0° \text{ C}) + (0.500 \text{ kg})(2.26\times10^6 \text{ J/kg}) = 1.30\times10^6 \text{ J}$$

The time required is

$$\Delta t = \frac{Q}{P_c} = \frac{1.30\times10^6 \text{ J}}{67.9 \text{ J/s}} = 1.91\times10^4 \text{ s} = 5.31 \text{ h} \quad \Diamond$$

---

51. An ideal gas is enclosed in a cylinder with a movable piston on top of it. The piston has a mass of 8000 g and an area of 5.00 cm^2 and is free to slide up and down, keeping the pressure of the gas constant. How much work is done as the temperature of 0.200 mol of the gas is raised from 20.0 °C to 300 °C?

Solution In an isobaric (constant pressure) process, $W = P\,\Delta V = P\left(\frac{nR}{P}\right)(T_h - T_c)$

Therefore, $W = nR\,\Delta T = (0.200 \text{ mol})(8.315 \text{ J/mol}\cdot\text{K})(280 \text{ K}) = 466 \text{ J} \quad \Diamond$

Chapter 18

This steam-driven locomotive runs from Durango to Silverton, Colorado. Before World War II, most trains were powered by steam; modern trains use electric or diesel engines. However, steam engines can still be found in some electric power plants and nuclear-powered ships. (Lois Moulton, Tony Stone Images)

HEAT ENGINES, ENTROPY, AND THE SECOND LAW OF THERMODYNAMICS

Chapter 18

HEAT ENGINES, ENTROPY, AND THE SECOND LAW OF THERMODYNAMICS

INTRODUCTION

The first law of thermodynamics, studied in Chapter 17, is a statement of conservation of energy, generalized to include heat as a form of energy transfer. This law tells us only that an increase in one form of energy must be accompanied by a decrease in some other form of energy. It places no restrictions on the types of energy conversions that can occur. Furthermore, it makes no distinction between heat and work. However, such a distinction exists. One manifestation of this difference is the fact that it is impossible to convert thermal energy to mechanical energy entirely and continuously. In nature, only certain types of energy conversions can take place. The second law of thermodynamics establishes which processes in nature can and which cannot occur.

From an engineering viewpoint, perhaps the most important application of the second law of thermodynamics is the limited efficiency of heat engines. The second law says that a machine capable of continuously converting thermal energy completely to other forms of energy in a cyclic process cannot be constructed.

NOTES FROM SELECTED CHAPTER SECTIONS

18.1 Heat Engines and the Second Law of Thermodynamics

A heat engine is a device that converts thermal energy to other useful forms, such as electrical and mechanical energy.

A heat engine carries some working substance through a cyclic process during which (1) heat is absorbed from a source at a high temperature, (2) work is done by the engine, and (3) heat is expelled by the engine to a source at a lower temperature.

The engine absorbs a quantity of heat, Q_h, from a hot reservoir, does work W, and then gives up heat Q_c to a cold reservoir. Because the working substance goes through a cycle, its initial and final internal energies are equal, so $\Delta U = 0$. Hence, from the first law we see that **the net work, W, done by a heat engine equals the net heat flowing into it.**

If the working substance is a gas, **the net work done for a cyclic process is the area enclosed by the curve representing the process on a *PV* diagram.**

The **thermal efficiency,** e, of a heat engine is the ratio of the net work done to the heat absorbed at the higher temperature during one cycle.

The **second law of thermodynamics** can be stated in several ways:

Clausius Statement: No thermodynamic process can occur whose only result is to transfer heat from a colder to a hotter body. Such a process is only possible if work is done on the system.

Kelvin-Planck Statement: It is impossible for a thermodynamic process to occur whose only final result is the complete conversion of heat extracted from a hot reservoir into work. That is, **it is impossible to construct a heat engine that, operating in a cycle, produces no other effect than the absorption of heat from a reservoir and the performance of an equal amount of work.** A heat engine must also release heat into a cold reservoir.

The Kelvin-Planck statement is equivalent to stating that **it is impossible to construct a perpetual motion machine of the second kind** (that is, a machine which violates the second law). Perpetual motion machines of the **first kind** would be those which violate the first law of thermodynamics, which requires that energy be conserved.

18.2 Reversible and Irreversible Processes

A process is **reversible** if the system passes from the initial to the final state through a succession of equilibrium states. Then, the process can be made to run in the opposite direction by an infinitesimally small change in conditions. Otherwise the process is **irreversible**.

18.3 The Carnot Engine

The ideal reversible cyclic process, called the **Carnot cycle,** is described in the *PV* diagram of Figure 18.1. The Carnot cycle consists of two adiabatic and two isothermal processes, all being reversible.

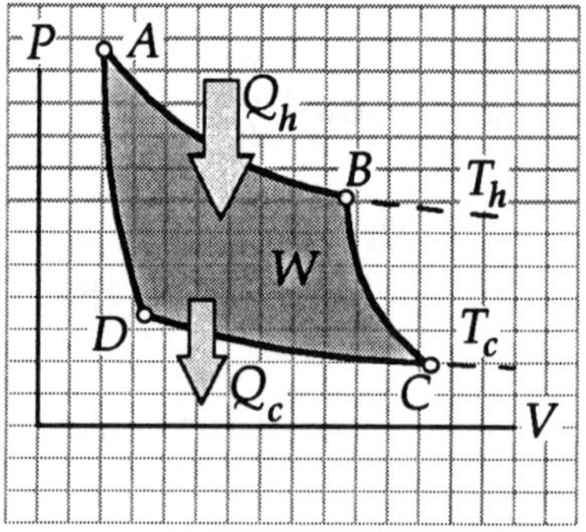

Figure 18.1

- The process $A \to B$ is an isotherm (constant T), during which time the gas expands at constant temperature T_h, and absorbs heat Q_h from the hot reservoir.
- The process $B \to C$ is an adiabatic expansion ($Q = 0$), during which time the gas expands and cools to a temperature T_c.
- The process $C \to D$ is a second isotherm, during which time the gas is compressed at constant temperature T_c, and gives up heat Q_c to the cold reservoir.
- The final process $D \to A$ is an adiabatic compression in which the gas temperature increases to a final temperature of T_h.

In practice, no working engine is 100% efficient, even when losses such as friction are neglected. One can determine the theoretical limit on the efficiency of a real engine by comparison with the ideal Carnot engine. A **reversible engine** is one which will operate with the same efficiency in the forward and reverse directions. The Carnot engine is one example of a reversible engine.

Carnot's theorems, which are consistent with the first and second laws of thermodynamics, can be stated as follows:

> **Theorem I.** No real (irreversible) engine can have an efficiency greater than that of a reversible engine operating between the same two temperatures.
>
> **Theorem II.** All reversible engines operating between T_h and T_c have the **same** efficiency given by Equation 18.2.

A schematic diagram of a heat engine is shown in Figure 18.2a, where Q_h is the heat extracted from the hot reservoir at temperature T_h, Q_c is the heat rejected to the cold reservoir at temperature T_c, and W is the work done by the engine.

18.6 Heat Pumps and Refrigerators

A refrigerator is a heat engine operating in reverse, as shown in Figure 18.2b. During one cycle of operation, the refrigerator absorbs heat Q_c from the cold reservoir, expels heat Q_h to the hot reservoir, and the work done on the system is $W = Q_h - Q_c$.

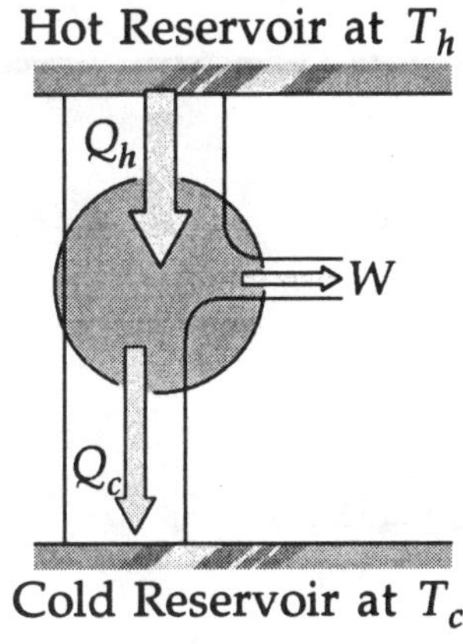

(a) Heat engine

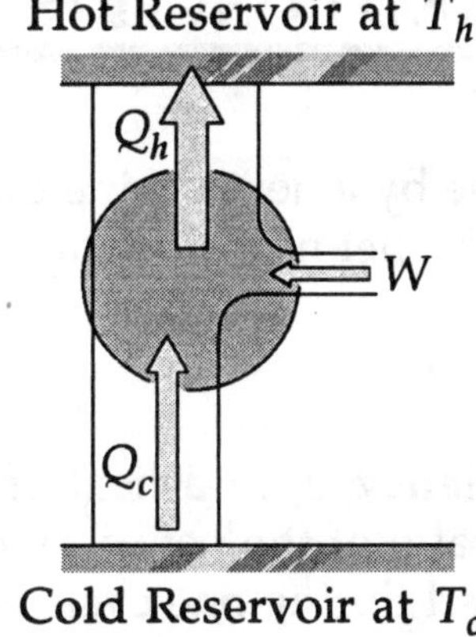

(b) Refrigerator (heat pump)

Figure 18.2

18.7 Entropy and Entropy Changes in Irreversible Processes

Entropy is a quantity used to measure the degree of **disorder** in a system. For example, the molecules of a gas in a container at a high temperature are in a more disordered state (higher entropy) than the same molecules at a lower temperature.

When heat is added to a system, dQ_r is **positive** and the entropy **increases.** When heat is removed, dQ_r is **negative** and the entropy **decreases.** Note that only **changes** in entropy are defined by Equation 18.6; therefore, the concept of entropy is most useful when a system undergoes a **change in its state.**

When using Equation 18.7 to calculate entropy changes, note that ΔS may be obtained even if the process is irreversible, since ΔS depends only on the initial and final equilibrium states, not on the path. In order to calculate ΔS for an irreversible process, you must devise a reversible process (or sequence of reversible processes) between the initial and final states, and compute dQ_r / T for the reversible process. The entropy change for the irreversible process is the **same** as that of the reversible process between the same initial and final equilibrium states.

The second law of thermodynamics can be stated in terms of entropy as follows: **The total entropy of an isolated system always increases in time if the system undergoes an irreversible process.** If an isolated system undergoes a **reversible** process, the total entropy **remains constant.**

EQUATIONS AND CONCEPTS

The net **work done** by a heat engine during one cycle equals the net heat flowing into the engine.

$$W = Q_h - Q_c \tag{18.1}$$

The **thermal efficiency**, e, of a heat engine is defined as the ratio of the net work done to the heat absorbed during one cycle of the process.

$$e = \frac{W}{Q_h} = 1 - \frac{Q_c}{Q_h} \tag{18.2}$$

No real engine operating between the temperatures T_c and T_h can be more efficient than an engine operating reversibly in a Carnot cycle between the same two temperatures.

$$e_c = 1 - \frac{T_c}{T_h} \tag{18.3}$$

The **coefficient of performance** of a refrigerator or a heat pump used in the cooling cycle is defined by the ratio of the heat absorbed, Q_c, to the work done. A good refrigerator has a high coefficient of performance.

$$\text{COP (heat pump)} = \frac{Q_h}{W} \tag{18.4}$$

$$\text{COP (refrigerator)} = \frac{Q_c}{W} \tag{18.5}$$

If a system changes from one equilibrium state to another, under a reversible, quasi-static process, and a quantity of heat dQ_r is added (or removed) at the absolute temperature T, the **change in entropy** is defined by the ratio dQ_r / T.

$$dS = \frac{dQ_r}{T} \tag{18.6}$$

The **change in entropy** of a system which undergoes a reversible process between the states i and f depends only on the properties of the initial and final equilibrium states.

$$\Delta S = \int_i^f \frac{dQ_r}{T} \quad \text{(reversible path)} \tag{18.7}$$

The change in entropy of a system for any arbitrary **reversible cycle is zero.**

$$\oint \frac{dQ_r}{T} = 0 \text{ (closed path)} \tag{18.8}$$

During a **free expansion** (irreversible, adiabatic process), the entropy of a gas **increases.**

$$\Delta S = nR \ln \frac{V_f}{V_i} \tag{18.9}$$

During the process of irreversible heat transfer between two masses, the entropy of the system will increase.

$$\Delta S = m_1 c_1 \ln \frac{T_f}{T_1} + m_2 c_2 \ln \frac{T_f}{T_2} \tag{18.11}$$

All isolated systems tend toward disorder and the increase in entropy is proportional to the probability, W, of occurrence of the event. Entropy is a measure of microscopic disorder.

$$S \equiv k_B \ln W \tag{18.14}$$

REVIEW CHECKLIST

▷ Understand the basic principle of the operation of a heat engine, and be able to define and discuss the **thermal efficiency** of a heat engine.

▷ State the second law of thermodynamics, and discuss the difference between reversible and irreversible processes. Discuss the importance of the first and second laws of thermodynamics as they apply to various forms of energy conversion and thermal pollution.

▷ Describe the processes which take place in an ideal heat engine taken through a **Carnot cycle**.

▷ Calculate the efficiency of a Carnot engine, and note that the efficiency of real heat engines is always less than the Carnot efficiency.

▷ Calculate entropy changes for reversible processes (such as one involving an ideal gas).

▷ Calculate entropy changes for irreversible processes, recognizing that the entropy change for an irreversible process is equivalent to that of a reversible process between the same two equilibrium states.

ANSWERS TO SELECTED CONCEPTUAL QUESTIONS

1. What are some factors that affect the efficiency of automobile engines?

Answer The most basic limit of efficiency for automobile engines is the Carnot cycle; the efficiency cannot **ever** exceed this ideal model. The efficiency is therefore affected by the maximum temperature that the engine block can stand, and the intake and exhaust pressures. However, the Carnot cycle assumes that the exhaust gases can be released over an infinite period of time. If you want your engine to act more quickly than "never," you must take additional losses. Other limits on efficiency are imposed by friction, the mass and acceleration of the engine parts, and the timing of the ignition.

□ □ □ □

2. A steam-driven turbine is one major component of an electric power plant. Why is it advantageous to have the temperature of the steam as high as possible?

Answer The most optimistic limit of efficiency in a steam engine is the ideal (Carnot) efficiency. This can be calculated from the high and low temperatures of the steam, as it expands:

$$e_c = \frac{T_h - T_c}{T_h} = 1 - \frac{T_c}{T_h}$$

The engine will be most efficient when the low temperature is extremely low, and the high temperature is extremely high. However, since the electric power plant is typically placed on the surface of the earth, there is a limit to how much the steam can expand, and how low the lower temperature can be. The only way to further increase the efficiency, then, is to raise the temperature of the hot steam, as high as possible.

□ □ □ □

3. Is it possible to construct a heat engine that creates no thermal pollution?

Answer Practically speaking, it is not possible to create a heat engine on earth that creates no thermal pollution, because there must be both a hot heat source, and a cold heat sink. The heat engine will warm the cold heat sink, and will cool down the heat source. If either of those two events is undesirable, then there will be thermal pollution.

That being said, there are some circumstances where the thermal pollution would be negligible. For example, suppose a satellite in space were to run a heat pump between its sunny side, and its dark side. The satellite would intercept some of the heat that gathered on one side, and would dump it to the dark side. Since neither of those effects would be particularly undesirable, it could be said that such a heat pump produced no thermal pollution.

It is also possible to make the heat source and sink totally self contained. In such a case, the engine would function until the sink became as hot as the source. However, building such an engine would probably be more wasteful than simply finding an acceptable form of thermal pollution.

□ □ □ □

4. Discuss the change in entropy of a gas that expands (a) at constant temperature and (b) adiabatically.

Answer If you remember, the change in entropy is calculated as the change in heat divided by the temperature:

$$dS = dQ / T$$

(a) In the case of constant temperature expansion, the change in entropy is a function of the starting and ending volumes of the gas:

$$\Delta S = \frac{1}{T}\int dQ = nR\ln\frac{V_2}{V_1}$$

(b) In the case of adiabatic expansion, $\Delta Q = 0$, so the term "adiabatic" also means that the change in entropy is zero.

□ □ □ □

18. The device in Figure Q18.18 of the textbook, called a thermoelectric converter, uses a series of semiconductor cells to convert thermal energy to electrical energy. In the photograph at the left, both legs of the device are at the same temperature, and no electrical energy is produced. However, when one leg is at a higher temperature than the other, as in the photograph on the right, electrical energy is produced as the device extracts energy from the hot reservoir and drives a small electric motor. (a) Why does the temperature differential produce electrical energy in this demonstration? (b) In what sense does this intriguing experiment demonstrate the second law of thermodynamics?

Answer (a) The semiconductor converter operates essentially like a thermocouple, which is a pair of wires of different metals, with a junction at each end. When the junctions are at different temperatures, a small voltage appears around the loop, so that the device can be used to measure temperature or (here) to drive a small motor.

(b) The second law's statement is that it is impossible for a cycling engine simply to extract heat from one reservoir, and convert it into work. This exactly describes the first situation, where both legs are in contact with a single reservoir, and the thermocouple fails to produce work. To convert heat to work, the device must transfer heat from a hot reservoir to a cold reservoir, as in the second situation.

□ □ □ □

SOLUTIONS TO SELECTED END-OF-CHAPTER PROBLEMS

3. A particular engine has a power output of 5.00 kW and an efficiency of 25.0%. If the engine expels 8000 J of thermal energy in each cycle, find (a) the heat absorbed in each cycle and (b) the time for each cycle.

Solution

(a) We have
$$e = \frac{W}{Q_h} = \frac{Q_h - Q_c}{Q_h} = 1 - \frac{Q_c}{Q_h},$$

so
$$Q_h = \frac{Q_c}{1-e} = \frac{8000 \text{ J}}{1-0.250} = 10.7 \text{ kJ} \quad \lozenge$$

(b) $W = Q_h - Q_c = 2667 \text{ J}$ and from $P = \frac{W}{t}$, we have

$$t = \frac{W}{P} = \frac{2667 \text{ J}}{5000 \text{ J/s}} = 0.533 \text{ s} \quad \lozenge$$

---

5. One of the most efficient engines ever built (its actual efficiency is 42.0%) operates between 430°C and 1870°C. (a) What is its maximum theoretical efficiency? (b) How much power does the engine deliver if it absorbs 1.40×10^5 J of thermal energy each second?

Solution $T_c = 430°\text{C} = 703 \text{ K}$ and $T_h = 1870°\text{C} = 2143 \text{ K}$

(a) $e_c = \frac{\Delta T}{T_h} = \frac{1440 \text{ K}}{2143 \text{ K}} = 0.672$ or 67.2% ◊

(b) $Q_h = 1.40 \times 10^5 \text{ J}$ $W = 0.420 Q_h = 5.88 \times 10^4 \text{ J}$

$$P = \frac{W}{t} = \frac{5.88 \times 10^4 \text{ J}}{1 \text{ s}} = 58.8 \text{ kW} \quad \lozenge$$

---

7. A steam engine is operated in a cold climate in which the exhaust temperature is 0 °C. (a) Calculate the theoretical maximum efficiency of the engine using an intake steam temperature of 100 °C. (b) If, instead, superheated steam at 200 °C is used, find the maximum possible efficiency.

Solution

(a) $e_c = \dfrac{\Delta T}{T_h} = \dfrac{100}{373} = 0.268 = 26.8\%$ ◊

(b) $e_c = \dfrac{\Delta T}{T_h} = \dfrac{200}{473} = 0.423 = 42.3\%$ ◊

It is important to remember that the temperatures to which Carnot's equation refers are absolute temperatures.

---

9. An ideal gas is taken through a Carnot cycle. The isothermal expansion occurs at 250 °C, and the isothermal compression takes place at 50.0 °C. If the gas absorbs 1200 J of heat during the isothermal expansion, find (a) the heat expelled to the cold reservoir in each cycle and (b) the net work done by the gas in each cycle.

Solution

(a) For a Carnot cycle, $e_c = 1 - \dfrac{T_c}{T_h}$

For any engine, $e = \dfrac{W}{Q_h} = 1 - \dfrac{Q_c}{Q_h}$

Therefore, for a Carnot engine, $1 - \dfrac{T_c}{T_h} = 1 - \dfrac{Q_c}{Q_h}$

$$Q_c = Q_h(T_c / T_h) = (1200\text{ J})\left(\frac{323\text{ K}}{523\text{ K}}\right) = 741\text{ J} \quad ◊$$

(b) The work is calculated as $W = Q_h - Q_c = 1200\text{ J} - 741\text{ J} = 459\text{ J}$

---

15. An ideal refrigerator or ideal heat pump is equivalent to a Carnot engine running in reverse. That is, heat Q_c is absorbed from a cold reservoir and heat Q_h is rejected to a hot reservoir. (a) Show that the work that must be supplied to run the refrigerator or pump is

$$W = \frac{T_h - T_c}{T_c} Q_c$$

(b) Show that the coefficient of performance of the ideal refrigerator is $\text{COP} = \dfrac{T_c}{T_h - T_c}$

Solution (a) For a complete cycle $\Delta U = 0$, and $W = Q_h - Q_c = Q_c\left(\dfrac{Q_h}{Q_c} - 1\right)$

We have already shown for a Carnot cycle (and only for a Carnot cycle) that

$$\frac{Q_h}{Q_c} = \frac{T_h}{T_c}$$

Therefore, $$W = \frac{T_h - T_c}{T_c} Q_c \quad \Diamond$$

(b) From Equation 18.5, $\text{COP} = \dfrac{Q_c}{W}$.

Using the result from part (a), this becomes $\text{COP} = \dfrac{T_c}{T_h - T_c}$ ◊

---

16. How much work is required, using an ideal Carnot refrigerator, to remove 1.00 J of thermal energy from helium gas at 4.00 K and reject this thermal energy to a room-temperature (293 K) environment?

Solution The coefficient of performance for an ideal Carnot refrigerator is:

$$\text{COP} = \frac{T_c}{\Delta T} = \frac{4.00}{289} = 0.0138$$

$$W = \frac{Q_c}{\text{COP}} = \frac{1.00\text{ J}}{0.0138} = 72.3\text{ J} \quad \Diamond$$

---

21. Calculate the change in entropy of 250 g of water heated slowly from 20.0 °C to 80.0 °C. (**Hint:** Note that $dQ = mcdT$.)

Solution To do the heating reversibly, put the water pot successively into contact with reservoirs at temperatures 20.0° C + δ, 20.0° C + 2δ, . . . 80.0 °C, where δ is some small increment. Then

$$\Delta S = \int_{\text{reversible path}} \frac{dQ}{T} = \int_{T_i}^{T_f} mc\frac{dT}{T}$$

Here T means the absolute temperature. We would ordinarily think of dT as the change in the Celsius temperature, but one Celsius degree of temperature change is the same size as one kelvin of change, so dT is also the change in absolute T. Then

$$\Delta S = mc \ln T\Big|_{T_i}^{T_f} = mc \ln\left(\frac{T_f}{T_i}\right)$$

$$\Delta S = (0.250 \text{ kg})(4186 \text{ J/kg}\cdot\text{K}) \ln\left(\frac{353 \text{ K}}{293 \text{ K}}\right) = 195 \text{ J/K} \quad \lozenge$$

---

23. A 1500-kg car is moving at 20.0 m/s. The driver brakes to a stop. The brakes cool off to the temperature of the surrounding air, which is nearly constant at 20.0 °C. What is the total entropy change?

Solution The original kinetic energy of the car,

$$K = \frac{1}{2}mv^2 = \frac{1}{2}(1500 \text{ kg})(20.0 \text{ m/s})^2 = 300 \text{ kJ}$$

becomes irreversibly 300 kJ of extra internal energy in the brakes, the car, and its surroundings. Since their total heat capacity is so large, their temperature ends up very nearly still at 20.0° C. To carry them reversibly to this same final state, imagine putting 300 kJ from a heater into the car and its environment. Then

$$\Delta S = \int \frac{dQ}{T} = \frac{1}{T}\int dQ = \frac{Q}{T} = \frac{300 \text{ kJ}}{293 \text{ K}}$$

$$\Delta S = 1.02 \text{ kJ/K} \quad \lozenge$$

---

27. One mole of H_2 gas is contained in the left-hand side of the container shown in Figure P18.27, which has equal volumes left and right. The right-hand side is evacuated. When the valve is opened, the gas streams into the right side. What is the final entropy change? Does the temperature of the gas change?

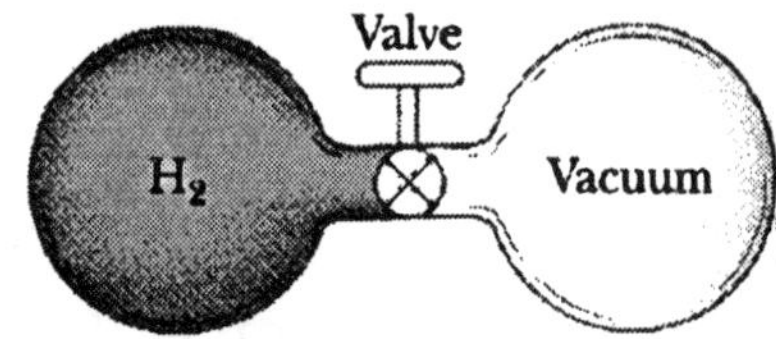

Figure P18.27

Solution

This is an example of free expansion; from Equation 18.9 we have

$$\Delta S = nR \ln\left(\frac{V_f}{V_i}\right) = (1.00 \text{ mole})\left(8.31 \frac{\text{J}}{\text{mole} \cdot \text{K}}\right) \ln\left(\frac{2}{1}\right)$$

$$\Delta S = 5.76 \text{ J/K} = 1.38 \text{ cal/K} \quad \lozenge$$

The gas is expanding into an evacuated region. Therefore, $W = 0$; and it expands so fast that heat has no time to flow; $Q = 0$. But $\Delta U = Q - W$, so in this case $\Delta U = 0$. For an ideal gas, the internal energy is a function of the temperature, so if $\Delta U = 0$, the temperature remains constant. ◊

---

31. Repeat the procedure used to construct Table 18.1 (a) for the case in which you draw three marbles from your bag rather than four and (b) for the case in which you draw five rather than four.

Solution:

The combinations are as follows:

(a)

Result	Possible combinations	Total
All Red	RRR	1
2R, 1G	RRG, RGR, GRR	3
1R, 2G	RGG, GRG, GGR	3
All Green	GGG	1

(b)

Result	Possible combinations	Total
All Red	RRRRR	1
4R, 1G	RRRRG, RRRGR, RRGRR, RGRRR, GRRRR	5
3R, 2G	RRRGG, RRGRG, RGRRG, GRRRG, RRGGR, RGRGR, GRRGR, RGGRR, GRGRR, GGRRR	10
2R, 3G	GGGRR, GGRGR, GRGGR, RGGGR, GGRRG, GRGRG, RGGRG, GRRGG, RGRGG, RRGGG	10
1R, 4G	GGGGR, GGGRG, GGRGG, GRGGG, RGGGG	5
All Green	GGGGG	1

---

37. A house loses heat through the exterior walls and roof at a rate of 5000 J/s = 5.00 kW when the interior temperature is 22.0 °C and the outside temperature is –5.00 °C. Calculate the electric power required to maintain the interior temperature at 22.0 °C for the following two cases. (a) The electric power is used in electric resistance heaters (which convert all of the electricity supplied into heat). (b) The electric power is used to drive an electric motor that operates the compressor of a heat pump (which has a coefficient of performance equal to 60.0% of the Carnot cycle value).

Solution

(a) $P_{\text{electric}} = \dfrac{\Delta E}{\Delta t}$ and if all of the electricity is converted into thermal energy, $\Delta E = \Delta Q$.

Therefore, $$P_{\text{electric}} = \frac{\Delta Q}{\Delta t} = 5000 \text{ W} \quad \Diamond$$

(b) For a heat pump, $$(\text{COP})_{\text{Carnot}} = \frac{T_h}{\Delta T} = \frac{295}{27} = 10.93$$

$$\text{Actual COP} = (0.600)(10.93) = 6.56 = \frac{Q_h}{W} = \frac{Q_h / t}{W / t}$$

Therefore, to bring 5000 W of heat into the house only requires input power

$$\frac{W}{t} = \frac{Q_h / t}{\text{COP}} = \frac{5000 \text{ W}}{6.56} = 763 \text{ W} \quad \Diamond$$

---

45. Figure P18.45 represents n mol of an ideal monatomic gas being taken through a cycle that consists of two isothermal processes at temperatures $3T_o$ and T_o and two constant-volume processes. For each cycle, determine, in terms of n, R, and T_o, (a) the net heat transferred to the gas and (b) the efficiency of an engine operating in this cycle.

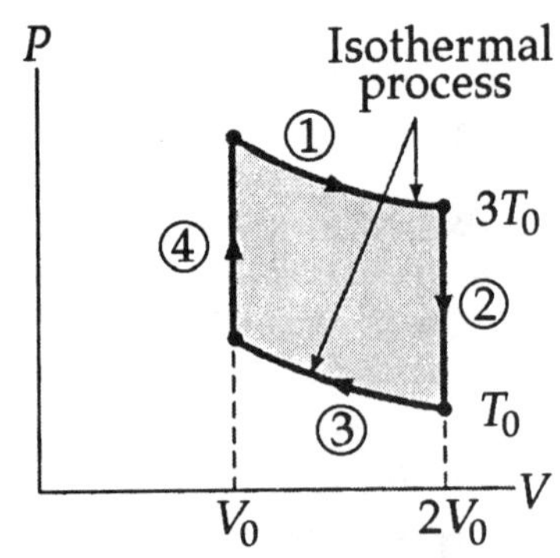

Figure P18.45

Solution

(a) For an isothermal process,

$$Q = nRT \ln\left(\frac{V_f}{V_i}\right)$$

Therefore, $Q_1 = nR(3T_o)\ln 2$ and $Q_3 = nR(T_o)\ln\left(\frac{1}{2}\right)$

The internal energy of a monatomic ideal gas is, from equation 16.17, $U = \frac{3}{2}nRT$. In the constant-volume processes,

$$Q_2 = \Delta U_2 = \tfrac{3}{2}nR(T_o - 3T_o) \qquad \text{and} \qquad Q_4 = \Delta U_4 = \tfrac{3}{2}nR(3T_o - T_o)$$

The net heat transferred is then

$$Q = Q_1 + Q_2 + Q_3 + Q_4 \qquad \text{or} \qquad Q = 2nRT_o \ln 2 \quad \Diamond$$

(b) Heat > 0 is the heat added to the system. Therefore,

$$Q_h = Q_1 + Q_4 = 3nRT_o(1 + \ln 2)$$

Since the change in temperature for the complete cycle is zero, $\Delta U = 0$ and $W = Q$.

Therefore, the efficiency is

$$e = \frac{W}{Q_h} = \frac{Q}{Q_h} = \frac{2\ln 2}{3(1+\ln 2)} = 0.273 \qquad \text{or} \qquad 27.3\% \quad \Diamond$$

Chapter 19

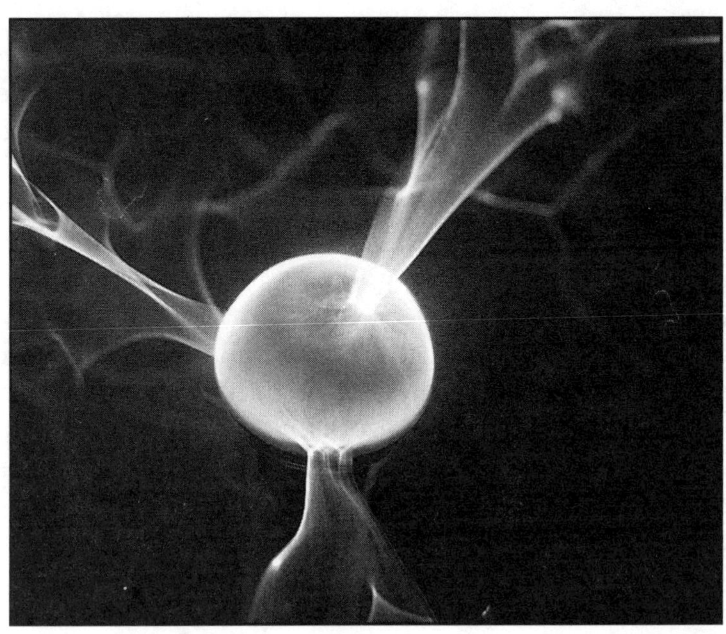

This metallic sphere is charged by a generator to a high voltage. The high concentration of charge formed on the sphere creates a strong electric field around the sphere. The charges then leak through the surrounding gas, producing the pink glow. Consider what would happen if a needle was charged in the same manner: would it also glow? (E.R. Degginger, H. Armstrong Roberts)

ELECTRIC FORCES AND ELECTRIC FIELDS

Chapter 19

ELECTRIC FORCES AND ELECTRIC FIELDS

INTRODUCTION

The electromagnetic force between charged particles is one of the fundamental forces of nature. In this chapter, we begin by describing some of the basic properties of electric forces. We then discuss Coulomb's law, which is the fundamental law of force between any two stationary charged particles. The concept of an electric field associated with a charge distribution is then introduced, and its effect on other charged particles is described. The method of using Coulomb's law to calculate electric fields of a given charge distribution is discussed, and several examples are given.

This chapter also describes an alternative procedure for calculating electric fields known as **Gauss's law**. This formulation is based on the fact that the fundamental electrostatic force between point charges is an inverse-square law. Although Gauss's law is a consequence of Coulomb's law, Gauss's law is much more convenient for calculating the electric field of highly symmetric charge distributions. Furthermore, Gauss's law serves as a guide for understanding more complicated problems.

NOTES FROM SELECTED CHAPTER SECTIONS

19.2 Properties of Electric Charges

Electric charge has the following important properties:

- There are two kinds of charges in nature, positive and negative, with the property that unlike charges attract one another and like charges repel one another.
- Charge is conserved.
- Charge is quantized.

19.3 Insulators and Conductors

Conductors are materials in which electric charges move freely under the influence of an electric field; **insulators** are materials that do not readily transport charge.

Chapter 19

19.4 Coulomb's Law

Experiments show that an **electric force** has the following properties:

- Two charged objects exert on each other equal-size forces in opposite directions.
- The force is inversely proportional to the square of the separation, r, between the two particles and is along the line joining them.
- It is proportional to the product of the magnitudes of the charges, $|q_1|$ and $|q_2|$, on the two particles.
- It is attractive if the charges are of opposite sign and repulsive if the charges have the same sign.

19.5 Electric Fields

An electric field exists at some point if a test charge at rest placed at that point experiences an electrical force.

The electric field vector **E** at some point in space is defined as the electric force **F** acting on a positive test charge placed at that point divided by the magnitude of the test charge q_o.

At any point the total electric field created by a group of discrete point charges equals the **vector sum** of the electric fields due to each of the charges individually.

19.6 Electric Field Lines

A convenient aid for visualizing electric field patterns is to draw lines pointing in the same direction as the electric field vector at any point. These lines, called electric field lines, are related to the electric field in any region of space in the following manner:

- The electric field vector **E** is **tangent** to the electric field line at each point.
- The number of lines per unit area through a surface perpendicular to the lines is proportional to the strength of the electric field in that region. Thus, **E** is large when the field lines are close together and small when they are far apart.

The rules for drawing electric field lines for any charge distribution are as follows:

- The lines must begin on positive charges and terminate on negative charges, or at infinity in the case of an excess of charge.
- The number of lines drawn leaving a positive charge or approaching a negative charge is proportional to the magnitude of the charge.
- No two field lines can cross.

19.8 Gauss's Law

Gauss's law states that the net electric flux through a closed gaussian surface is equal to the net charge inside the surface divided by ϵ_0.

The gaussian surface should be chosen so that it has the same symmetry as the charge distribution.

19.10 Conductors in Electrostatic Equilibrium

A conductor in electrostatic equilibrium has the following properties:

- The electric field is zero everywhere inside the conductor.
- Any excess charge on an isolated conductor resides entirely on its surface.
- The electric field just outside a charged conductor is perpendicular to the conductor's surface and has a magnitude σ/ϵ_0, where σ is the charge per unit area at that point.
- On an irregularly shaped conductor, charge tends to accumulate at locations where the radius of curvature of the surface is the smallest, that is, at sharp points.

EQUATIONS AND CONCEPTS

The **magnitude** of the **electrostatic force** between two stationary point charges, q_1 and q_2, separated by a distance r is given by Coulomb's law.

$$F = k_e \frac{|q_1||q_2|}{r^2} \qquad (19.1)$$

where $k_e = \frac{1}{4\pi\epsilon_o}$

$$\epsilon_o = 8.8542 \times 10^{-12}\ \mathrm{C^2/N \cdot m^2}$$

In calculations, an approximate value for k_e may be used.

$$k_e = 8.99 \times 10^9 \frac{\mathrm{N \cdot m^2}}{\mathrm{C^2}}$$

The direction of the electrostatic force on each charge is determined from the experimental observation that like sign charges experience forces of mutual repulsion and unlike sign charges attract each other. By virtue of Newton's third law, the magnitude of the force on each of the two charges is the same regardless of the relative magnitude of the values of q_1 and q_2.

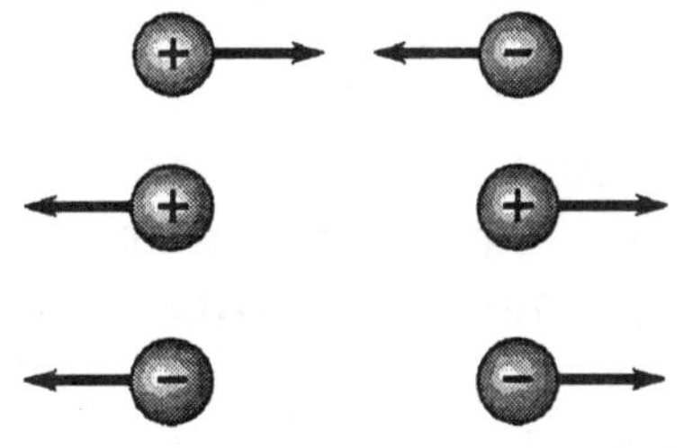

The electric force between two charges can be expressed in vector form. $\mathbf{F}_{21}$ is the force **on** q_2 **due to** q_1 and $\hat{\mathbf{r}}$ is a unit vector directed from q_1 to q_2. Coulomb's law applies exactly to point charges or particles.

$$\mathbf{F}_{21} = k_e \frac{q_1 q_2}{r^2}\hat{\mathbf{r}} \qquad (19.2)$$

In cases where there are more than two charges present, the resultant force on any one charge is the vector sum of the forces exerted on that charge by the remaining individual charges present. The principle of superposition applies.

$$\mathbf{F}_i = \sum_{\substack{i,j=1 \\ j \neq i}}^{N} \mathbf{F}_{ij}$$

The **electric field** at any point in space is defined as the ratio of electric force per charge exerted on a small positive test charge placed at the point where the field is to be determined.

$$\mathbf{E} \equiv \frac{\mathbf{F}}{q_o} \qquad (19.3)$$

The above definition (Equation 19.3) together with Coulomb's law leads to an expression for calculating the **electric field a distance** r **from a point charge,** q. In this case the unit vector $\hat{\mathbf{r}}$ is directed away from q and toward the point P where the field is to be calculated. The direction of the electric field is radially outward from a positive point charge and radially inward toward a negative point charge.

$$\mathbf{E} = k_e \frac{q}{r^2} \hat{\mathbf{r}} \qquad (19.4)$$

The superposition principle holds when the electric field at a point is due to a number of point charges.

$$\mathbf{E} = k_e \sum_i \frac{q_i}{r_i^2} \hat{\mathbf{r}}_i \qquad (19.5)$$

(vector sum)

Chapter 19

Electric field lines are a convenient graphical representation of electric field patterns. These lines are drawn so that the electric field vector, **E**, is tangent to the electric field lines at each point. Also, the number of lines per unit area through a surface perpendicular to the lines is proportional to the strength or magnitude of the electric field over the region. In every case, electric field lines must begin on positive charges and terminate on negative charges or at infinity; the number of lines leaving or approaching a charge is proportional to the magnitude of the charge; and no two field lines can cross.

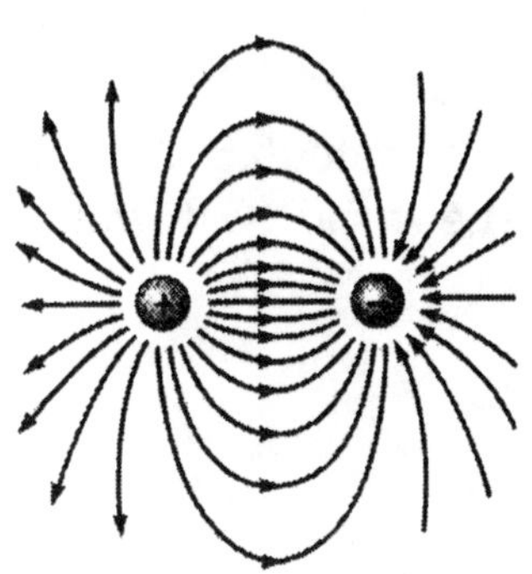

When the electric field is due to a **continuous charge distribution**, the contribution to the field by each element of charge must be integrated over the total line, surface, or volume which contains the charge.

$$\mathbf{E} = k_e \int_{\substack{\text{All}\\\text{charge}}} \frac{dq}{r^2}\hat{\mathbf{r}} \qquad (19.6)$$

In order to perform the integration described above, it is convenient to represent a charge increment dq as the product of an element of length, area, or volume and the **charge density over that region. Note**: For those cases in which the charge is not uniformly distributed, the densities λ, σ, and ρ must be stated as functions of position.

For an element of length dx

$$dq = \lambda\, dx$$

For an element of area dA

$$dq = \sigma\, dA$$

For an element of volume dV

$$dq = \rho\, dV$$

For uniform charge distributions the volume charge density (ρ), the surface charge density (σ), and the linear charge density (λ) can be calculated from the total charge and the total volume, area, or length.

$$\rho \equiv \frac{Q}{V} \qquad (19.7)$$

$$\sigma \equiv \frac{Q}{A} \qquad (19.8)$$

$$\lambda \equiv \frac{Q}{\ell} \qquad (19.9)$$

The **electric flux** is a measure of the number of electric field lines that penetrate some surface. For a **plane surface** in a **uniform field**, the flux depends on the angle between the normal to the surface and the direction of the field.

$$\Phi = EA\cos\theta \qquad (19.11)$$

In the case of a **surface of arbitrary shape** in the region of a **nonuniform** field, the flux is calculated by integrating the normal component of the field over the surface in question.

$$\Phi = \int_{\text{surface}} \mathbf{E}\cdot d\mathbf{A} \qquad (19.12)$$

Gauss's law states that when Φ (Eq. 19.12) is evaluated over a **closed** surface (gaussian surface), the result equals **the net charge enclosed by the surface** divided by the constant ϵ_0. The symbol $\oint$ in this equation indicates that the integral must be evaluated over a **closed** surface.

$$\Phi_c = \oint_{\substack{\text{Closed}\\ \text{Surface}}} \mathbf{E}\cdot d\mathbf{A} = \frac{q_{in}}{\epsilon_0} \qquad (19.15)$$

The electric field just outside the surface of a charged conductor in equilibrium can be expressed in terms of the surface charge density on the conductor. Just outside the conductor, the field is normal to the surface.

$$E_n = \frac{\sigma}{\epsilon_0} \qquad (19.18)$$

SUGGESTIONS, SKILLS, AND STRATEGIES

Problem-Solving Hints for Electric Forces and Fields:

- Units: When performing calculations that involve the use of the Coulomb constant k_e that appears in Coulomb's law, charges must be in coulombs and distances in meters. If they are given in other units, you must convert them to SI.

- Applying Coulomb's law to point charges: It is important to use the superposition principle properly when dealing with a collection of interacting point charges. When several charges are present, the resultant force on any one of them is found by finding the individual force that every other charge exerts on it and then finding the vector sum of all these forces. The magnitude of the force that any charged object exerts on another is given by Coulomb's law, and the direction of the force is found by noting that the forces are repulsive between like charges and attractive between unlike charges.

- Calculating the electric field of point charges: Remember that the superposition principle can also be applied to electric fields, which are also vector quantities. To find the total electric field at a given point, first calculate the electric field at the point due to each individual charge. The resultant field at the point is the vector sum of the fields due to the individual charges.

- To evaluate the electric field of a continuous charge distribution, it is convenient to employ the concept of charge density. Charge density can be written in different ways: charge per unit volume, ρ; charge per unit area, σ; or charge per unit length, λ. The total charge distribution is then subdivided into a small element of volume dV, area dA, or length dx. Each element contains an increment of charge dq (equal to ρdV, σdA, or λdx). If the charge is **nonuniformly** distributed over the region, then the charge densities must be written as functions of position. For example, if the charge density along a line or long bar is proportional to the distance from one end of the bar, then the linear charge density could be written as $\lambda = bx$ and the charge increment dq becomes $dq = bx\, dx$.

- Symmetry: Whenever dealing with either a distribution of point charges or a continuous charge distribution, take advantage of any symmetry in the system to simplify your calculations.

Problem-Solving Hints for Gauss's Law

Gauss's law is a very powerful theorem which relates any charge distribution to the resulting electric field at any point in the vicinity of the charge. In this chapter you should learn how to apply Gauss's law to those cases in which the charge distribution has a sufficiently high degree of symmetry. As you review Examples 19.7 through 19.10 of the text, observe how each of the following steps have been included in the application of the equation $\oint \mathbf{E} \cdot d\mathbf{A} = \frac{q}{\epsilon_0}$ to that particular situation.

- The gaussian surface should be chosen to have the **same symmetry as the charge distribution.**

- The dimensions of the surface must be such that the surface includes the point where the electric field is to be calculated.

- From the symmetry of the charge distribution, you should be able to correctly describe the direction of the electric field vector, **E**, relative to the direction of an element of surface area vector, $d\mathbf{A}$, over each region of the gaussian surface.

- From the symmetry of the charge distribution, you should also be able to identify one or more portions of the closed surface (and in some cases the entire surface) over which the magnitude of **E** remains constant.

- Write $\mathbf{E} \cdot d\mathbf{A}$ as $E\, dA \cos\theta$, and divide the surface into separate regions such that over each region:

  $E(dA)(\cos\theta)$ will equal 0 when $\mathbf{E} \perp d\mathbf{A}$ or when $\mathbf{E} = 0$. An example of the first case may be seen at the top of the accompanying figure; the second case will typically be true of the area inside the surface of a conductor.

  $E(dA)(\cos\theta)$ will equal $E\, dA$ when $\mathbf{E} \parallel d\mathbf{A}$. An example of this is shown in the figure to the right, over the curved portion of the cylindrical gaussian surface.

  $E(dA)(\cos\theta) = -E(dA)$ when **E** and $d\mathbf{A}$ are oppositely directed.

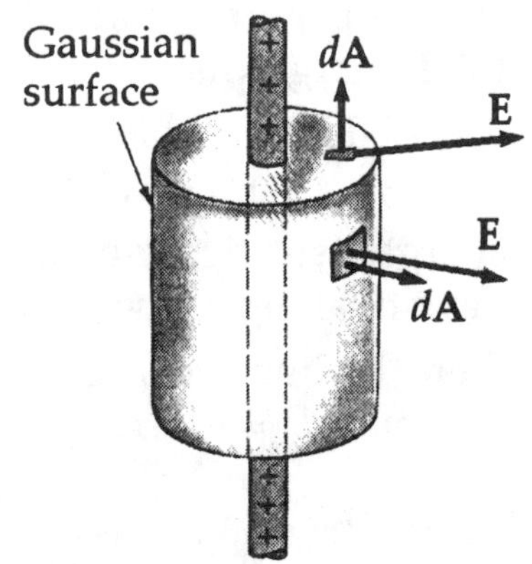

- If the gaussian surface has been chosen and subdivided so that the magnitude of **E** is **constant** over those regions where $\mathbf{E}\cdot d\mathbf{A} = E\,dA$, then over each of those regions

$$\int \mathbf{E}\cdot d\mathbf{A} = E\int dA = E\,(\text{area of region})$$

- The total charge enclosed by the gaussian surface is that portion of the charge **inside** the gaussian surface.

- Once the left and right sides of Gauss's law have been evaluated, you can calculate the electric field on the gaussian surface, assuming the charge distribution is given in the problem. Conversely, if the electric field is known, you can calculate the charge distribution that produces the field.

REVIEW CHECKLIST

▷ Describe the fundamental properties of electric charge and the nature of electrostatic forces between charged bodies.

▷ Use Coulomb's law to determine the net electrostatic force on a point electric charge due to a known distribution of a finite number of point charges.

▷ Calculate the electric field **E** (magnitude and direction) at a specified location in the vicinity of a group of point charges.

▷ Calculate the electric field due to a continuous charge distribution. The charge may be distributed uniformly or nonuniformly along a line, over a surface, or throughout a volume.

▷ Calculate the **electric** flux through a surface; in particular, find the net electric flux through a **closed** surface.

▷ Understand that a gaussian surface must be a real or imaginary **closed** surface within a conductor, an insulator, or in space. Also remember that the net electric flux through a closed gaussian surface is equal to the net charge enclosed by the surface divided by the constant ϵ_0.

▷ Use Gauss's law to evaluate the electric field at points in the vicinity of charge distributions which exhibit spherical, cylindrical, or planar symmetry.

ANSWERS TO SELECTED CONCEPTUAL QUESTIONS

2. A balloon is negatively charged by rubbing and then clings to a wall. Does this mean that the wall is positively charged? Why does the balloon eventually fall?

Answer No. The balloon induces charge of opposite sign in the wall, causing it to be attracted. The balloon eventually falls since its charge slowly diminishes as it leaks to ground. Some of the charge could also be lost due to ions of opposite sign in the surrounding atmosphere which would tend to neutralize the charge.

□ □ □ □

3. A charged comb often attracts small bits of dry paper that then fly away when they touch the comb. Explain.

Answer When the comb is nearby, charges separate on the paper, resulting in the paper being attracted. After contact, charges from the comb are transferred to the paper so that it has the same type charge as the comb. It is thus repelled.

□ □ □ □

5. Would life be different if the electron were positively charged and the proton were negatively charged? Does the choice of signs have any bearing on physical and chemical interactions? Explain.

Answer No, life would not be different. The character and effect of electric forces is defined by (1) the fact that there are only two types of electric charge—positive and negative, and (2) the fact that opposite charges attract, while like charges repel. The choice of signs is completely arbitrary.

As a related exercise, you might consider what would happen in a world where there were three types of electric charge, or in a world where opposite charges repelled, and like charges attracted.

□ □ □ □

12. A common demonstration involves charging a balloon, which is an insulator, by rubbing it on your head, and touching the balloon to a ceiling or wall, which is also an insulator. The electrical attraction between the charged balloon and the neutral wall results in the balloon sticking to the wall. Imagine now that we have two infinitely large sheets of insulating material. One is charged and the other neutral. If these are brought into contact, will there be an attractive force between them, as there was for the balloon and the wall?

Answer There will **not** be an attractive force. There are two factors to consider in the attractive force between a balloon and a wall, or between any pair of charged and neutral objects. The first factor is that the molecules in the wall will orient themselves with their negative ends toward the balloon, and their positive ends pointing away from the balloon. The second factor to consider is that the balloon is of finite, curved dimensions, and thus the field is not uniform. Therefore, the negative charges in the wall are in a stronger field (due to the positive charges on the balloon) than are the positive charges. As a result, the attractive force on the negative charges is stronger than the repulsive force on the positive charges, and the result is an overall attractive force.

Now, consider the infinite sheets brought into contact. The polarization of the molecules in the neutral sheet will indeed occur, as in the wall. But the electric field from the charged sheet is **uniform**, and therefore is independent of the distance from the sheet. Thus, both the negative and positive charges in the neutral sheet will experience the same electric field and the same magnitude of electric force. The attractive force on the negative charges will cancel with the repulsive force on the positive charges, and there will be no net force.

□ □ □ □

13. If the total charge inside a closed surface is known but the distribution of the charge is unspecified, can you use Gauss's law to find the electric field? Explain.

Answer **No.** If we wish to use Gauss's law to find the electric field, we must be able to bring the electric field out of the integral. This **can** be done, in some cases—when the field is constant, for example. However, since we do not know the charge distribution, we cannot claim that the field is constant, and thus cannot find the electric field.

To illustrate this point, consider a sphere that contained a net charge of 100 μC. The charges could be located near the center, or they could all be grouped at the northernmost point within the sphere. In either case, the net electric flux would be the same, but the electric field would vary greatly.

□ □ □ □

15. A person is placed in a large, hollow, metallic sphere that is insulated from ground. If a large charge is placed on the sphere, will the person be harmed on touching the inside of the sphere? Explain what will happen if the person also has an initial charge the sign of which is opposite that of the charge on the sphere.

Answer The metallic sphere is a good conductor, so any excess charge on the sphere will reside on the outside of the sphere. From Gauss's Law, we know that the field **inside** the sphere will then be zero. As a result, when the person touches the inside of the sphere, no charge will be exchanged between the person and the sphere, and the person will not be harmed.

What happens, then, if the person has an initial charge? Regardless of the sign of the person's initial charge, the charges in the conducting surface will redistribute themselves to maintain a net zero charge within the **conducting metal**. Thus, if the person has a 5.00 μC charge on his skin, exactly –5.00 μC will gather on the inner surface of the sphere, so that the electric field inside the metal will be zero. When the person touches the metallic sphere then, he will receive a shock due to the charge on his own skin.

□ □ □ □

16. Imagine two electric dipoles in empty space. Each dipole has zero net charge. Is there an electric force between the dipoles—that is, can two objects with **zero net charge** exert electric forces on each other? If so, is the force attractive or repulsive?

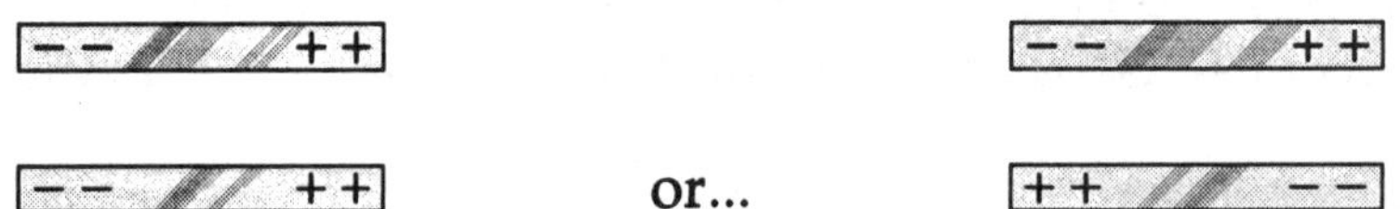

Answer At first thought, it might seem that neutral objects could not exert electric forces. But there is a charge structure within the dipoles which makes things a little more complicated. Imagine that the dipoles are oriented as in the diagram to the left. In such a case, there is a repulsive force between the left-hand side of the dipoles and another repulsive force between the positive charges at the right. There are also attractive forces between the top of each dipole and the unlike charge at the bottom of the other. The separation distance between these charges is larger, however, than that between the like charges at top and bottom. As a result, the attractive force is smaller than the repulsive force, giving a net repulsive force between the dipoles.

If the dipoles are oriented as in the diagram to the right, then the attractive and repulsive forces in the above discussion are reversed, and there is a net attractive force between the dipoles.

□ □ □ □

SOLUTIONS TO SELECTED END-OF-CHAPTER PROBLEMS

3. Richard Feynman once said that if two persons stood at arm's length from each other and each person had 1% more electrons than protons, the force of repulsion between them would be enough to lift a "weight" equal to that of the entire Earth. Carry out an order-of-magnitude calculation to substantiate this assertion.

Solution

Suppose each person has mass 70 kg. In terms of elementary charges, each consists of precisely equal numbers of protons and electrons and a nearly equal number of neutrons. The electrons comprise very little of the mass, so we find the number of protons-and-neutrons in each person:

$$70 \text{ kg}\left(\frac{1 \text{ u}}{1.66\times10^{-27} \text{ kg}}\right)=4\times10^{28} \text{ u}$$

Of these, nearly one half, 2×10^{28}, are protons, and 1% of this is 2×10^{26}, constituting a charge of $(2 \times 10^{26})(1.60 \times 10^{-19} \text{ C}) = 3 \times 10^{7} \text{ C}$. Thus, Feynman's force is

$$F=\frac{k_e q_1 q_2}{r^2}=\frac{\left(8.99\times10^{9} \text{ N}\cdot\text{m}^2/\text{C}^2\right)\left(3\times10^{7} \text{ C}\right)^2}{(0.5 \text{ m})^2}\approx10^{26} \text{ N}$$

where we have used a half-meter arm's length.

The mass of the Earth in a gravitational field of magnitude 9.80 m/s^2 would weigh

$$w=mg=(6\times10^{24} \text{ kg})(10 \text{ m}/\text{s}^2)\approx10^{26} \text{ N}$$

Thus, the forces are of the same order of magnitude. ◊

7. The electrons in a particle beam each have a kinetic energy K. What are the magnitude and direction of the electric field that will stop these electrons in a distance d?

Solution In order to stop the negatively charged electron, the field must be oriented in the direction of the electron's motion, because the direction of the field is defined to be the direction of the force on a positive charge, and is opposite the direction of the force on a negative charge. ◊

The electron is moving with a kinetic energy of K, so work equal to K must be performed. The magnitude of the electric force is: $F = eE$

This force acts over a distance d. Therefore, the work is: $W = K = eEd$

and the electric field must be: $E = \frac{K}{ed}$ ◊

Related Calculation: Perform the above calculation, with $K = 1.60 \times 10^{-17}$ J, and $d = 10.0$ cm.

$$E = \frac{K}{ed} = \frac{1.60 \times 10^{-17}\ \text{J}}{(1.60 \times 10^{-19}\ \text{C})(0.100\ \text{m})} = 1.00\ \text{kN/C} \quad ◊$$

---

11. A continuous line of charge lies along the x axis, extending from $x = +x_0$ to positive infinity. The line carries a uniform linear charge density λ_0. What are the magnitude and direction of the electric field at the origin?

Solution A segment of the line between x and $x + dx$ has charge $\lambda_0\, dx$ and creates an electric field at the origin of

$$d\mathbf{E} = \left(k_e \frac{dq}{r^2}\right)\hat{\mathbf{r}} = \left(k_e \lambda_0 \frac{dx}{x^2}\right)(-\mathbf{i})$$

The total field at the origin is

$$\mathbf{E} = \int_{\text{All charge}} d\mathbf{E} = \int_{x_0}^{\infty} k_e \lambda_0 (-\mathbf{i}) x^{-2}\, dx$$

$$\mathbf{E} = (-k_e \lambda_0 \mathbf{i}) \frac{x^{-1}}{-1}\Bigg]_{x_0}^{\infty} = (k_e \lambda_0 \mathbf{i})\left(\frac{1}{\infty} - \frac{1}{x_0}\right)$$

$$\mathbf{E} = \left(k_e \frac{\lambda_0}{x_0}\right)(-\mathbf{i}) \quad ◊$$

---

13. Three equal positive charges q are at the corners of an equilateral triangle of sides a as in Figure P19.13. (a) At what point in the plane of the charges (other than infinity) is the electric field zero? (b) What are the magnitude and direction of the electric field at P due to the two charges at the base?

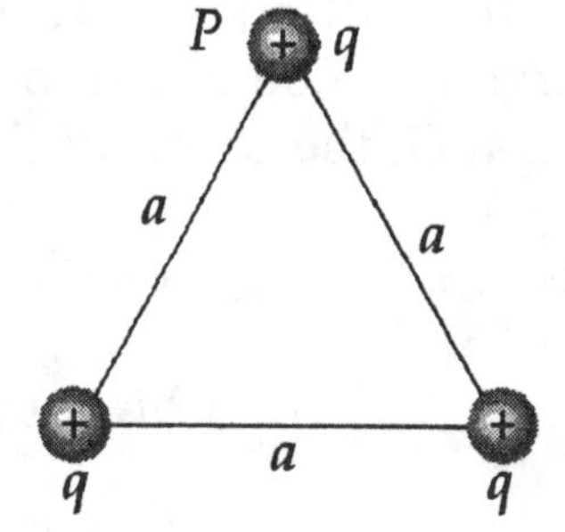

Figure P19.13

Solution

(a) The electric field has the general appearance shown. It is zero at the center, where (by symmetry) one can see that the three charges individually produce fields that cancel out.

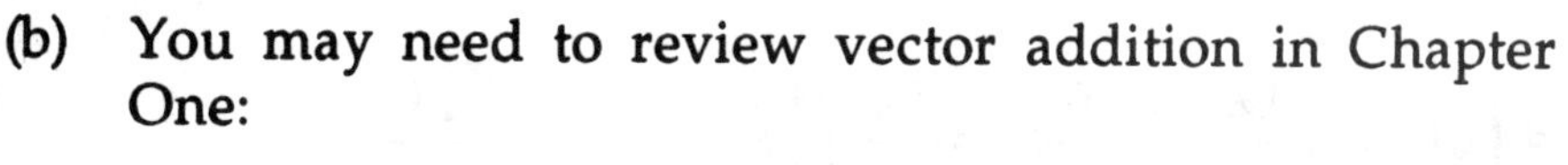

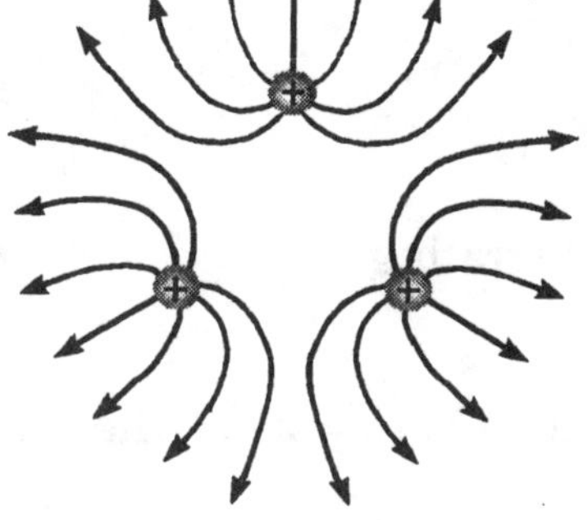

(b) You may need to review vector addition in Chapter One:

$$\mathbf{E} = k_e \sum_i \frac{q_i}{r_i^2} \hat{\mathbf{r}}_i$$

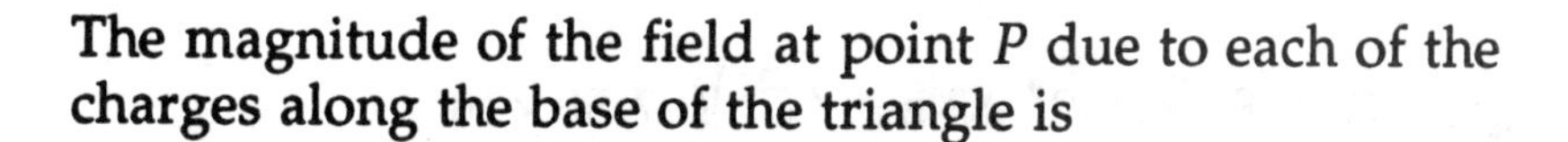

The magnitude of the field at point P due to each of the charges along the base of the triangle is

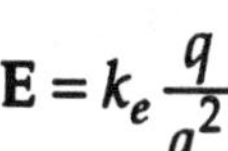

$$\mathbf{E} = k_e \frac{q}{a^2}$$

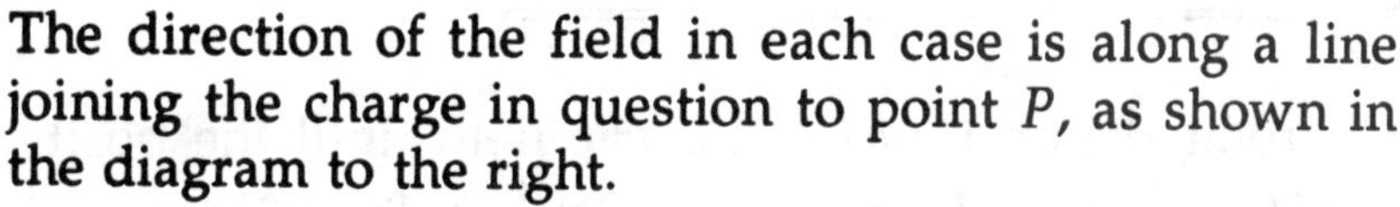

The direction of the field in each case is along a line joining the charge in question to point P, as shown in the diagram to the right.

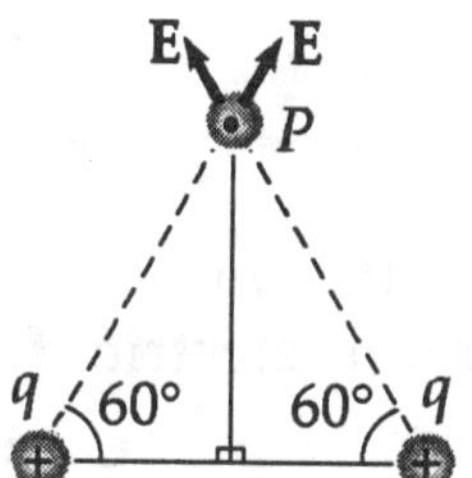

The x components add to zero, leaving

$$\mathbf{E} = \frac{k_e q}{a^2} \sin 60° \ \mathbf{j} + \frac{k_e q}{a^2} \sin 60° \ \mathbf{j} = \sqrt{3} \frac{k_e q}{a^2} \ \mathbf{j} \quad \Diamond$$

17. A uniformly charged insulating rod of length 14.0 cm is bent into the shape of a semicircle as in Figure P19.17. If the rod has a total charge of –7.50 μC, find the magnitude and direction of the electric field at O, the center of the semicircle.

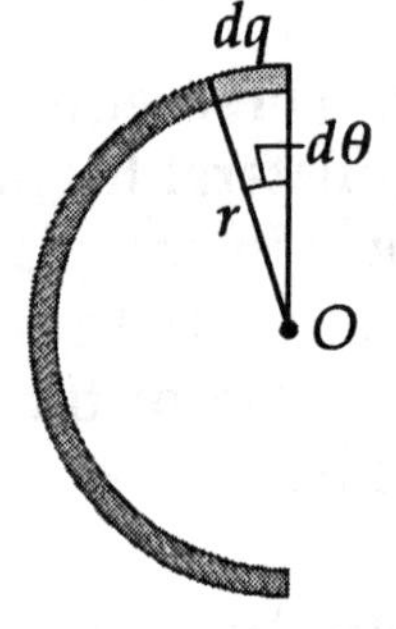

Figure P19.17 (modified)

Solution Let λ be the charge per unit length.

Then, $dq = \lambda ds = \lambda r d\theta$

and $dE = \frac{kdq}{r^2}$

In component form, $E_y = 0$ (from symmetry)

$dE_x = dE \cos\theta$

Integrating, $$E_x = \int dE_x = \int \frac{k\lambda r \cos\theta}{r^2} d\theta = \frac{k\lambda}{r}\int_{-\pi/2}^{\pi/2} \cos\theta \, d\theta = \frac{2k\lambda}{r}$$

But $Q_{\text{total}} = \lambda \ell$. where $\ell = 0.140$ m, and $r = \ell/\pi$.

Thus, $$E_x = \frac{2\pi kQ}{\ell^2} = \frac{(2\pi)(8.99\times10^9 \text{ N}\cdot\text{m}^2/\text{C}^2)(-7.50\times10^{-6}\text{ C})}{(0.140 \text{ m})^2}$$

$$\mathbf{E} = (-2.16\times10^7 \text{ N/C})\mathbf{i} \quad \lozenge$$

---

19. A 40.0-cm-diameter loop is rotated in a uniform electric field until the position of maximum electric flux is found. The flux in this position is measured to be $5.20\times10^5 \text{ N}\cdot\text{m}^2/\text{C}$. What is the electric field strength?

Solution We calculate the flux as $\Phi = \mathbf{E}\cdot\mathbf{A} = EA\cos\theta$

The maximum value of the flux occurs when $\theta = 0$.

Therefore, we can calculate the field strength at this point as $E = \frac{\Phi_{\text{max}}}{A} = \frac{\Phi_{\text{max}}}{\pi r^2}$

$$E = \frac{5.20\times10^5}{\pi(0.200 \text{ m})^2} = 4.14\times10^6 \text{ N/C} \quad \lozenge$$

---

23. A point charge Q is located just above the center of the flat face of a hemisphere of radius R as in Figure P19.23. What is the electric flux (a) through the curved surface and (b) through the flat face?

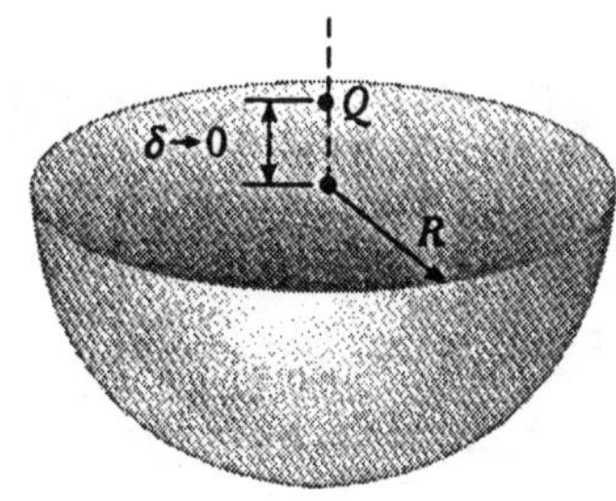

Figure P19.23

Solution

With δ very small, all points on the hemisphere are nearly at distance R from the charge, so the field everywhere on the curved surface is k_eQ/R^2 radially outward.

(a) The flux over the curved surface is this field strength times the area of half a sphere:

$$\Phi_{\text{curved}} = \left(k_e \frac{Q}{R^2}\right)\left(\tfrac{1}{2}\right)\left(4\pi R^2\right) = \frac{1}{4\pi\epsilon_0} Q(2\pi) = \frac{Q}{2\epsilon_0} \quad \Diamond$$

(b) The closed surface encloses zero charge, so Gauss's law gives:

$$\Phi_{\text{curved}} + \Phi_{\text{flat}} = 0$$

$$\Phi_{\text{flat}} = -\Phi_{\text{curved}} = \frac{-Q}{2\epsilon_0} \quad \Diamond$$

---

29. Consider a long cylindrical charge distribution of radius R with a uniform charge density ρ. Find the electric field at distance r from the axis where $r < R$.

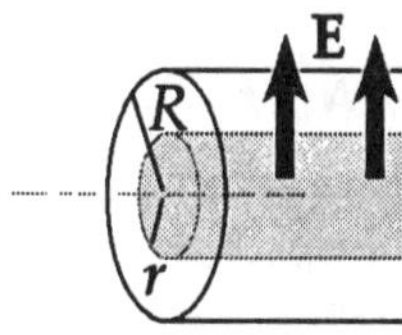

Solution If ρ is positive, the field must everywhere be radially outward. Choose as the gaussian surface a cylinder of length L and radius r, contained inside the charged rod. Its volume is $\pi r^2 L$ and it encloses charge $\rho\pi r^2 L$. The circular end caps have no electric flux through them; there $\mathbf{E}\cdot d\mathbf{A} = 0$. The curved surface has $\mathbf{E}\cdot d\mathbf{A} = E\,dA \cos 0°$, and E must be the same strength everywhere over the curved surface.

Then $\oint \mathbf{E}\cdot d\mathbf{A} = \frac{q}{\epsilon_0}$ becomes $E \int_{\text{Curved Surface}} dA = \frac{\rho\pi r^2 L}{\epsilon_0}$

Noting that $2\pi rL$ is the lateral surface area of the cylinder, $E(2\pi r L) = \frac{\rho\pi r^2 L}{\epsilon_0}$

Thus, $\mathbf{E} = \frac{\rho r}{2\epsilon_0}$ radially away from the axis ◊

---

31. A long, straight metal rod has a radius of 5.00 cm and a charge per unit length of 30.0 nC/m. Find the electric field (a) 3.00 cm, (b) 10.0 cm, and (c) 100 cm from the axis of the rod, where distances are measured perpendicular to the rod.

Solution

(a) Inside the conductor, $E = 0$ ◊

(b) Outside the conductor, $E = 2\frac{k_e \lambda}{r}$ (Eq. 19.16)

(See Example 19.9, p. 546)

At $r = 0.10$ m,

$$E = 2\frac{(8.99\times10^9\ \text{N}\cdot\text{m}^2/\text{C}^2)(30.0\times10^{-9}\ \text{C/m})}{0.100\ \text{m}}$$

giving $E = 5.39\times10^3\ \text{N/C}$ ◊

(c) At $r = 1.00$ m,

$$E = 2\frac{(8.99\times10^9\ \text{N}\cdot\text{m}^2/\text{C}^2)(30.0\times10^{-9}\ \text{C/m})}{1.00\ \text{m}}$$

$E = 539\ \text{N/C}$ ◊

---

41. A charged cork ball of mass 1.00 g is suspended on a light string in the presence of a uniform electric field, as in Figure P19.41. When **E** = (3.00**i** + 5.00**j**) × 10^5 N/C, the ball is in equilibrium at θ = 37.0°. Find (a) the charge on the ball and (b) the tension in the string.

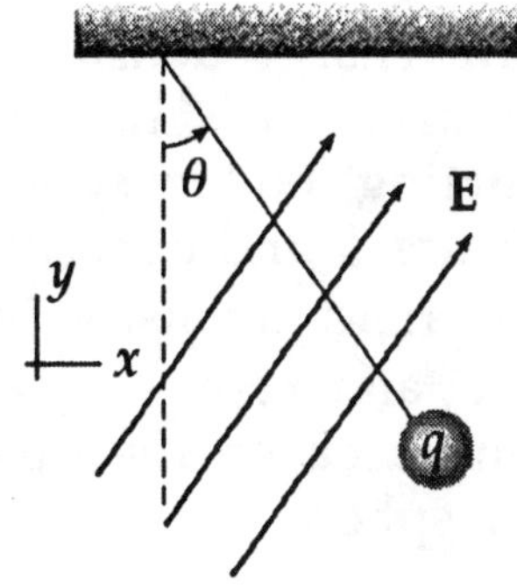

Figure P19.41

Solution

(a) $E_x = 3.00 \times 10^5\ \text{N/C}$ and $E_y = 5.00 \times 10^5\ \text{N/C}$

Applying Newton's Second Law in the x and y directions,

(1) $\Sigma F_x = qE_x - T \sin 37.0° = 0$

(2) $\Sigma F_y = qE_y + T \cos 37.0° - mg = 0$

Substituting T from Equation (1) into Equation (2):

$$q = \frac{mg}{\left(E_y + \dfrac{E_x}{\tan 37.0°}\right)}$$

Evaluating,

$$q = \frac{(1.00 \times 10^{-3}\ \text{kg})(9.80\ \text{m/s}^2)}{\left(5.00 + \dfrac{3.00}{\tan 37.0°}\right) \times 10^5\ \text{N/C}} = 1.09 \times 10^{-8}\ \text{C} \quad \Diamond$$

(b) From Equation (1),

$$T = \frac{qE_x}{\sin 37.0°} = 5.44 \times 10^{-3}\ \text{N} \quad \Diamond$$

43. A solid, **insulating** sphere of radius a has a uniform charge density of ρ and a total charge of Q. Concentric with this sphere is an **uncharged, conducting** hollow sphere the inner and outer radii of which are b and c, as in Figure P19.43. (a) Find the electric field intensity in the regions $r<a$, $a<r<b$, $b<r<c$, and $r>c$. (b) Determine the induced charge per unit area on the inner and outer surfaces of the hollow sphere.

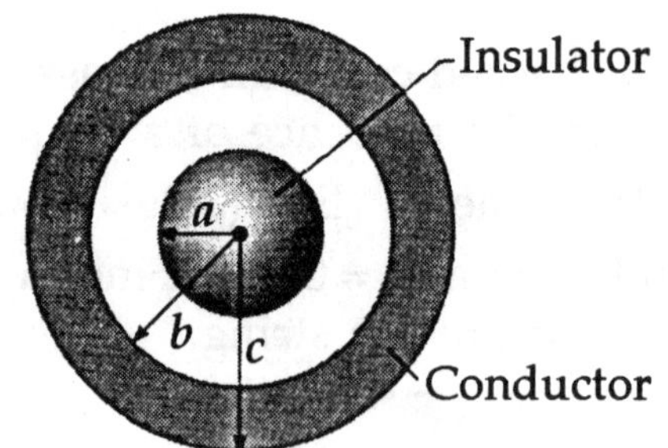

Figure P19.43

Solution

(a) Choose as the gaussian surface a concentric sphere of radius r. The electric field will be perpendicular to its surface, and will be uniform in strength over its surface.

The sphere of radius $r < a$ encloses charge $\rho \frac{4}{3} \pi r^3$

so $\Phi = \frac{q}{\epsilon_0}$ becomes $E \cdot 4\pi r^2 = \frac{\rho \frac{4}{3} \pi r^3}{\epsilon_0}$, and $E = \frac{\rho r}{3\epsilon_0}$ ◊

For $a < r < b$, we have $E(4\pi r^2) = \frac{4}{3}\rho \frac{\pi a^3}{\epsilon_0} = \frac{Q}{\epsilon_0}$ and $E = \frac{\rho a^3}{3\epsilon_0 r^2} = \frac{Q}{4\pi\epsilon_0 r^2}$ ◊

For $b < r < c$, we must have $E = 0$ ◊ because any nonzero field would be moving charges in the metal. Free charges did move in the metal to deposit charge $-Q_b$ on its inner surface, at radius b, leaving charge $+Q_c$ on its outer surface, at radius c. Since the shell as a whole is neutral, $Q_c - Q_b = 0$.

For $r > c$, $\Phi = \frac{q}{\epsilon_0}$ becomes $E(4\pi r^2) = \frac{Q + Q_c - Q_b}{\epsilon_0}$, and $E = \frac{Q}{4\pi\epsilon_0 r^2}$ ◊

(b) For a gaussian surface of radius $b < r < c$, we have $0 = \frac{Q - Q_b}{\epsilon_0}$

so $Q_b = Q$ and the charge density on the inner surface is $\frac{-Q_b}{A} = \frac{-Q}{4\pi b^2}$ ◊

Then $Q_c = Q_b = Q$, and the charge density on the outer surface is $+\frac{Q}{4\pi c^2}$ ◊

---

47. Repeat the calculations for Problem 46 when both sheets have **positive** uniform charge densities of value σ.

Solution The new, modified problem states: "Two infinite, nonconducting sheets of charge are parallel to each other as in Figure P19.46. Both sheets have a uniform surface charge density $+\sigma$. Calculate the value of the electric field at points (a) to the left of, (b) in between, and (c) to the right of the two sheets."

For each sheet, the magnitude of the field at any point is $|\mathbf{E}| = \dfrac{\sigma}{2\epsilon_0}$

(a) At point to the left of the two parallel sheets

$$\mathbf{E} = E_1(-\mathbf{i}) + E_2(-\mathbf{i}) = 2E(-\mathbf{i})$$

$$\mathbf{E} = -\frac{\sigma}{\epsilon_0}\mathbf{i} \quad \Diamond$$

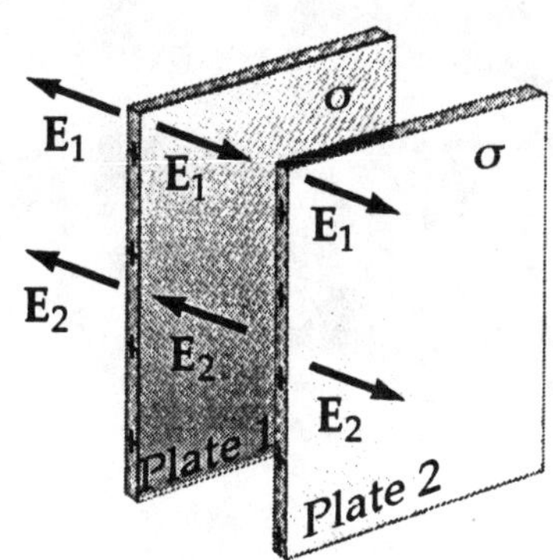

Figure P19.46
(modified)

(b) At point between the two sheets

$$\mathbf{E} = E_1\mathbf{i} + E_2(-\mathbf{i}) = 0$$

$$\mathbf{E} = 0 \quad \Diamond$$

(c) At point to the right of the two parallel sheets

$$\mathbf{E} = E_1\mathbf{i} + E_2\mathbf{i} = 2E\mathbf{i}$$

$$\mathbf{E} = \frac{\sigma}{\epsilon_0}\mathbf{i} \quad \Diamond$$

Chapter 20

Jennifer is holding on to a charged sphere that reaches a potential of about 100 000 volts. The device that generates this high potential is called a Van de Graaff generator. Why do you suppose Jennifer's hair stands on end like the needles of a porcupine? Why is it important that she stand on a pedestal insulated from the ground? (*Courtesy of Henry Leap and Jim Lehman*)

ELECTRIC POTENTIAL AND CAPACITANCE

Chapter 20

ELECTRIC POTENTIAL AND CAPACITANCE

INTRODUCTION

The concept of potential energy was first introduced in Chapter 7 of the text, in connection with such conservative forces as the force of gravity and the elastic force of a spring. By using the law of energy conservation, we were often able to avoid working directly with forces when solving various mechanical problems. In this chapter we see that the energy concept is also of great value in the study of electricity. Since the electrostatic force given by Coulomb's law is conservative, electrostatic phenomena can be described conveniently in terms of an electrical potential energy. This idea enables us to define a scalar quantity called **electric potential.** Because the potential is a scalar function of position, it offers a simpler way of describing electrostatic phenomena than does the electric field.

This chapter is also concerned with the properties of capacitors, devices that store charge. Capacitors are commonly used in a variety of electrical circuits. For instance, they are used (1) to tune the frequency of radio receivers, (2) as filters in power supplies, (3) to eliminate sparking in automobile ignition systems, and (4) as energy-storing devices in electronic flash units.

A capacitor basically consists of two conductors separated by an insulator. We shall see that the capacitance of a given device depends on its geometry and on the material separating the charged conductors, called a **dielectric.**

NOTES FROM SELECTED CHAPTER SECTIONS

20.1 Potential Difference and Electric Potential

The potential difference between two points $V_B - V_A$ equals the work per unit charge that an **external agent** must perform in order to move a test charge, q_0, from point A to point B **without** a change in kinetic energy.

The electric potential at an arbitrary point is the work required per unit charge to bring a **positive test charge from infinity to that point.**

Chapter 20

20.2 Potential Differences in a Uniform Electric Field

Electric field lines always point in the direction of decreasing electric potential. A positive electric charge loses electric potential energy when it moves in the direction of the electric field. An equipotential surface is any surface consisting of a continuous distribution of points having the same electric potential. Equipotential surfaces are perpendicular to electric field lines.

20.3 Electric Potential and Electric Potential Energy Due to Point Charges

Equipotential surfaces for a point charge are a family of spheres concentric with the charge.

20.4 Obtaining E From the Electric Potential

If the electric potential (which is a scalar) is known as a function of coordinates (x, y, z), the components of the electric field (a vector quantity) can be obtained by taking the negative derivative of the potential with respect to the coordinates.

20.6 Electric Potential of a Charged Conductor

The surface of any charged conductor in equilibrium is an equipotential surface. Also (**since the electric field is zero inside the conductor**), the potential is constant everywhere inside the conductor and equal to its value at the surface.

20.7 Capacitance

The capacitance of a capacitor depends on the physical characteristics of the device (size, shape, and separation of plates and the nature of the dielectric medium filling the space between the plates). Since the potential difference between the plates is proportional to the quantity of charge on each plate, the value of the capacitance is independent of the charge on the capacitor.

20.8 Combinations of Capacitors

When two or more unequal capacitors are connected in **series,** they carry the same charge, but the potential differences are not the same. Their capacitances add as reciprocals, and the equivalent capacitance of the combination is always **less** than the smallest individual capacitor.

When two or more capacitors are connected in **parallel**, the potential difference across each is the same. The charge on each capacitor is proportional to its capacitance, hence the capacitances add directly to give the equivalent capacitance of the parallel combination.

20.9 Energy Stored in a Charged Capacitor

The electrostatic potential energy stored in a charged capacitor equals the work done in the charging process—moving charges from one conductor at a lower potential to another conductor at a higher potential.

20.10 Capacitors with Dielectrics

A dielectric is a nonconducting material characterized by a dimensionless parameter—the dielectric constant, κ. In general, the use of a dielectric has the following effects:

- increases the capacitance
- increases the maximum operating voltage
- provides mechanical support for the two conductors

EQUATIONS AND CONCEPTS

The **potential difference** between two points A and B in an electric field, $\Delta V = V_B - V_A$, can be found by integrating $\mathbf{E} \cdot d\mathbf{s}$ along **any path** from A to B.

$$\Delta V = -\int_A^B \mathbf{E} \cdot d\mathbf{s} \qquad (20.3)$$

If the field is **uniform**, the potential difference depends only on the displacement d in the direction parallel to $\mathbf{E}$.

$$\Delta V = -E\int_A^B ds = -Ed \qquad (20.6)$$

The **change in potential energy,** ΔU, of a charge in moving from point A to point B in an electric field depends on the sign and magnitude of the charge as well as on the change in potential, ΔV.

$$\Delta U = q_o \, \Delta V = -q_o E d \tag{20.7}$$

In the special case where the electric field is uniform, the change in potential energy is proportional to the distance the charge moves along a direction parallel to the electric field. Note that a positive charge loses electric potential energy when it moves in the direction of the electric field.

The **electric potential** at a point in the vicinity of several point charges is calculated in a manner which assumes that the potential is zero at infinity.

$$V = k_e \sum_i \frac{q_i}{r_i} \tag{20.12}$$

The **potential energy of a pair of charges** separated by a distance r represents the work required to assemble the charges from an infinite separation. Hence, the negative of the potential energy equals the minimum work required to separate them by an infinite distance. The electric potential energy associated with a system of two charged particles is positive if the two charges have the same sign, and negative if they are of opposite sign.

$$U = k_e \frac{q_1 q_2}{r_{12}} \tag{20.13}$$

If there are more than two charged particles in the system, the **total potential energy** is found by calculating U for each pair of charges and summing the terms algebraically.

$$U = \frac{k_e}{2} \sum_{i=1}^{N} \sum_{j=1}^{N} \frac{q_i q_j}{r_{ij}}$$

when $i \neq j$

If the scalar electric potential function throughout a region of space is known, then the vector electric field can be calculated from the potential function. The **components of the electric field** in rectangular coordinates are given in terms of partial derivatives of the potential.

$$E_x = -\frac{\partial V}{\partial x} \qquad (20.15)$$

$$E_y = -\frac{\partial V}{\partial y}$$

$$E_z = -\frac{\partial V}{\partial z}$$

If the charge density has spherical symmetry, where the charge density depends only on the radial distance, r, then the electric field is radial.

$$\mathbf{E} = -\frac{dV}{dr}\hat{\mathbf{r}} \qquad (20.16)$$

The **potential** (relative to zero at infinity) **for a continuous charge distribution** can be calculated by integrating the contribution due to a charge element dq over the line, surface, or volume which contains all the charge. Here, as in the case of a continuous charge distribution, it is convenient to represent dq in terms of the appropriate charge density.

$$V = k_e \int_{\substack{\text{All}\\\text{Charge}}} \frac{dq}{r} \qquad (20.18)$$

The **capacitance** of a capacitor is defined as the ratio of the charge on either conductor (or plate) to the magnitude of the potential difference between the conductors.

$$C \equiv \frac{Q}{\Delta V} \qquad (20.19)$$

The **capacitance of an air-filled parallel-plate capacitor** is proportional to the area of the plates and inversely proportional to the separation of the plates.

$$C = \frac{\epsilon_0 A}{d} \qquad (20.21)$$

When the region between the plates of a capacitor is completely filled by a material of dielectric constant κ, the capacitance increases by the factor κ.

$$C = \kappa \frac{\epsilon_o A}{d} \qquad (20.33)$$

The **equivalent capacitance** of a **parallel combination** of capacitors is larger than any individual capacitor in the group.

$$C_{eq} = C_1 + C_2 + C_3 + \ldots \qquad (20.25)$$

The **equivalent capacitance** of a **series combination** of capacitors is smaller than the smallest capacitor in the group.

$$\frac{1}{C_{eq}} = \frac{1}{C_1} + \frac{1}{C_2} + \frac{1}{C_3} + \ldots \qquad (20.28)$$

In the special case of only **two capacitors in series,** the equivalent capacitance is equal to the ratio of the product to the sum of their capacitance.

$$C_{eq} = \frac{C_1 C_2}{C_1 + C_2}$$

The **electrostatic energy** stored in the electrostatic field of a charged capacitor equals the work done (by a battery or other source) in charging the capacitor from $q = 0$ to $q = Q$.

$$U = \frac{Q^2}{2C} = \tfrac{1}{2} Q \Delta V = \tfrac{1}{2} C \Delta V^2 \qquad (20.29)$$

The **energy density,** energy per unit volume, at any point in an electrostatic field is proportional to the square of the electric field intensity at that point.

$$u_E = \tfrac{1}{2} \epsilon_o E^2 \qquad (20.31)$$

SUGGESTIONS, SKILLS, AND STRATEGIES

The vector expression giving the components of the electric field **E** over a region can be obtained from the scalar function which describes the electric potential, V, over the region by using Equation 20.15 and similar expressions for the y and z components:

$$E_x = -\frac{\partial V}{\partial x} \qquad E_y = -\frac{\partial V}{\partial y} \qquad E_z = -\frac{\partial V}{\partial z}$$

The derivatives in the above expressions are called **partial derivatives.** This means that when the derivative is taken with respect to any one coordinate, any other coordinates which appear in the expression for the potential function are treated as constants.

Since the electrostatic force is a conservative force, the work done by the electrostatic force in moving a charge q from an initial point A to a final point B depends only on the location of the two points and is independent of the path taken between A and B. When calculating potential differences using the equation

$$V_B - V_A = -\int_A^B \mathbf{E}\cdot d\mathbf{s} \qquad (20.3)$$

any path between A and B may be chosen to evaluate the integral; therefore you should select a path for which the evaluation of the "line integral" in Equation 20.3 will be as convenient as possible. For example; where A, B, and C are constants, if **E** is in the form

$$\mathbf{E} = Ax\mathbf{i} + By\mathbf{j} + Cz\mathbf{k}$$

The potential is integrated as $\displaystyle\int_A^B \mathbf{E}\cdot d\mathbf{s} = \int_A^B (Ax\,dx + By\,dy + Cz\,dz)$

and Equation 20.3 becomes

$$V_B - V_A = -\left(A\int_{x_A}^{x_B} x\,dx + B\int_{y_A}^{y_B} y\,dy + C\int_{z_A}^{z_B} z\,dz\right)$$

$$V_B - V_A = \frac{A}{2}\left(x_A^2 - x_B^2\right) + \frac{B}{2}\left(y_A^2 - y_B^2\right) + \frac{C}{2}\left(z_A^2 - z_B^2\right)$$

Chapter 20

Problem-Solving Hints for Electric Potentials:

- When working problems involving electric potential, remember that potential is a **scalar quantity** (rather than a vector quantity like the electric field), so there are no components to worry about. Therefore, when using the superposition principle to evaluate the electric potential at a point due to a system of point charges, you simply take the algebraic sum of the potentials due to each charge. However, you must keep track of signs. The potential created by each positive charge ($V = k_e q / r$) is positive, while the potential for each negative charge is negative.

- Just as in mechanics, only **changes** in electric potential are significant, hence the point where you choose the potential to be zero is arbitrary. When dealing with point charges or a finite-sized charge distribution, we usually define $V = 0$ to be at a point infinitely far from the charges. However, if the charge distribution itself extends to infinity, some other nearby point must be selected as the reference point.

- The electric potential at some point P due to a continuous distribution of charge can be evaluated by dividing the charge distribution into infinitesimal elements of charge dq located at a distance r from the point P. You then treat this element as a point charge, so that the potential at P due to the element is $dV = k_e dq / r$. The total potential at P is obtained by integrating dV over the entire charge distribution. In performing the integration for most problems, it is necessary to express dq and r in terms of a single variable. In order to simplify the integration, it is important to give careful consideration of the geometry involved in the problem.

- Another method that can be used to obtain the potential due to a finite continuous charge distribution is to start with the definition of the potential difference given by Equation 20.3. If **E** is known or can be obtained easily (say from Gauss's law), then the line integral of $\mathbf{E}\cdot d\mathbf{s}$ can be evaluated. An example of this method is given in Ex. 20.6.

- Once you know the electric potential at a point, it is possible to obtain the electric field at that point by remembering that **the electric field is equal to the negative of the derivative of the potential with respect to some coordinate.** Example 20.5 illustrates how to use this procedure.

- Until now, we have been using the symbols V to represent the electric potential at some point and ΔV to represent the potential difference between two points. In descriptions of electrical devices, however, it is common practice to use the symbol V to represent the potential difference across the device. Hence, in this book both symbols will be used to denote potential differences, depending on the circumstances.

  In practice, a variety of phrases are used to describe the potential difference between two points, the most common being "voltage." A voltage **applied** to a device or **across** a device has the same meaning as the potential difference across the device. For example, if we say that the voltage across a certain capacitor is 12 volts, we mean that the potential difference between the capacitor's plates is 12 volts.

Problem-Solving Hints for Capacitance:

- When analyzing a series-parallel combination of capacitors to determine the equivalent capacitance, you should make a sequence of circuit diagrams which show the successive steps in the simplification of the circuit; combine at each step those capacitors which are in simple-parallel or simple-series relationship to each other and use appropriate equations for series or parallel capacitors at each step of the simplification. At each step, you know two of the three quantities: Q, V, and C. You will be able to determine the remaining quantity using the relation $Q = CV$.

- When calculating capacitance, be careful with your choice of units. To calculate capacitance in farads, make sure that distances are in meters and use the SI value of ϵ_o. When checking consistency of units, remember that the units for electric fields are newtons per coulomb (N/C) or the equivalent volts per meter (V/m).

- When two or more unequal capacitors are connected in series, they carry the same charge, but their potential differences are not the same. The capacitances add as reciprocals, and the equivalent capacitance of the combination is always less than the smallest individual capacitor.

- When two or more capacitors are connected in parallel, the potential differences across them are the same. The charge on each capacitor is proportional to its capacitance; hence, the capacitances add directly to give the equivalent capacitance of the parallel combination.

- A dielectric increases capacitance by the factor κ (the dielectric constant) because induced surface charges on the dielectric reduce the electric field inside the material from E to E/κ.

- Be careful about problems in which you may be connecting or disconnecting a battery to a capacitor. It is important to note whether modifications to the capacitor are being made while the capacitor is connected to the battery or after it is disconnected. If the capacitor remains connected to the battery, the voltage across the capacitor necessarily remains the same (equal to the battery voltage), and the charge is proportional to the capacitance, **however it may be modified** (say, by insertion of a dielectric). On the other hand, if you disconnect the capacitor from the battery before making any modifications to the capacitor, then its charge remains the same. In this case, as you vary the capacitance, the voltage across the plates changes in inverse proportion to capacitance, according to $V = Q/C$.

REVIEW CHECKLIST

▷ Understand that each point in the vicinity of a charge distribution can be characterized by a scalar quantity called the electric potential, V. The values of this potential function over the region (a scalar) are related to the values of the electrostatic field over the region (a vector field).

▷ Calculate the electric potential difference between any two points in a uniform **electric field**, and the electric potential difference between any two points in the vicinity of a **group of point charges**.

▷ Calculate the electric **potential energy** associated with a group of point charges.

▷ Obtain an expression for the electric field (a **vector** quantity) over a region of space if the scalar electric potential function for the region is known.

▷ Calculate the work done by an external force in moving a charge q between any two points in an electric field when (a) an expression giving the field as a function of position is known, or when (b) the charge distribution (either point charges or a continuous distribution of charge) giving rise to the field is known.

▷ Determine the equivalent capacitance of a network of capacitors in series-parallel combination and calculate the final charge on each capacitor and the potential difference across each when a known potential is applied across the combination.

▷ Calculate the capacitance, potential difference, and stored energy of a capacitor which is partially or completely filled with a **dielectric**.

Chapter 20

ANSWERS TO SELECTED CONCEPTUAL QUESTIONS

2. Give a physical explanation of the fact that the potential energy of a pair of like charges is positive whereas the potential energy of a pair of unlike charges is negative.

Answer You may remember from the chapter on gravitational potential energy that potential energy of a system is defined to be positive when positive work must have been performed by an external agent. For example, a flag has a positive potential energy relative to the ground, since positive work must be done by an external force in order to raise it from the ground to the top of the pole.

When assembling like charges from an infinite separation, it takes work to move them closer together to some distance r; therefore energy is being stored, and the potential energy is positive.

When assembling unlike charges from an infinite separation, the charges tend to accelerate towards each other, and thus energy is released as they approach a separation of distance r. Therefore, the potential energy of a pair of unlike charges is negative.

□ □ □ □

10. If the potential difference across a capacitor is doubled, by what factor does the energy stored change?

Answer Since $U = CV^2/2$, doubling V will quadruple the stored energy.

□ □ □ □

12. If you were asked to design a capacitor where small size and large capacitance were required, what factors would be important in your design?

Answer You should use a dielectric filled capacitor whose dielectric constant is very large. Furthermore, you should make the dielectric as thin as possible, keeping in mind that dielectric breakdown must also be considered.

□ □ □ □

SOLUTIONS TO SELECTED END-OF-CHAPTER PROBLEMS

3. (a) Calculate the speed of a proton that is accelerated from rest through a potentia difference of 120 V. (b) Calculate the speed of an electron that is accelerated through the same potential difference.

Solution

(a) Energy is conserved as the proton moves from high to low potential; we take it as from 120 V to ground:

$$K_i + U_i + \Delta K_{nc} = K_f + U_f$$

(Review this work-energy theory of motion from Chapter 7 to the full extent necessary for you.)

$$0 + qV + 0 = \tfrac{1}{2}mv^2 + 0$$

$$(1.60 \times 10^{-19}\ \text{C})(120\ \text{V})\left(\frac{1\ \text{J}}{1\ \text{V}\cdot\text{C}}\right) = \tfrac{1}{2}(1.67 \times 10^{-27}\ \text{kg})v^2$$

$$v = 1.52 \times 10^5\ \text{m/s} \quad \Diamond$$

(b) The electron will gain speed in moving the other way, from $V_i = 0$ to $V_f = 120$ V:

$$K_i + U_i + \Delta K_{nc} = K_f + U_f$$

$$0 + 0 + 0 = \tfrac{1}{2}mv^2 + qV$$

$$0 = \tfrac{1}{2}(9.11 \times 10^{-31}\ \text{kg})v^2 + (-1.60 \times 10^{-19}\ \text{C})(120\ \text{J/C})$$

$$v = 6.49 \times 10^6\ \text{m/s} \quad \Diamond$$

This is less than one tenth the speed of light, so we need not use the relativistic kinetic energy formula.

---

13. The three charges in Figure P20.13 are at the vertices of an isosceles triangle. Calculate the electric potential at the midpoint of the base, taking $q = 7.00\ \mu C$.

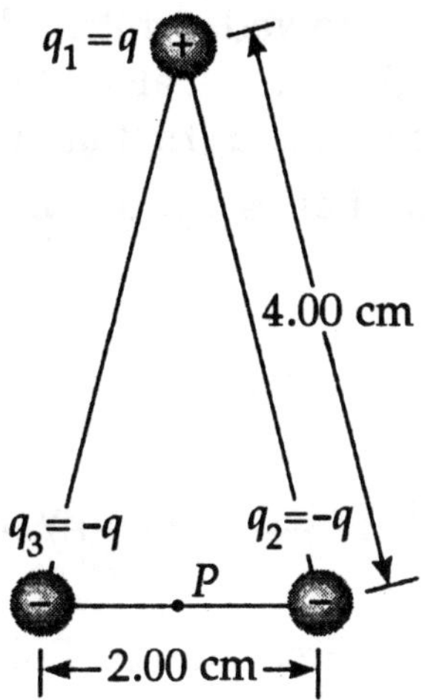

Figure P20.13

Solution

Let $\quad q_1 = q$

and $\quad q_2 = q_3 = -q$

The charges are at distances of

$$r_1 = \sqrt{(0.0400\ \text{m})^2 - (0.0100\ \text{m})^2} = 3.87 \times 10^{-2}\ \text{m}$$

and $r_2 = r_3 = 0.0100\ \text{m}$

The voltage at point P is

$$V_p = \frac{k_e q_1}{r_1} + \frac{k_e q_2}{r_2} + \frac{k_e q_3}{r_3} = (k_e)(q)\left(\frac{1}{r_1} + \frac{-1}{r_2} + \frac{-1}{r_3}\right)$$

so
$$V_p = \left(8.99 \times 10^9\ \frac{\text{N}\cdot\text{m}^2}{\text{C}^2}\right)(7.00 \times 10^{-6}\ \text{C})\left(\frac{1}{0.0387\ \text{m}} - \frac{1}{0.0100\ \text{m}} - \frac{1}{0.0100\ \text{m}}\right)$$

and $\quad V_p = -11.0 \times 10^6\ \text{V}$ ◊

Related Calculation Calculate the electric field vector at the same point due to the three charges.

The separate fields of the two negative charges are in opposite directions and add to zero:

$$\mathbf{E}_P = \frac{k_e q_1}{r_1^2}\hat{\mathbf{r}}_1 = \frac{(8.99 \times 10^9\ \text{N}\cdot\text{m}^2/\text{C}^2)(7 \times 10^{-6}\ \text{C})}{((0.0400\ \text{m})^2 - (0.0100\ \text{m})^2)}\ \text{down}$$

$$\mathbf{E}_P = (42.0 \times 10^6\ \text{N/C})(-\mathbf{j})$$

15. The Bohr model of the hydrogen atom holds that the electron can exist onl in certain allowed orbits. The radius of each Bohr orbit is $r = n^2(0.0529 \text{ nm})$ wher $n = 1, 2, 3, \ldots$. Calculate the electric potential energy of a hydrogen atom when th electron is in the (a) first allowed orbit, $n = 1$; (b) second allowed orbit, $n = 2$; and (c) whe the electron has escaped from the atom, $r = \infty$. Express your answers in electron volts.

Solution

The electric potential energy is given by $U = k_e \frac{q_1 q_2}{r}$

(a) For the first allowed Bohr orbit,

$$U = (8.99 \times 10^9 \text{ N} \cdot \text{m}^2/\text{C}^2)\frac{(-1.60 \times 10^{-19} \text{ C})(1.60 \times 10^{-19} \text{ C})}{(0.0529 \times 10^{-9} \text{ m})}$$

$$U = -4.35 \times 10^{-18} \text{ J} = \frac{-4.35 \times 10^{-18} \text{ J}}{1.60 \times 10^{-19} \text{ J/eV}} = -27.2 \text{ eV} \quad \lozenge$$

(b) For the second allowed orbit,

$$U = (8.99 \times 10^9 \text{ N} \cdot \text{m}^2/\text{C}^2)\frac{(-1.60 \times 10^{-19} \text{ C})(1.60 \times 10^{-19} \text{ C})}{2^2(0.0529 \times 10^{-9} \text{ m})}$$

$$U = -1.088 \times 10^{-18} \text{ J} = -6.80 \text{ eV} \quad \lozenge$$

(c) When the electron is at $r = \infty$,

$$U = (8.99 \times 10^9 \text{ N} \cdot \text{m}^2/\text{C}^2)\frac{(-1.60 \times 10^{-19} \text{ C})(1.60 \times 10^{-19} \text{ C})}{\infty \text{ m}}$$

$$U = 0 \text{ J} \quad \lozenge$$

17. Show that the amount of work required to assemble four identical point charges of magnitude Q at the corners of a square of side s is $5.41k_e Q^2/s$.

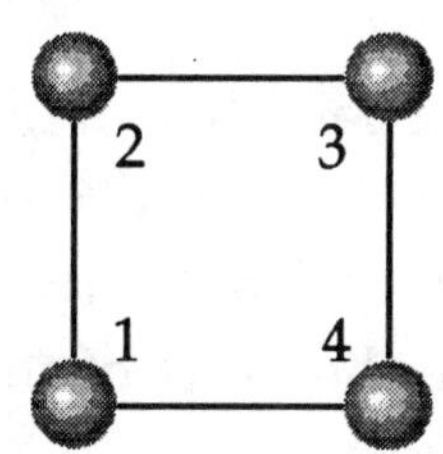

Solution The work required equals the sum of the potential energies of the four charges. We must add up $U = qV$ contributions for all pairs:

$$U = q_1V_2 + q_1V_3 + q_1V_4 + q_2V_3 + q_2V_4 + q_3V_4$$

$$U = \frac{q_1k_eq_2}{r_{12}} + \frac{q_1k_eq_3}{r_{13}} + \frac{q_1k_eq_4}{r_{14}} + \frac{q_2k_eq_3}{r_{23}} + \frac{q_2k_eq_4}{r_{24}} + \frac{q_3k_eq_4}{r_{34}}$$

$$U = \frac{Qk_eQ}{s} + \frac{Qk_eQ}{s\sqrt{2}} + \frac{Qk_eQ}{s} + \frac{Qk_eQ}{s} + \frac{Qk_eQ}{s\sqrt{2}} + \frac{Qk_eQ}{s}$$

Evaluating, $$U = \frac{k_eQ^2}{s}\left(4 + \frac{2}{\sqrt{2}}\right) = 5.41k_e\frac{Q^2}{s} \quad \lozenge$$

---

21. Over a certain region of space, the electric potential is $V = 5x - 3x^2y + 2yz^2$. Find the expressions for x, y, and z components of the electric field over this region. What is the magnitude of the field at the point P, which has coordinates (1.00, 0, -2.00) m?

Solution First, we find the x, y, and z components of the field; then, we evaluate them at point P. (We assume that V is given in volts, as a function of distances in meters.)

$E_x = -\frac{\partial V}{\partial x} = -5 + 6xy \quad \lozenge$ At point P, $E_x = -5 + 6(1.00\text{ m})(0\text{ m}) = -5\text{ N/C}$

$E_y = -\frac{\partial V}{\partial y} = 3x^2 - 2z^2 \quad \lozenge$ At point P, $E_y = 3(1.00\text{ m})^2 - 2(-2.00\text{ m})^2 = -5.00\text{ N/C}$

$E_z = -\frac{\partial V}{\partial z} = -4yz \quad \lozenge$ At point P, $E_z = -4(0\text{ m})(-2.00\text{ m}) = 0\text{ N/C}$

At P, the field's magnitude $E = \sqrt{(-5.00\text{ N/C})^2 + (-5.00\text{ N/C})^2 + 0^2} = 7.07\text{ N/C} \quad \lozenge$

---

25. A rod of length L (Fig. P20.25) lies along the x axis with its left end at the origin and has a nonuniform charge density $\lambda = \alpha x$ (where α is a positive constant). (a) What are the units of the α? (b) Calculate the electric potential at A.

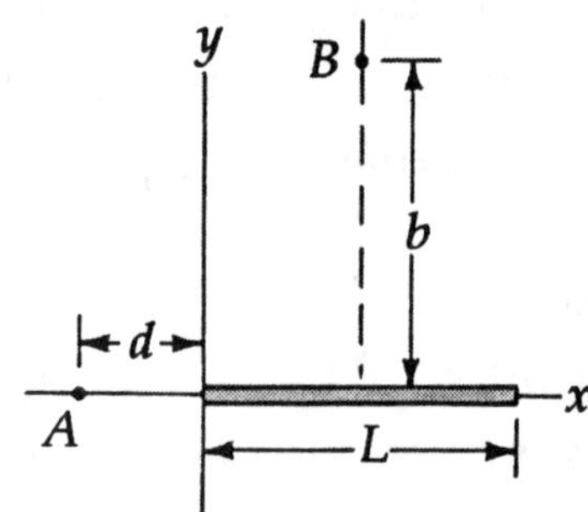

Figure P20.25

Solution

(a) As a linear charge density, λ has units of C/m.
So $\alpha = \lambda/x$ must have units of C/m². ◊

(b) Consider a small segment of the rod at location x and of length dx. The amount of charge on it is $\lambda dx = (\alpha x)dx$. Its distance from A is $d + x$, so its contribution to the electric potential at A is

$$dV = k_e \frac{dq}{r} = k_e \alpha x \frac{dx}{(d+x)}$$

We must integrate all these contributions for the whole rod, from $x = 0$ to $x = L$:

$$V = \int_{\text{all } q} dV = \int_0^L \frac{k_e \alpha x\, dx}{d+x}$$

To perform the integral, make a change of variables to

$$u = d + x, \quad du = dx, \quad u \text{ (at } x = 0) = d, \quad \text{and} \quad u \text{ (at } x = L) = d + L$$

$$V = \int_d^{d+L} \frac{k_e \alpha (u-d)du}{u} = k_e \alpha \int_d^{d+L} du - k_e \alpha d \int_d^{d+L} \left(\frac{1}{u}\right) du$$

[Keep track of symbols: the **unknown** is V. The values k_e, α, d, and L are **known** and constant. And x and u are variables, and will not appear in the answer.]

$$V = k_e \alpha u \Big|_d^{d+L} - k_e \alpha d \ln u \Big|_d^{d+L} = k_e \alpha (d + L - d) - k_e \alpha d(\ln(d+L) - \ln d)$$

$$V = k_e \alpha L - k_e \alpha d \ln\left(\frac{d+L}{d}\right) \quad ◊$$

We have the answer when the unknown is expressed in terms of the d, L, and α mentioned in the problem and the universal constant k_e.

33. An air-filled capacitor consists of two parallel plates, each with area 7.60 cm^2, separated by a distance of 1.80 mm. If a 20.0-V potential difference is applied to these plates, calculate (a) the electric field between the plates, (b) the surface charge density, (c) the capacitance, and (d) the charge on each plate.

Solution

(a) The potential difference between two points in a uniform electric field is $V = Ed$, so:

$$E = \frac{V}{d} = \frac{20.0\ \text{V}}{1.80 \times 10^{-3}\ \text{m}}$$

$$E = 1.11 \times 10^{4}\ \text{V/m} \quad \Diamond$$

(b) The electric field between capacitor plates is $E = \frac{\sigma}{\epsilon_0}$, so $\sigma = \epsilon_0 E$:

$$\sigma = \left(8.85 \times 10^{-12}\ \frac{\text{C}^2}{\text{N} \cdot \text{m}^2}\right)(1.11 \times 10^{4}\ \text{V/m})$$

$$\sigma = 9.83 \times 10^{-8}\ \text{C/m}^2 = 98.3\ \text{nC/m}^2 \quad \Diamond$$

(c) For a parallel-plate capacitor, $C = \frac{\epsilon_0 A}{d}$:

$$C = \frac{\left(8.85 \times 10^{-12}\ \frac{\text{C}^2}{\text{N} \cdot \text{m}^2}\right)(7.60 \times 10^{-4}\ \text{m}^2)}{1.80 \times 10^{-3}\ \text{m}}$$

$$C = 3.74 \times 10^{-12}\ \text{F} = 3.74\ \text{pF} \quad \Diamond$$

(d) The charge on each plate is $Q = CV$:

$$Q = (3.74 \times 10^{-12}\ \text{F})(20.0\ \text{V})$$

$$Q = 7.47 \times 10^{-11}\ \text{C} = 74.7\ \text{pC} \quad \Diamond$$

39. **Four capacitors are connected as shown in Figure P20.39. (a) Find the equivalent capacitance between points *a* and *b*. (b) Calculate the charge on each capacitor if $\Delta V_{ab} = 15.0$ V.**

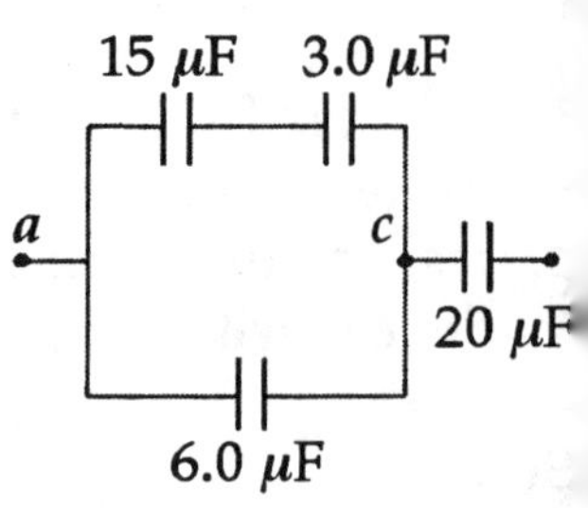

Figure P20.39

(a) We successively simplify the circuit, proceeding from the given diagram through solution figures (a) - (c).

First, the 15.0 μF and 3.00 μF in series are equivalent to

$$\frac{1}{\left(\dfrac{1}{15.0\ \mu\text{F}}+\dfrac{1}{3.00\ \mu\text{F}}\right)} = 2.50\ \mu\text{F}$$

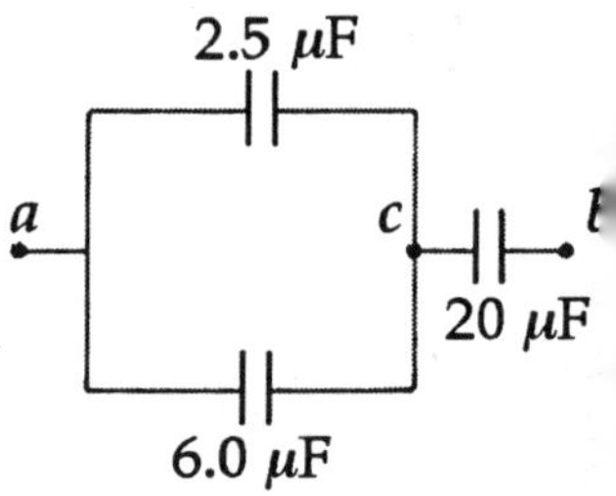

(a)

Next, 2.50 μF combines in parallel with 6.00 μF, creating an equivalent capacitance of 8.50 μF.

At last, 8.50 μF and 20.0 μF are in series, equivalent to

$$\frac{1}{\left(\dfrac{1}{8.50\ \mu F}+\dfrac{1}{20.0\ \mu F}\right)} = 5.96\ \mu F \quad \lozenge$$

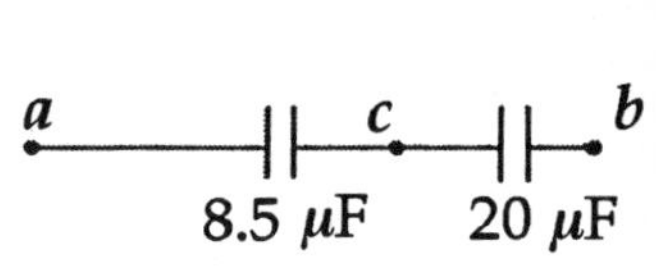

(b)

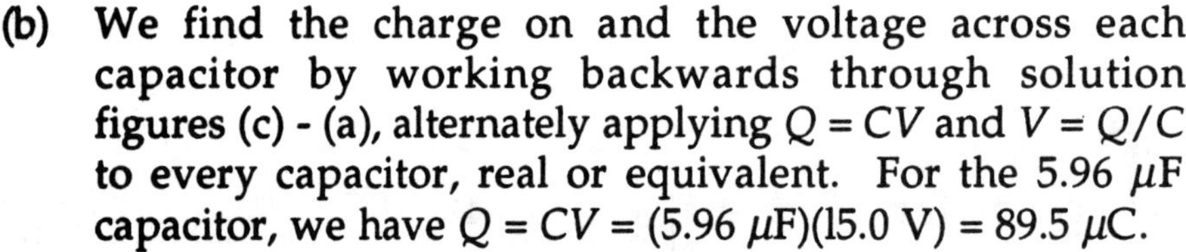

(b) We find the charge on and the voltage across each capacitor by working backwards through solution figures (c) - (a), alternately applying $Q = CV$ and $V = Q/C$ to every capacitor, real or equivalent. For the 5.96 μF capacitor, we have $Q = CV = (5.96\ \mu\text{F})(15.0\ \text{V}) = 89.5\ \mu\text{C}$.

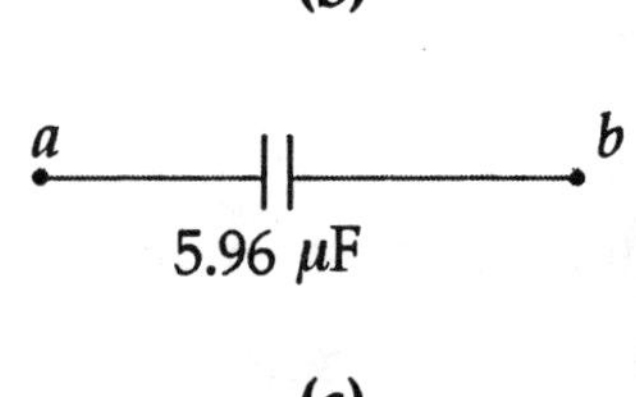

(c)

Thus, if *a* is higher in potential than *b*, just 89.5 μC flows to the right past *a* and past *b* to charge the capacitors in each picture. In (b) we have, for the 8.5 μF capacitor,

$$V_{ac} = \frac{Q}{C} = \frac{89.5\ \mu\text{C}}{8.50\ \mu\text{F}} = 10.5\ \text{V}$$

and for the 20.0 μF in (b), (a), and the original circuit, we have $Q_{20} = 89.5\ \mu\text{C}$ ◊

$$V_{cb} = \frac{Q}{C} = \frac{89.5\ \mu\text{C}}{20.0\ \mu\text{F}} = 4.47\ \text{V}$$

Next, (a) is equivalent to (b), so $V_{cb} = 4.47$ V and $V_{ac} = 10.5$ V.

For the 2.50 μF, $V = 10.5$ V and $Q = CV = (2.50\ \mu\text{F})(10.5\ \text{V}) = 26.3\ \mu\text{C}$

For the 6.00 μF, $V = 10.5$ V and $Q_6 = CV = (6.00\ \mu\text{F})(10.5\ \text{V}) = 63.2\ \mu\text{C} = Q_6$ ◊

Now, 26.3 μC having flowed in the upper parallel branch in (a), back in the original circuit we have

$$Q_{15} = 26.3\ \mu\text{C} \quad \Diamond \qquad \text{and} \qquad Q_3 = 26.3\ \mu\text{C} \quad \Diamond$$

Related Calculation An exam problem might also ask for the voltage across each:

$$V_{15} = \frac{Q}{C} = \frac{26.3\ \mu\text{C}}{15.0\ \mu\text{F}} = 1.75\ \text{V} \qquad \text{and} \qquad V_3 = \frac{Q}{C} = \frac{26.3\ \mu\text{C}}{3.00\ \mu\text{F}} = 8.77\ \text{V} \quad \Diamond$$

---

41. Consider the circuit shown in Figure P20.41, where $C_1 = 6.00\ \mu$F, $C_2 = 3.00\ \mu$F, and $V = 20.0$ V. Capacitor C_1 is first charged by the closing of switch S_1. Switch S_1 is then opened, and the charged capacitor is connected to the uncharged capacitor by the closing of S_2. Calculate the initial charge acquired by C_1 and the final charge on each.

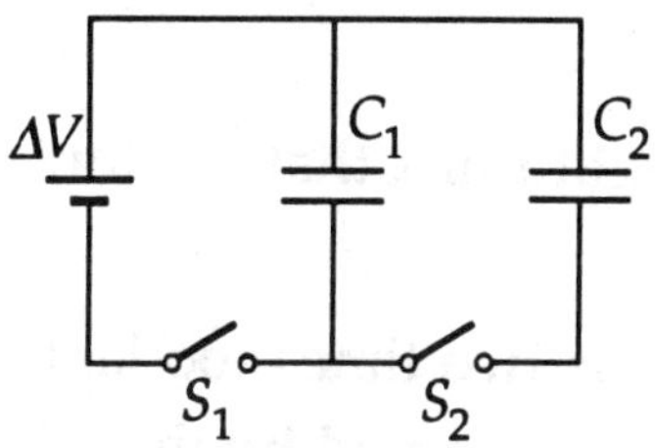

Figure P20.41

Solution When S_1 is closed, the charge on C_1 will be

$$Q_1 = C_1V_1 = (6.00\ \mu\text{F})(20.0\ \text{V}) = 120\ \mu\text{C} \quad \Diamond$$

When S_1 is opened and S_2 is closed, the total charge will remain constant and be shared by the two capacitors:

$$Q_1' = 120\ \mu\text{C} - Q_2'$$

The potential across the two capacitors will be equal.

$$V' = \frac{Q_1'}{C_1} = \frac{Q_2'}{C_2} \qquad \text{or} \qquad \frac{120\ \mu\text{C} - Q_2'}{6.00\ \mu\text{F}} = \frac{Q_2'}{3.00\ \mu\text{F}}$$

and

$$Q_2' = 40.0\ \mu\text{C} \quad \Diamond$$

$$Q_1' = 120\ \mu\text{C} - 40.0\ \mu\text{C} = 80.0\ \mu\text{C} \quad \Diamond$$

---

47. A parallel-plate capacitor has a charge Q and plates of area A. Show that the force exerted on each plate by the other is $F = Q^2/2\,\epsilon_0 A$. (**Hint:** Let $C = \epsilon_0 A/x$ for an arbitrary plate separation x; then require that the work done in separating the two charged plates be $W = \int F\,dx$.)

Solution

The electric field in the space between the plates is

$$E = \frac{\sigma}{\epsilon_0} = \frac{Q}{A\epsilon_0}.$$

You might think that the force on one plate is $F = QE = \dfrac{Q^2}{A\epsilon_0}$,

but this is two times too large, because neither plate exerts a force on itself. The force **on** one plate is exerted **by** the other, through its electric field $\mathbf{E} = \sigma/2\,\epsilon_0 = Q/2A\,\epsilon_0$. The force on each plate is:

$$F = (Q_{self})(E_{other}) = Q^2/2A\,\epsilon_0.$$

To prove this, we follow the hint, and calculate that the work done in separating the plates, which equals the potential energy stored in the charged capacitor:

$$U = \frac{1}{2}\frac{Q^2}{C} = \int F\,dx$$

From the fundamental theorem of calculus, $dU = F\,dx$, and

$$F = \frac{d}{dx}U = \frac{d}{dx}\left(\frac{Q^2}{2C}\right) = \frac{1}{2}\frac{d}{dx}\left(\frac{Q^2}{\epsilon_0 A/x}\right)$$

Solving, $$F = \frac{1}{2}\frac{d}{dx}\left(\frac{Q^2 x}{\epsilon_0 A}\right) = \frac{1}{2}\left(\frac{Q^2}{\epsilon_0 A}\right) \quad \Diamond$$

61. A parallel-plate capacitor is constructed using a dielectric material the dielectric constant of which is 3.00 and the dielectric strength of which is 2.00×10^8 V/m. The desired capacitance is 0.250 μF, and the capacitor must withstand a maximum potential difference of 4000 V. Find the minimum area of the capacitor plates.

Solution $\kappa = 3.00,$ $E_{max} = 2.00 \times 10^8 \text{ V/m} = \dfrac{V_{max}}{d}$ so $d = \dfrac{V_{max}}{E_{max}}$

For $C = \dfrac{\kappa \epsilon_0 A}{d} = 0.250 \times 10^{-6}$ F,

$$A = \frac{Cd}{\kappa \epsilon_0} = \frac{CV_{max}}{\kappa \epsilon_0 E_{max}}$$

$$A = \frac{(0.250 \times 10^{-6}\text{ F})(4000\text{ V})}{3.00(8.85 \times 10^{-12}\text{ F/m})(2.00 \times 10^8\text{ V/m})} = 0.188\text{ m}^2 \quad \lozenge$$

---

63. A conducting slab of a thickness d and area A is inserted into the space between the plates of a parallel-plate capacitor with spacing s and surface area A, as in Figure P20.63. What is the capacitance of the system?

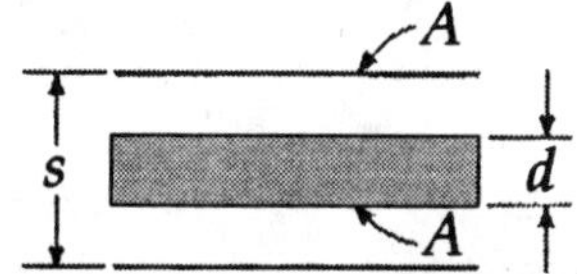

Figure P20.63

Solution If the capacitor is charged with charge Q, free charges will move across the slab to neutralize the electric field inside it, with the top and bottom faces of the slab then carrying charges $+Q$ and $-Q$. Then the capacitor with slab is electrically equivalent to two capacitors in series. Call x the upper gap, so $s - d - x$ is the distance between the lower two surfaces.

The upper capacitor has $C_1 = \dfrac{\epsilon_0 A}{x}$ and the lower has $C_2 = \dfrac{\epsilon_0 A}{s - d - x}$

So the combination has
$$C = \frac{1}{1/C_1 + 1/C_2} = \frac{1}{\dfrac{x}{\epsilon_0 A} + \dfrac{s-d-x}{\epsilon_0 A}} = \frac{\epsilon_0 A}{s-d} \quad \lozenge$$

Chapter 21

This is a carbon filament incandescent lamp. The resistance of such a lamp is typically 10 Ω, *though its value changes with temperature.* (Courtesy of Central Scientific Co.)

CURRENT AND DIRECT CURRENT CIRCUITS

Chapter 21

CURRENT AND DIRECT CURRENT CIRCUITS

INTRODUCTION

Thus far our discussion of electrical phenomena has been confined to charges at rest, or electrostatics. We now consider situations involving electric charges in motion. The term **electric current,** or simply **current,** is used to describe the rate of flow of charge through some region of space.

In this chapter we first define current and current density. A microscopic description of current is given, and some of the factors that contribute to the resistance to the flow of charge in conductors are discussed. Mechanisms responsible for the electrical resistance of various materials depend on the composition of the material and on temperature. A classical model is used to describe electrical conduction in metals, and some of the limitations of this model are pointed out.

In this chapter we also analyze some simple circuits whose elements include batteries, resistors, and capacitors in various combinations. Such analysis is simplified by the use of two rules known as **Kirchhoff's rules,** which follow from the laws of conservation of energy and conservation of charge. Most of the circuits are assumed to be in **steady state,** which means that the currents are constant in magnitude and direction. We close the chapter with a discussion of circuits containing resistors and capacitors, in which current varies with time.

NOTES FROM SELECTED CHAPTER SECTIONS

21.1 Electric Current

The direction of conventional current is designated as the direction of motion of positive charge. In an ordinary metal conductor, the direction of current will be **opposite** the **direction of flow of electrons** (which are the charge carriers in this case).

21.2 Resistance and Ohm's Law

For ohmic materials, the ratio of the current density to the electric field (that gives rise to the current) is equal to a constant σ, the conductivity of the material. The reciprocal of the conductivity is called the resistivity, ρ. Each ohmic material has a characteristic resistivity which depends only on the properties of the specific material and is a function of temperature.

Chapter 21

21.4 A Model for Electrical Conduction

In the classical model of electronic conduction in a metal, electrons are treated like molecules in a gas and, in the absence of an electric field, have a **zero average velocity**.

Under the influence of an electric field, the electrons move along a direction opposite the direction of the applied field with a **drift velocity** which is proportional to the average time between collisions with atoms of the metal. The current is proportional to the magnitude of the drift velocity and to the number of free electrons per unit volume.

21.7 Resistors in Series and in Parallel

The **current** must be the same for each of a group of resistors connected in **series.**

The **potential difference** must be the same across each of a group of resistors in **parallel.**

21.8 Kirchhoff's Rules and Simple DC Circuits

- The sum of the currents entering any junction must equal the sum of the currents leaving that junction. A junction is defined to be any point in the circuit where the current can split. (A separate instance of this rule can be applied at every junction but one.)
- The algebraic sum of the changes in potential around any closed circuit loop must be **zero.**

The first rule is a statement of **conservation of charge**; the second rule follows from the **conservation of energy.**

21.9 *RC* Circuits

Consider an uncharged capacitor in series with a resistor, a battery, and a switch. In the charging process, charges are transferred from one plate of the capacitor to the other moving along a path **through the resistor, battery, and switch.** The charges **do not move across the gap between the plates of the capacitor.**

The battery does work on the charges to increase their electrostatic potential energy as they move from one plate to the other.

Chapter 21

EQUATIONS AND CONCEPTS

Under the action of an electric field, electric charges will move through gases, liquids, and solid conductors. **Electric current,** I, is defined as the rate at which charge moves through a cross section of the conductor.

$$I \equiv \frac{dQ}{dt} \qquad (21.2)$$

The direction of the current is in the direction of the flow of positive charges. The SI unit of current is the **ampere** (A).

$$1\text{ A} = 1\text{ C/s} \qquad (21.3)$$

The current in a conductor can be related to the number of mobile charge carriers per unit volume, n; the quantity of charge associated with each carrier, q; and the **drift velocity,** v_d, of the carriers.

$$I = nqv_dA \qquad (21.4)$$

The **current density,** J, in a conductor is a vector quantity which is proportional to the electric field in the conductor.

$$J \equiv \frac{I}{A} \qquad (21.5)$$

or

$$\mathbf{J} = nq\mathbf{v}_d \qquad (21.6)$$

For many practical applications, a more useful form of Ohm's law relates the potential difference across a conductor and the current in the conductor to a composite of several physical characteristics of the conductor called the **resistance,** R.

$$R \equiv \frac{\Delta V}{I} \qquad (21.7)$$

The **resistance** of a given conductor of uniform cross section depends on the length, cross-sectional area, and a characteristic property of the material of which the conductor is made. The parameter, ρ, is the **resistivity** of the material of which the conductor is made. The resistivity is the inverse of the **conductivity** and has units of ohm-meters. The unit of resistance is the ohm (Ω).

$$R = \rho\frac{\ell}{A} \qquad (21.9)$$

$$1\,\Omega = 1\,\text{V/A}$$

The **resistivity** and therefore the resistance of a conductor vary with temperature in an approximately linear manner. In these expressions, α is the **temperature coefficient of resistivity** and T_0 is a stated reference temperature (usually 20°C).

$$\rho = \rho_0[1 + \alpha(T - T_0)] \qquad (21.11)$$

$$R = R_0[1 + \alpha(T - T_0)] \qquad (21.13)$$

Power will be supplied to a resistor or other current-carrying devices when a potential difference is maintained between the terminals of the circuit element. The quantities can be related in an equation called Joule's law and the SI unit of power is the watt (W). When the device obeys Ohm's law, the power dissipated can be expressed in alternative forms.

$$P = I(\Delta V) \qquad (21.21)$$

$$P = I^2R = \frac{(\Delta V)^2}{R} \qquad (21.22)$$

The average time between collisions with atoms of a metal is an important parameter in the description of the classical model of electrical conduction in metals. This characteristic time is denoted by τ and can be related to the drift velocity (see Eq. 21.4) or the resistivity (see Eq. 21.11) associated with the conductor. In these equations, m and q represent the mass and charge of the electron, E is the magnitude of the applied electric field, and n is the number of free electrons per unit volume.

$$\mathbf{v}_d = \frac{q\mathbf{E}}{m}\tau \qquad (21.16)$$

$$\rho = \frac{m}{nq^2\tau} \qquad (21.19)$$

When a battery is providing a current to an external circuit, the **terminal voltage** of the battery will be less than the emf due to **internal resistance** of the battery.

$$\Delta V = \mathcal{E} - Ir \qquad (21.24)$$

The **current,** I, delivered by a battery in a simple **dc** circuit, depends on the value of the **emf** of the source, $\mathcal{E}$; the total **load resistance** in the circuit, R; and the **internal resistance** of the source, r.

$$\mathcal{E} = IR + Ir \qquad (21.25)$$

The total or equivalent resistance of a series combination of resistors is equal to the sum of the resistances of the individual resistors.

$$R_{eq} = R_1 + R_2 + R_3 + \ldots \qquad (21.27)$$

(Series combination)

A group of resistors connected in parallel has an equivalent resistance which is less than the smallest individual value of resistance in the group.

$$\frac{1}{R_{eq}} = \frac{1}{R_1} + \frac{1}{R_2} + \frac{1}{R_3} + \ldots \qquad (21.29)$$

(Parallel combination)

Resistors in series are connected so that they have only one common circuit point per pair; there is a common current through each resistor in the group.

Resistors in parallel are connected so that each resistor in the group has two circuit points in common with each of the other resistors; there is a common potential difference across each resistor in the group.

Many circuits which contain several resistors can be reduced to an equivalent single-loop circuit by successive step-by-step combinations of groups of resistors in series and parallel.

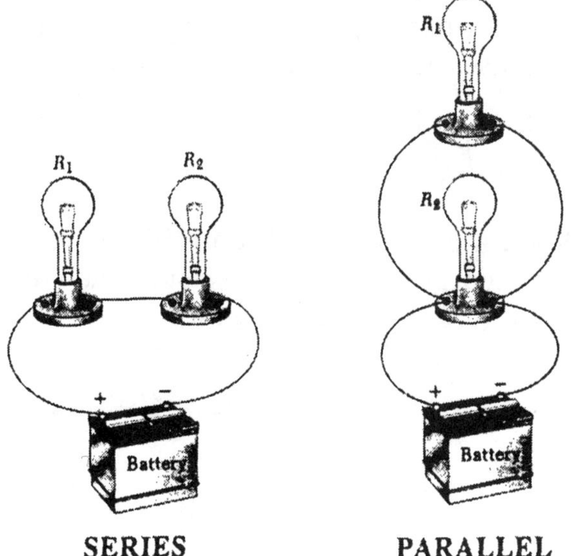

In the most general case, however, successive reduction is not possible and you must solve a true multiloop circuit by use of Kirchhoff's rules. Review the procedure suggested in the next section to apply Kirchhoff's rules.

When a potential difference is suddenly applied across an uncharged capacitor, the **current** in the circuit and the **charge** on the capacitor are functions of time. The **instantaneous values** of I and q depend on the capacitance and on the resistance in the circuit.

$$I(t) = \frac{\mathcal{E}}{R} e^{-t/RC} \tag{21.34}$$

$$q(t) = Q\left[1 - e^{-t/RC}\right] \tag{21.33}$$

When a battery is used to charge a capacitor in series with a resistor, a quantity τ, called the time constant of the circuit, is used to describe the manner in which the charge on the capacitor varies with time. The charge on the capacitor increases from zero to 63% of its maximum value in a time interval equal to one time constant. Also, during one time constant, the charging current decreases from its initial maximum value of $I_0 = \dfrac{\mathcal{E}}{R}$ to 37% of I_0.

$$\tau = RC$$

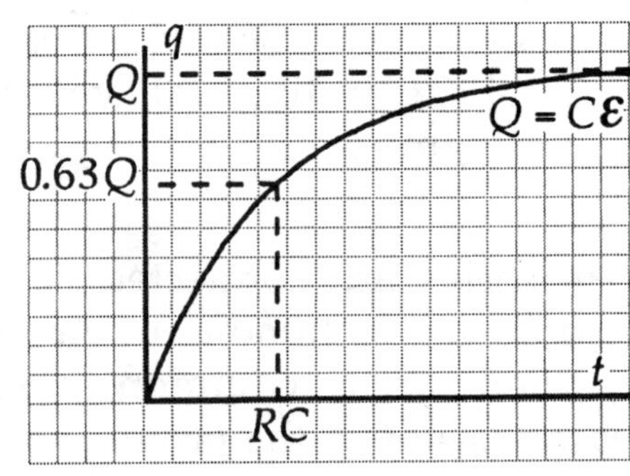

When a capacitor with an initial charge Q is discharged through a resistor, the **charge** and **current** decrease exponentially with time.

$$q(t) = Qe^{-t/RC} \tag{21.36}$$

$$I(t) = I_0 e^{-t/RC} \tag{21.37}$$

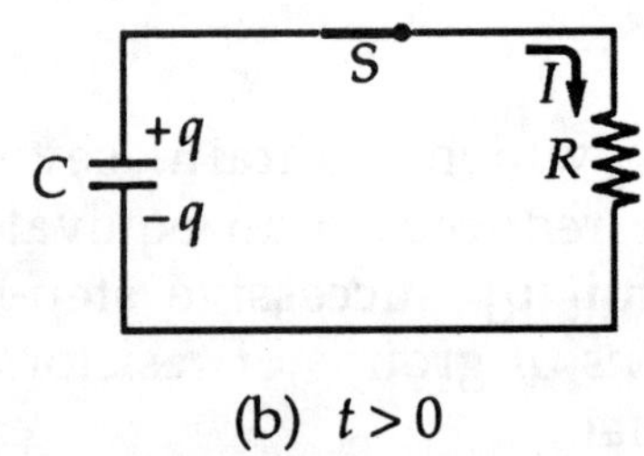

(a) $t < 0$ (b) $t > 0$

Chapter 21

SUGGESTIONS, SKILLS, AND STRATEGIES

A problem-solving strategy for resistors:

- Be careful with your choice of units. To calculate the resistance of a device in ohms, make sure that distances are in meters and use the SI value of ρ.

- When two or more unequal resistors are connected in **series,** they carry the same current, but the potential differences across them are not the same. The resistors add directly to give the equivalent resistance of the series combination.

- When two or more unequal resistors are connected in **parallel,** the potential differences across them are the same. Since the current is inversely proportional to the resistance, the currents through them are not the same. The equivalent resistance of a parallel combination of resistors is found through reciprocal addition, and the equivalent resistor is always **less** than the smallest individual resistor.

- A complicated circuit consisting of resistors can often be reduced to a simple circuit containing only one resistor. To do so, repeatedly examine the circuit and replace any resistors that are in series or in parallel, using the procedures outlined above. Sketch the new circuit after each set of changes has been made. Continue this process until a single equivalent resistance is found.

- If the current through, or the potential difference across, a resistor in the complicated circuit is to be identified, start with the final equivalent circuit found in the last step, and gradually work your way back through the circuits, using $V = IR$ to find the voltage drop across each equivalent resistor.

A strategy for using Kirchhoff's rules:

- First, draw the circuit diagram and assign labels and symbols to all the known and unknown quantities. You must assign **directions** to the currents in each part of the circuit. Do not be alarmed if you guess the direction of a current incorrectly; the resulting value will be negative, but **its magnitude will be correct**. Although the assignment of current directions is arbitrary, you must stick with your guess throughout as you apply Kirchhoff's rules.

- Apply the junction rule to any junction in the circuit. The junction rule may be applied as many times as a new current (one not used in a previous application) appears in the resulting equation. In general, the number of times the junction rule can be used is one fewer than the number of junction points in the circuit.

- Now apply Kirchhoff's loop rule to as many loops in the circuit as are needed to solve for the unknowns. Remember you must have as many equations as there are unknowns (I's, R's, and $\mathcal{E}$'s). In order to apply this rule, you must correctly identify the change in potential as you cross each element in traversing the closed loop. Watch out for signs!

The following figure illustrates convenient rules which you may use to determine the increase or decrease in potential as the current in a loop crosses a resistor or seat of emf. Notice that the potential **decreases** (changes by $-IR$) when the resistor is traversed **in the direction of the current**. There is an **increase** in potential of $+IR$ if the direction of travel is **opposite** the direction of current. If a seat of emf is traversed **in** the direction of the emf (from – to + on the battery), the potential **increases** by $\mathcal{E}$. If the direction of travel is from + to –, the potential **decreases** by $\mathcal{E}$ (changes by $-\mathcal{E}$).

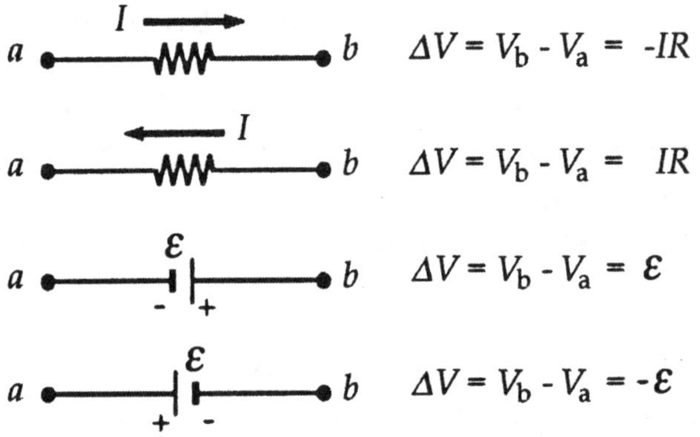

$\Delta V = V_b - V_a = -IR$

$\Delta V = V_b - V_a = IR$

$\Delta V = V_b - V_a = \mathcal{E}$

$\Delta V = V_b - V_a = -\mathcal{E}$

- Finally, you must solve the equations simultaneously for the unknown quantities. Be careful in your algebraic steps, and check your numerical answers for consistency.

As an illustration of the use of Kirchhoff's rules, consider a three-loop circuit which has the **general form** shown in the following figure on the left. In this illustration, the actual circuit elements, R's and $\mathcal{E}$'s are not shown but assumed known. There are six possible different values of I in the circuit; therefore you will need six independent equations to solve for the six values of I. There are four junction points in the circuit (at points a, d, f, and h). The first rule applied at **any three** of these points will yield three equations. The circuit can be thought of as a group of three "blocks" or meshes as shown in the following figure on the right. Kirchhoff's second law, when applied to each of these loops (*abcda*, *ahfga*, and *defhd*), will yield three additional equations. You can then solve the total of six equations simultaneously for the six values of I_1, I_2, I_3, I_4, I_5, and I_6. You can, of course, expect that the sum of the changes in potential difference around **any other closed loop** in the circuit will be zero (for example, *abcdefga* or *ahfedcba*); however the equations found by applying Kirchhoff's second rule to these additional loops **will not be independent** of the six equations found previously.

Chapter 21

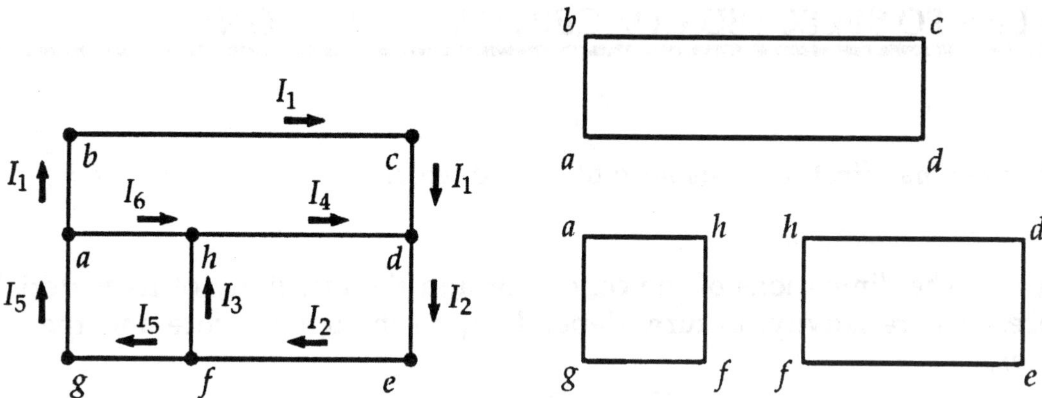

REVIEW CHECKLIST

▷ Define the term electric current in terms of rate of charge flow, and its corresponding unit of measure, the ampere. Calculate electron drift velocity, and the quantity of charge passing a point in a given time interval in a specified current-carrying conductor.

▷ Make calculations of the variation of resistance with temperature, which involves the concept of the temperature coefficient of resistivity.

▷ Use Joule's law to calculate the power dissipated in a resistor.

▷ Calculate the equivalent resistance of a group of resistors in parallel, series, or series-parallel combination.

▷ Apply Kirchhoff's rules to solve multiloop circuits; that is, find the currents and the potential difference between any two points.

▷ Calculate the charging (discharging) current $I(t)$ and the accumulated (residual) charge $q(t)$ during charging (and discharging) of a capacitor in an RC circuit.

ANSWERS TO SELECTED CONCEPTUAL QUESTIONS

2. What factors affect the resistance of a conductor?

Answer The dimensions of the conductor and the resistivity of its material affect its resistance. The resistivity, in turn, depends upon the temperature and the presence of impurities.

□ □ □ □

3. Two wires A and B of circular cross section are made of the same metal and have equal lengths, but the resistance of wire A is three times greater than that of wire B. What is the ratio of their cross-sectional areas? How do their radii compare?

Answer Since $R = \rho L / \text{Area}$, the ratio of resistances is given by

$$R_A / R_B = \text{Area}_A / \text{Area}_B.$$

Hence, the ratio of their areas is three. That is, the area of wire B is three times that of wire A. From the ratio of the areas, we can calculate that the radius of wire B is $\sqrt{3}$ times the radius of wire A.

□ □ □ □

4. Use the atomic theory of matter to explain why the resistance of a material should increase as its temperature increases.

Answer As the temperature increases, the amplitude of atomic vibrations increases. This makes it more likely that the electrons will be scattered by atomic vibrations, and makes it more difficult for charges to move inside the conductor.

□ □ □ □

6. What would happen to the drift velocity of the electrons in a wire and to the current in the wire if the electrons could move freely without resistance through the wire?

Answer Suppose in a normal metal we could proceed to a limit of zero resistance by lengthening the average time between collisions. The classical model of conduction then suggests that a constant applied voltage would cause constant acceleration of the free electrons, and a current steadily increasing in time.

On the other hand, we can actually switch to zero resistance by substituting a superconducting wire for the normal metal. In this case, the drift velocity of electrons is established by vibrations of atoms in the crystal lattice; the maximum current is limited; and it becomes impossible to apply a voltage.

□ □ □ □

7. If charges flow very slowly through a metal, why does it not require several hours for a light to come on when you throw a switch?

Answer Individual electrons move with a small average velocity through the conductor, but as soon as the voltage is applied, electrons all along the conductor start to move. Actually, the current does not flow "immediately," but is limited by the speed of light.

□ □ □ □

12. Why is it possible for a bird to sit on a high-voltage wire without being electrocuted?

Answer The bird is resting on a wire of a fixed potential. In order to be electrocuted, a potential difference is required. There is no potential difference between the bird's feet.

It could also be argued that a small amount of current does flow through the bird's feet, according to the rule of parallel resistors. However, since the resistance of the bird's feet is much higher than that of the wire, the amount of current that flows through the bird is still not enough to harm it.

□ □ □ □

22. A student claims that a second lightbulb in series is less bright than the first, because the first bulb uses up some of the current. How would you respond to this statement?

Answer Current is not something which is "used up" in a circuit. The current through a series circuit depends on all the devices in the circuit, and the same amount of current must flow through each device, in order to satisfy conservation of electrical charge.

□ □ □ □

SOLUTIONS TO SELECTED END-OF-CHAPTER PROBLEMS

3. Suppose that the current through a conductor decreases exponentially with time according to

$$I(t) = I_0 e^{-t/\tau},$$

where I_0 is the initial current (at $t = 0$), and τ is a constant having dimensions of time. Consider a fixed observation point within the conductor. (a) How much charge passes this point between $t = 0$ and $t = \tau$? (b) How much charge passes this point between $t = 0$ and $t = 10\tau$? (c) How much charge passes this point between $t = 0$ and $t = \infty$?

Solution

From $I = \frac{dQ}{dt}$, we have $$dQ = I\,dt$$

From this, we derive a general integral $$Q = \int dQ = \int I\,dt$$

In all three cases, define an end-time, T. $$Q = \int_0^T I_0\, e^{-t/\tau} dt$$

Integrating from time $t = 0$ to time $t = T$, $$Q = \int_0^T (-I_0\,\tau) e^{-t/\tau}\left(-\frac{dt}{\tau}\right)$$

Setting $Q = 0$ at $t = 0$, $$Q = I_0\,\tau\left(1 - e^{-T/\tau}\right)$$

(a) If $T = \tau$ $$Q = -I_0 \tau (e^{-1} - e^0)$$

$$Q = I_0 \tau (e^0 - e^{-1}) = 0.6321\, I_0 \tau \quad \Diamond$$

(b) If $T = 10\tau$, $$Q = I_0\,\tau\left(1 - e^{-10}\right) = 0.99995 I_0\,\tau \quad \Diamond$$

(c) If $T = \infty$, $$Q = I_0\,\tau\left(1 - e^{-\infty}\right) = I_0\,\tau \quad \Diamond$$

. A 0.900-V potential difference is maintained across a 1.50-m length of tungsten wire that has a cross-sectional area of 0.600 mm². What is the current in the wire?

Solution From Ohm's law, $I = \frac{V}{R}$ where $R = \frac{\rho L}{A}$. Therefore,

$$I = \frac{VA}{\rho L} = \frac{(0.900\text{ V})(6.00 \times 10^{-7}\text{ m}^2)}{(5.6 \times 10^{-8}\ \Omega \cdot \text{m})(1.50\text{ m})} = 6.43\text{ A} \quad \lozenge$$

---

15. If the drift velocity of free electrons in a copper wire is $7.84 \times 10^{-4}\text{ m/s}$, calculate the electric field in the conductor.

Solution The electron density in copper, from Example 21.1, is $8.48 \times 10^{28}/\text{m}^3$. The current density in this wire is

$$J = nqv_d = (8.48 \times 10^{28}/\text{m}^3)(1.60 \times 10^{-19}\text{ C})(7.84 \times 10^{-4}\text{ m/s})$$

$$J = 1.06 \times 10^7\text{ A/m}^2$$

From the footnote on page 608 of the text, $J = \sigma E = E/\rho$

$$E = \rho J = (1.7 \times 10^{-8}\ \Omega \cdot \text{m})(1.06 \times 10^7\text{ A/m}^2) = 0.181\text{ V/m} \quad \lozenge$$

---

19. What is the required resistance of an immersion heater that will increase the temperature of 1.50 kg of water from 10.0 °C to 50.0 °C in 10.0 min while operating at 110 V?

Solution From conservation of energy, $E_{(\text{thermal})} = E_{(\text{electrical})}$

$$E_{(\text{thermal})} = mc\Delta T \quad \text{and} \quad E_{(\text{electrical})} = \left(\frac{V^2}{R}\right)t$$

Therefore, since $c = 4186\text{ J/kg}\cdot{}^\circ\text{C}$

$$R = \frac{V^2 t}{cm\,\Delta T} = \frac{(110\text{ V})^2(600\text{ s})}{(4186\text{ J/kg}\cdot{}^\circ\text{C})(1.50\text{ kg})(40.0\ {}^\circ\text{C})} = 28.9\ \Omega \quad \lozenge$$

---

21. Suppose that a voltage surge produces 140 V for a moment. By what percentage wil the output of a 120-V, 100-W light bulb increase assuming its resistance does not change?

Solution We find the resistance:

$$P_1 = V_1 I_1$$

$$I_1 = \frac{P_1}{V_1} = \frac{100\text{ W}}{120\text{ V}} = 0.833\text{ A}$$

$$R = \frac{V_1}{I_1} = \frac{120\text{ V}}{0.833\text{ A}} = 144\ \Omega$$

Now the current is larger,

$$I_2 = \frac{V_2}{R} = \frac{140\text{ V}}{144\ \Omega} = 0.972\text{ A}$$

and the power is much larger:

$$P_2 = I_2 V_2 = (0.972\text{ A})(140\text{ V}) = 136\text{ W}$$

The percentage increase is

$$\frac{136\text{ W} - 100\text{ W}}{100\text{ W}} = 0.361 = 36.1\% \quad \Diamond$$

---

25. A battery has an emf of 15.0 V. The terminal voltage of the battery is 11.6 V when it is delivering 20.0 W of power to an external load resistor R. (a) What is the value of R? (b) What is the internal resistance of the battery?

Solution

(a) Combining Joule's law, $P = VI$, and Ohm's law, $V = IR$, gives

$$R = \frac{V^2}{P} = \frac{(11.6\text{ V})^2}{20.0\text{ W}} = 6.73\ \Omega \quad \Diamond$$

(b) $\mathcal{E} = IR + Ir$

$r = \dfrac{\mathcal{E} - IR}{I}$ where $I = \dfrac{V}{R}$

Therefore,

$$r = \frac{(\mathcal{E} - V)R}{V} = \frac{(15.0\text{ V} - 11.6\text{ V})(6.73\ \Omega)}{11.6\text{ V}} = 1.97\ \Omega \quad \Diamond$$

---

29. Consider the circuit shown in Figure P21.29. Find (a) the current in the 20.0-Ω resistor and (b) the potential difference between points *a* and *b*.

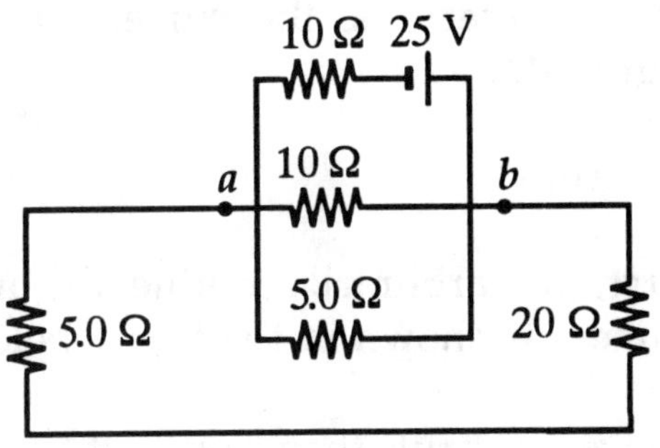

Figure P21.29

Solution If we turn the given diagram on its side, we find that it is the same as figure (a). The 20.0-Ω and 5.00-Ω resistors are in series, so the first reduction is as shown in (b). In addition, since the 10.0-Ω, 5.00-Ω, and 25.0-Ω resistors are then in parallel, we can solve for their equivalent resistance as:

$$R_{eq} = \frac{1}{\left(\frac{1}{10.0\ \Omega} + \frac{1}{5.00\ \Omega} + \frac{1}{25.0\ \Omega}\right)} = 2.94\ \Omega.$$

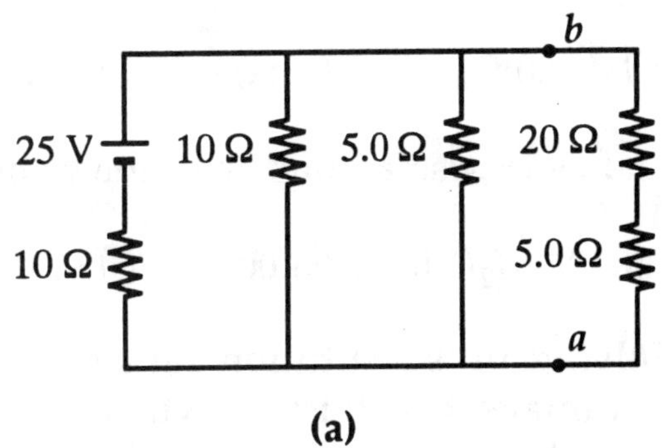

(a)

This is shown in figure (c), which in turn reduces to the circuit shown in (d).

Next, we work backwards through the diagrams, applying $I = V/R$ and $V = IR$. The 12.94-Ω resistor is connected across 25.0-V, so the current through the voltage source in every diagram is

$$I = \frac{V}{R} = \frac{25.0\ \text{V}}{12.94\ \Omega} = 1.93\ \text{A}$$

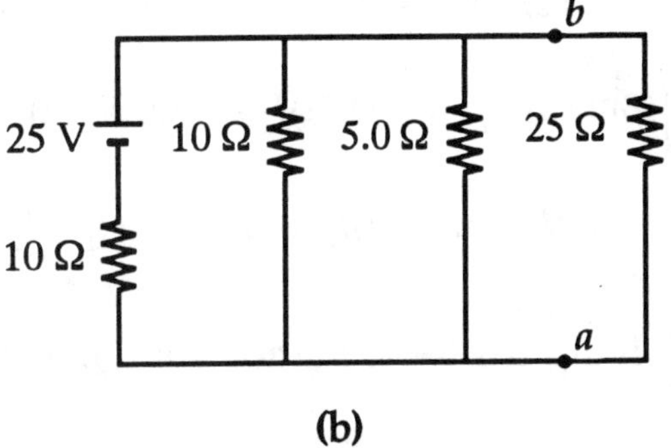

(b)

In figure (c), this 1.93 A goes through the 2.94-Ω equivalent resistor to give a voltage drop of:

$$V = IR = (1.93\ \text{A})(2.94\ \Omega) = 5.68\ \text{V}$$

From figure (b), we see that this voltage drop is the same across V_{ab}, the 10-Ω resistor, and the 5.00-Ω resistor.

(b) Therefore, $V_{ab} = 5.68\ \text{V}$ ◊

Since the current through the 20-Ω resistor is also the current through the 25-Ω line *ab*,

(a) $$I = \frac{V_{ab}}{R_{ab}} = \frac{5.68\ \text{V}}{25\ \Omega} = 0.227\ \text{A}$$ ◊

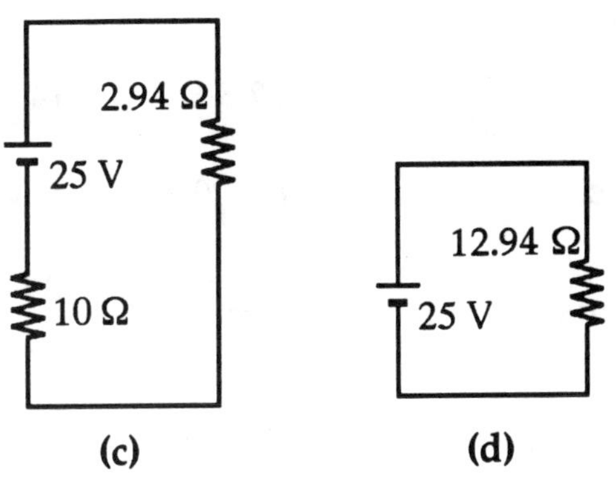

(c) (d)

33. Determine the current in each branch of the circuit in Figure P21.33.

3.0 Ω
5.0 Ω
1.0 Ω
1.0 Ω
8.0 Ω
4.0 V
12.0 V

Figure P21.33

Solution

First, we arbitrarily define the initial current directions and names, as shown in the figure below.

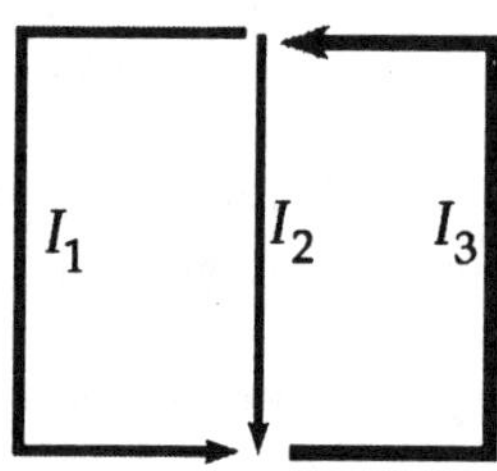

The current rule then says that $\qquad I_3 = I_1 + I_2. \qquad (1)$

By the voltage rule, clockwise around the left-hand loop,

$$+I_1(8.00\ \Omega) - I_2(5.00\ \Omega) - I_2(1.00\ \Omega) - 4.00\ \text{V} = 0 \qquad (2)$$

and clockwise around the right-hand loop,

$$4.00\ \text{V} + I_2(1.00\ \Omega + 5.00\ \Omega) + I_3(3.00\ \Omega + 1.00\ \Omega) - 12.0\ \text{V} = 0 \qquad (3)$$

Solving by substitution rather than by determinants has the advantage that (just as when a cat has kittens) the answers, after the first, come out much more easily. Thus we substitute $(I_1 + I_2)$ for I_3, and reduce our three equations to:

$$\left\{\begin{array}{l}(8.00\ \Omega)I_1 - (6.00\ \Omega)I_2 - 4.00\ \text{V} = 0 \\ \\ 4.00\ \text{V} + (6.00\ \Omega)I_2 + (4.00\ \Omega)(I_1 + I_2) - 12.0\ \text{V} = 0\end{array}\right\} \quad \text{or} \quad \left\{\begin{array}{l}I_2 = \dfrac{(8.00\ \Omega)I_1 - 4.00\ \text{V}}{6.00\ \Omega} \\ \\ -8.0\ \text{V} + (4.0\ \Omega)I_1 + (10.0\ \Omega)I_2 = 0\end{array}\right\}$$

Solving the top equation for I_2, and then substituting I_2 into the equation below it,

$$-8.00\ \text{V} + (4.00\ \Omega)I_1 + \frac{10.0}{6.00}\left((8.00\ \Omega)I_1 - 4.00\ \text{V}\right) = 0$$

$$(17.3\ \Omega)I_1 - (14.7\ \text{V}) = 0,$$

and $\qquad I_1 = 0.846$ A down the 8 Ω resistor ◊

Thus, $\qquad I_2 = \dfrac{(8.00\ \Omega)(0.846\ \text{A}) - 4.00\ \text{V}}{6.00\ \Omega} = 0.462\ \text{A}$ down in middle branch ◊

$I_3 = 0.846\ \text{A} + 0.462\ \text{A} = 1.31\ \text{A}$ up in the right-hand branch ◊

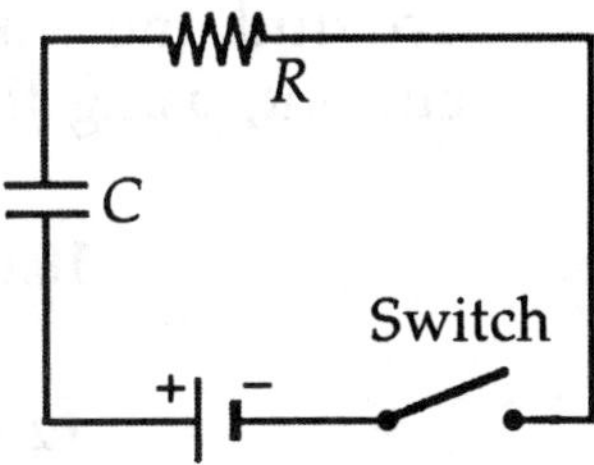

Figure 21.25

39. Consider a series RC circuit (Figure 21.25) for which $R = 1.00\ \text{M}\Omega$, $C = 5.00\ \mu\text{F}$, and $\mathcal{E} = 30.0$ V. Find (a) the time constant of the circuit and (b) the maximum charge on the capacitor after the switch is closed. (c) If the switch is closed at $t = 0$, find the current in the resistor 10.0 s later.

Solution

(a) $\tau = RC = (1.00\times 10^6\ \Omega)(5.00\times 10^{-6}\ \text{F}) = 5.00\ \Omega\cdot\text{F} = 5.00\ \text{s}$ ◊

(b) After a long time, the capacitor is "charged to thirty volts," separating charges

$$Q = CV = (5.00\times 10^{-6})(30.0\ \text{V}) = 150\ \mu\text{C} \quad ◊$$

(c) $I = I_0 e^{-t/\tau}$ where $I_0 = \dfrac{\mathcal{E}}{R}$ and $\tau = RC$

$$I = \frac{\mathcal{E}}{R}e^{-t/RC} = \left(\frac{30.0\ \text{V}}{1.00\times 10^6\ \Omega}\right)e^{-10.0\ \text{s}/5.00\ \text{s}}$$

$I = 4.06\times 10^{-6}\ \text{A} = 4.06\ \mu\text{A}$ ◊

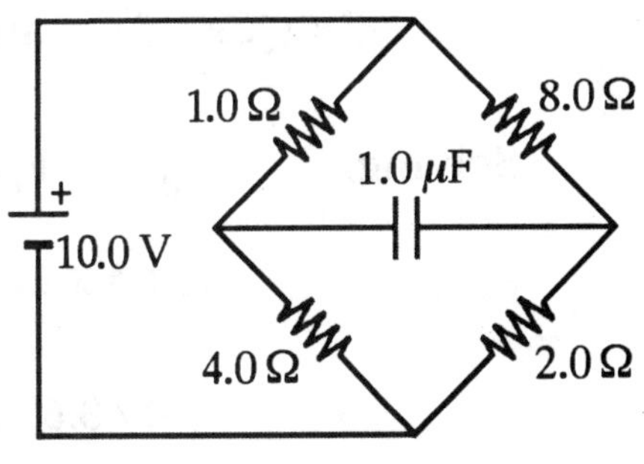

Figure P21.41

41. The circuit in Figure P21.41 has been connected for a long time. (a) What is the voltage across the capacitor? (b) If the battery is disconnected, how long does it take the capacitor to discharge to one tenth of its initial voltage?

Solution

(a) After a long time the capacitor branch will carry negligible current. The current flow is as shown in Figure (a).

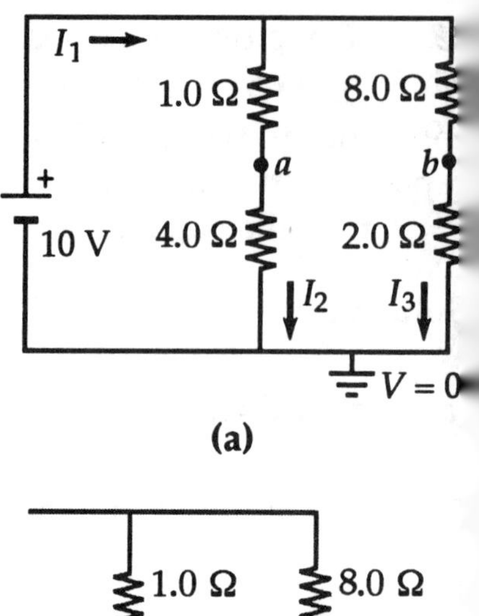

(a)

To find the voltage at point a, we first find the current, using the voltage rule:

$$10.0\ \text{V} - (1.00\ \Omega) I_2 - (4.00\ \Omega) I_2 = 0$$

$$I_2 = 2.00\ \text{A}$$

$$V_a = (4.00\ \Omega) I_2 = 8.00\ \text{V}$$

Similarly, $10.0\ \text{V} - (8.00\ \Omega) I_3 - (2.00\ \Omega) I_3 = 0$

$$I_3 = 1.00\ \text{A}$$

At point b, $V_b = (2.00\ \Omega) I_3 = 2.00\ \text{V}$

Thus, the voltage across the capacitor is

$$V_a - V_b = 8.00\ \text{V} - 2.00\ \text{V} = 6.00\ \text{V} \quad \lozenge$$

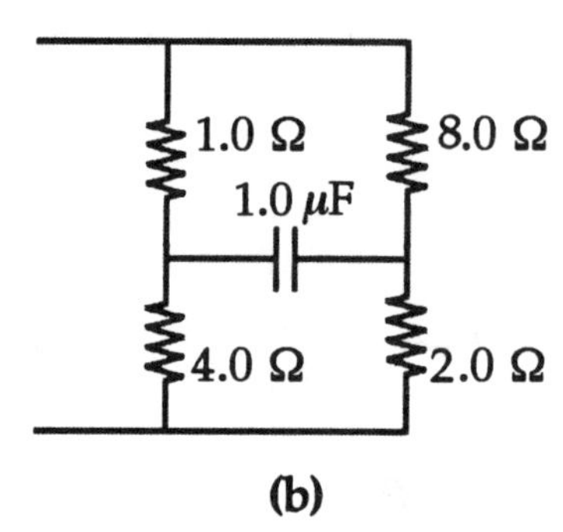

(b)

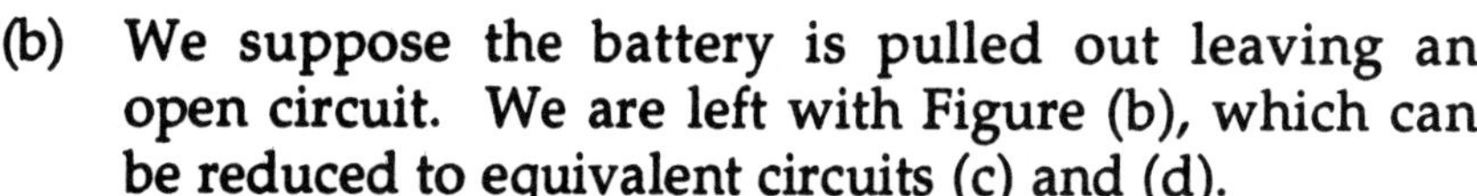

(b) We suppose the battery is pulled out leaving an open circuit. We are left with Figure (b), which can be reduced to equivalent circuits (c) and (d).

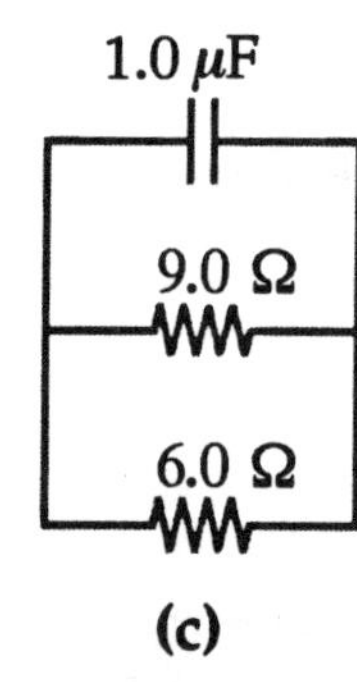

(c)

From (d), we see that the capacitor discharges through a 3.60 Ω equivalent resistance. According to $q = Qe^{-t/RC}$, we calculate that

$$qC = QCe^{-t/RC} \quad \text{and} \quad V = V_o e^{-t/RC}$$

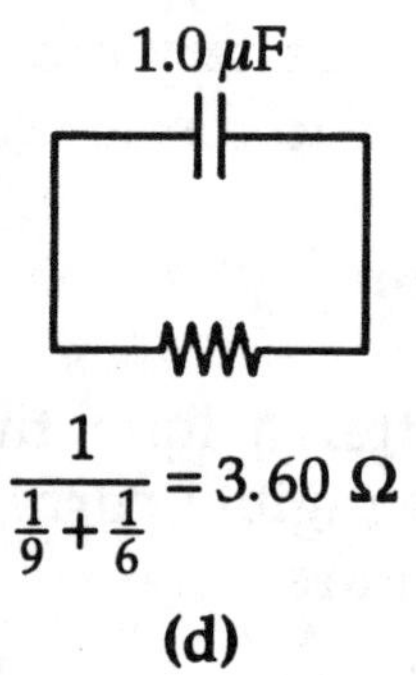

$$\frac{1}{\frac{1}{9}+\frac{1}{6}} = 3.60\ \Omega$$

(d)

Solving, $\frac{1}{10} V_o = V_o e^{-t/(3.60\ \Omega)(1.00\ \mu\text{F})}$

$$e^{-t/3.60\ \mu\text{s}} = 0.100$$

$$(-t / 3.60\ \mu\text{s}) = \ln\ 0.100 = -2.30$$

$$\frac{t}{3.60\ \mu\text{s}} = 2.30$$

$$t = (2.30)(3.60\ \mu\text{s}) = 8.29\ \mu\text{s} \quad \lozenge$$

49. An electric heater is rated at 1500 W, a toaster at 750 W, and an electric grill at 1000 W. The three appliances are connected to a common 120-V circuit. (a) How much current does each draw? (b) Is a circuit fused at 25.0-A circuit sufficient in this situation? Explain.

Solution

(a) Heater: $$I = \frac{P}{V} = \frac{1500\text{ W}}{120\text{ V}}\left(\frac{1\text{ J/s}}{1\text{ W}}\right)\left(\frac{1\text{ V}}{1\text{ J/C}}\right)\left(\frac{1\text{ A}}{1\text{ C/s}}\right) = 12.5\text{ A} \quad \Diamond$$

Toaster: $$I = \frac{750\text{ W}}{120\text{ V}} = 6.25\text{ A} \quad \Diamond$$

Grill: $$I = \frac{1000\text{ W}}{120\text{ V}} = 8.33\text{ A} \quad \Diamond$$

(b) Together in parallel they pass current $12.5 + 6.25 + 8.33\text{ A} = 27.1\text{ A}$, so 25-A wiring cannot feed all of them energy at once. ◊

---

51. An experiment is conducted to measure the electrical resistivity of Nichrome in the form of wires with different lengths and cross-sectional areas. For one set of measurements, a student uses #30 gauge wire, which has a cross-sectional area of $7.3 \times 10^{-8}\text{ m}^2$. The voltage across the wire and the current in the wire are measured with a voltmeter and ammeter, respectively. For each of the measurements given in the table below taken on wires of three different lengths, calculate the resistance of the wires and the corresponding values of the resistivity. What is the average value of the resistivity, and how does it compare with the value given in Table 21.1?

L (m)	ΔV (V)	I (A)	R (Ω)	ρ (Ω·m)
0.540	5.22	0.500		
1.028	5.82	0.276		
1.543	5.94	0.187		

Solution

Find each resistance from $R = \frac{\Delta V}{I}$, as in $\frac{5.22\text{ V}}{0.500\text{ A}} = 10.4\ \Omega$.

Find each resistivity from $R = \frac{\Delta V}{I} = \frac{\rho L}{A}$

$$\rho = \frac{\Delta VA}{IL} = \frac{(5.22\text{ V})\left(7.3\times10^{-8}\text{ m}^2\right)}{(0.500\text{ A})(0.540\text{ m})}$$

$$\rho = 1.41\times10^{-6}\ \Omega\cdot\text{m}$$

To obtain

L (m)	R (Ω)	ρ (Ω·m)
0.540	10.4	1.41×10^{-6}
1.028	21.1	1.50×10^{-6}
1.543	31.8	1.50×10^{-6}

Thus the average resistivity is $\rho = 1.47\times10^{-6}\ \Omega\cdot\text{m}$ ◊

This differs from the tabulated $1.50\times10^{-6}\ \Omega\cdot\text{m}$ by 2%. The difference is accounted for by the experimental uncertainty, which we may estimate as

$$\frac{1.47-1.41}{1.47} = 4\%$$

Chapter 22

The white arc in this photograph indicates the circular path followed by an electron beam moving in a magnetic field. The vessel contains gas at very low pressure, and the beam is made visible as the electrons collide with the gas atoms, which in turn emit visible light. The magnetic field is produced by two coils (not shown). The apparatus can be used to measure the charge-to-mass ratio for the electron. (Courtesy of CENCO)

MAGNETISM

Chapter 22

MAGNETISM

INTRODUCTION

Many historians of science believe that the compass, which uses a magnetic needle was used in China as early as the 13th century B.C., its invention being of Arab or Indian origin. The early Greeks knew about magnetism as early as 800 B.C. They discovered that certain stones, now called magnetite (Fe_3O_4), attract pieces of iron. In 1269 Pierre de Maricourt mapped out the directions taken by a needle when it was placed at various points on the surface of a spherical natural magnet. He found that the directions formed lines that encircled the sphere and passed through two points diametrically opposite to each other, which he called the **poles** of the magnet. Subsequent experiments showed that every magnet, regardless of its shape, has two poles, called **north** and **south poles**, which exhibit forces on each other in a manner analogous to electric charges. That is, like poles repel each other and unlike poles attract each other.

This chapter deals with the origin of the magnetic field, namely, moving charges or electric currents. We show how to use the Biot-Savart law to calculate the magnetic field produced at a point by a current element. Using this formalism and the superposition principle, we then calculate the total magnetic field due to a current loop. Next, we show how to determine the force between two current-carrying conductors, a calculation that leads to the definition of the ampere. We also introduce Ampère's law, which is very useful for calculating the magnetic field of highly symmetric configurations carrying steady currents. We apply Ampère's law to determine the magnetic field for a long current-carrying wire, a solenoid, and a toroid.

NOTES FROM SELECTED CHAPTER SECTIONS

22.2 The Magnetic Field

Particles with charge q, moving with speed v in a magnetic field **B**, experience a magnetic force **F**:

- The magnetic force is proportional to the charge q and speed v of the particle.
- The magnitude and direction of the magnetic force depend on the velocity of the particle and on the magnitude and direction of the magnetic field.
- When a charged particle moves in a direction **parallel** to the magnetic field vector, the magnetic force **F** on the charge is **zero**.

- The magnetic force acts in a direction perpendicular to both **v** and **B**; that is, **F** is perpendicular to the plane formed by **v** and **B**.
- The magnetic force on a positive charge is in the direction opposite to the force on a negative charge moving in the same direction.
- If the velocity vector makes an angle θ with the magnetic field, the magnitude of the magnetic force is proportional to $\sin\theta$.

There are several important differences between electric and magnetic forces:

- The electric force is always in the direction of the electric field, whereas the magnetic force is perpendicular to the magnetic field.
- The electric force acts on a charged particle independent of the particle's velocity, whereas the magnetic force acts on a charged particle only when the particle is in motion.
- The electric force does work in displacing a charged particle, whereas the magnetic force associated with a steady magnetic field does **no** work when a particle is displaced.

22.5 The Biot-Savart Law

The **Biot-Savart law** says that if a wire carries a steady current I, the magnetic field $d\mathbf{B}$ at a point P associated with an element $d\mathbf{s}$ has the following properties:

- The vector $d\mathbf{B}$ is perpendicular both to $d\mathbf{s}$ (which is in the current's direction) and to the unit vector $\hat{\mathbf{r}}$ (directed from the current element to the point P.)
- The magnitude of $d\mathbf{B}$ is inversely proportional to r^2, where r is the distance from the element to the point P.
- The magnitude of $d\mathbf{B}$ is proportional to the current and to the length of the element, $d\mathbf{s}$.
- The magnitude of $d\mathbf{B}$ is proportional to $\sin\theta$, where θ is the angle between the vectors $d\mathbf{s}$ and $\hat{\mathbf{r}}$.

22.6 The Magnetic Force Between Two Parallel Conductors

Parallel conductors carrying currents in the **same direction attract** each other, whereas parallel conductors carrying currents in **opposite directions repel** each other.

The force between two parallel wires each carrying a current is used to define the ampere as follows:

If two long, parallel wires 1 m apart in a vacuum carry the same current and the force per unit length on each wire is 2×10^{-7} N/m, then the current is defined to be 1 A.

22.7 Ampère's Law

The direction of the magnetic field due to a current in a conductor is given by the right-hand rule:

If the wire is grasped in the right hand with the thumb in the direction of the current, the fingers will wrap (or curl) in the direction of **B**.

Ampère's law is valid only for **steady** currents and is useful only in those cases where the current configuration has a **high degree** of **symmetry**.

22.9 Magnetism in Matter

In order to describe the magnetic properties of materials, it is convenient to classify the material into three categories: paramagnetic, ferromagnetic, and diamagnetic. **Paramagnetic and ferromagnetic materials** are those that have atoms with permanent magnetic dipole moments.

EQUATIONS AND CONCEPTS

The magnetic field (or magnetic induction) at some point in space is defined in terms of the **magnetic force** exerted on a moving positive electric charge at that point. The SI unit of the magnetic field is the tesla (T) or weber per square meter (Wb/m^2).

$$\mathbf{F} = q\mathbf{v} \times \mathbf{B} \tag{22.1}$$

$$1\,\mathrm{T} = 1\,\frac{\mathrm{N \cdot s}}{\mathrm{C \cdot m}}$$

The magnetic force will be of maximum magnitude when the charge moves along a direction perpendicular to the direction of the magnetic field. In general, the velocity vector may be directed along some direction other than 90.0° relative to the magnetic field. In this case, the magnetic force on the moving charge is less than its maximum value.

$$F = qvB\sin\theta \tag{22.2}$$

Equation 22.2 can be written in a form which serves to define the magnitude of the magnetic field.

$$B \equiv \frac{F}{qv\sin\theta}$$

The cgs unit of magnetic field is the gauss (G).

$$1\text{ T} = 10^4\text{ G}$$

In order to determine the direction of the magnetic force, apply the right-hand rule:

> Hold your right hand with your fingers curling, first in the direction of **v**, and then in the direction of **B**, as shown. The magnetic force on a positive charge points in the direction of your thumb. If the charge is negative, then the direction of the force is reversed.

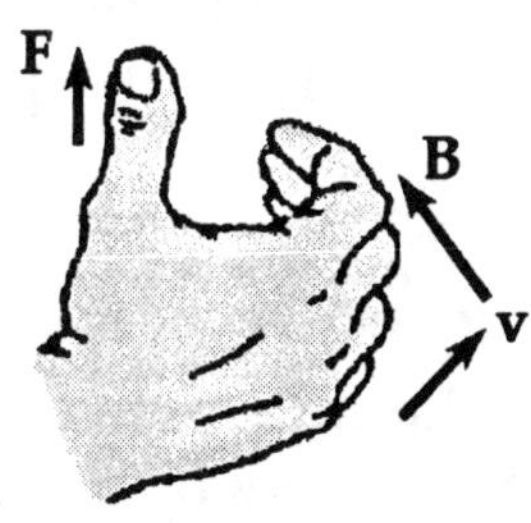

If a **straight** wire carrying a current is placed in an external magnetic field, a **magnetic force** will be exerted on the wire. The magnetic force on a wire of arbitrary shape is found by integrating over the length of the wire. In these equations the direction of $\boldsymbol{\ell}$ and $d\mathbf{s}$ is that of the current.

$$\mathbf{F} = I\boldsymbol{\ell} \times \mathbf{B} \qquad (22.3)$$

$$\mathbf{F} = I\int_a^b d\mathbf{s} \times \mathbf{B} \qquad (22.5)$$

The magnitude of the magnetic force on the conductor depends on the angle between the direction of the conductor and the direction of the field.

$$F = BIL\sin\theta$$

The magnetic force will be maximum when the conductor is directed perpendicular to the magnetic field.

$$F_{\max} = BIL$$

When a closed conducting loop carrying a current is placed in an external magnetic field, there is a **net torque** exerted on the loop. In Equation 22.7, the area vector **A** is directed perpendicular to the area of the loop with a sense given by the right-hand rule. The magnitude of **A** is numerically equal to the area of the loop.

$$\tau = I\mathbf{A} \times \mathbf{B} \quad (22.7)$$

The magnitude of the torque will depend on the angle between the direction of the magnetic field and the direction of the normal (or perpendicular) to the plane of the loop.

$$\tau = IAB\sin\theta$$

The magnitude of the torque will be maximum when the magnetic field is parallel to the plane of the loop.

$$\tau_{max} = IAB$$

The direction of rotation of the loop is such that the normal to the plane of the loop turns into a direction parallel to the magnetic field.

The torque on a current loop can also be expressed in terms of the magnetic dipole moment of the loop.

$$\tau = \mu \times \mathbf{B} \quad (22.9)$$

$$\text{where} \quad \mu = I\mathbf{A} \quad (22.8)$$

The **Biot-Savart law** gives the magnetic field at a point in space due to an element of conductor $d\mathbf{s}$ which carries a current I and is at a distance r away from the point.

$$d\mathbf{B} = \frac{\mu_0}{4\pi}\frac{I d\mathbf{s} \times \hat{\mathbf{r}}}{r^2} \quad (22.13)$$

The permeability of free space is a constant.

$$\mu_o = 4\pi \times 10^{-7}\ \text{T·m/A}$$

The **total magnetic field** is found by integrating the Biot-Savart law expression over the entire length of the conductor.

$$\mathbf{B} = \frac{\mu_0 I}{4\pi}\int \frac{d\mathbf{s} \times \hat{\mathbf{r}}}{r^2}$$

Chapter 22

The magnetic field due to several important geometric arrangements of a current-carrying conductor can be calculated by use of the Biot-Savart law:

B at a distance a from a **long straight conductor**, carrying a current I.

$$B = \frac{\mu_0 I}{2\pi a}$$

B at the center of an **arc of radius** R **which subtends an angle** θ **(in radians) at the center of the arc.**

$$B = \frac{\mu_0 I}{4\pi R}\theta$$

B_x on the axis of a **circular loop** of radius R and at a **distance** x **from the plane** of the loop.

$$B_x = \frac{\mu_0 I R^2}{2\left(x^2 + R^2\right)^{3/2}} \qquad (22.16)$$

Substituting $x = 0$ gives the magnetic field along the axis of the loop.

$$B = \frac{\mu_0 I}{2R} \qquad (22.17)$$

The magnitude of the **magnetic force per unit length** between very long parallel conductors depends on the distance a between the conductors and the magnitudes of the two currents.

$$\frac{F_1}{\ell} = \frac{\mu_0 I_1 I_2}{2\pi a} \qquad (22.20)$$

If the parallel currents I_1 and I_2 are in the same direction, the force between conductors will be one of attraction. Parallel conductors carrying currents in opposite directions will repel each other. In any case, the magnitude of the forces on the two conductors will be **equal.**

Ampère's law represents a relationship between the integral of the tangential component of the magnetic field around any closed path and the total current which threads the closed path.

$$\oint \mathbf{B} \cdot d\mathbf{s} = \mu_0 I \qquad (22.22)$$

B inside a toroid having N turns and at a distance r from the center of the toroid.

$$B = \frac{\mu_0 N I}{2\pi r} \quad (22.24)$$

B near the center of a **solenoid** of n turns per unit length.

$$B = \mu_0 n I \quad (22.25)$$

The direction of the magnetic field due to a current in a long wire is determined by using the right-hand rule for **B**:

> Hold the conductor in the right hand with the thumb pointing in the direction of the conventional current. The fingers will then wrap around the wire in the direction of the magnetic field lines. The magnetic field is tangent to the circular field lines at every point in the region around the conductor.

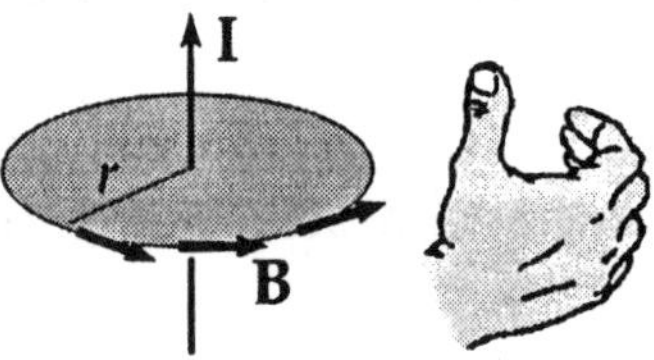

The direction of the magnetic field at the center of a current loop is perpendicular to the plane of the loop and directed in the sense given by the right-hand rule for **B**.

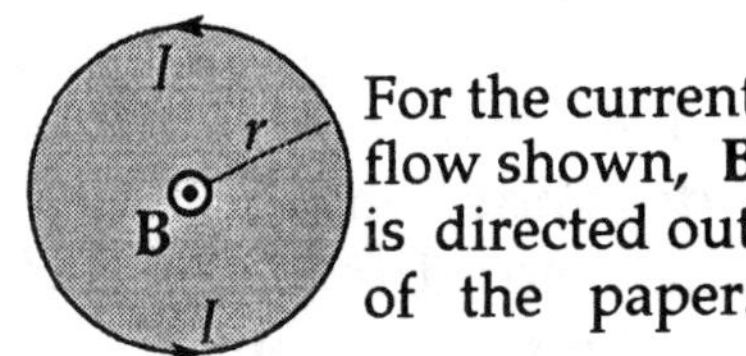

Within a solenoid, the magnetic field is parallel to the axis of the solenoid and pointing in a sense determined by applying right-hand rule B to one of the coils.

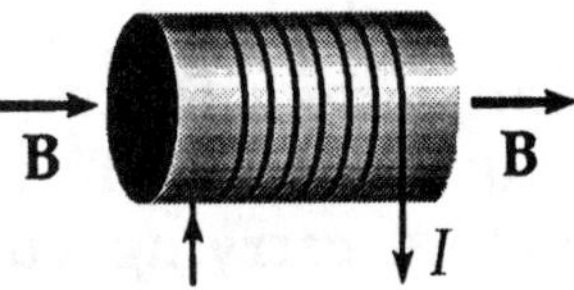

The magnetic moment μ of an orbiting electron is proportional to its orbital angular momentum L.

$$\mu = \left(\frac{e}{2m}\right)L \quad (22.26)$$

$$L = 0,\ \hbar,\ 2\hbar,\ 3\hbar,\ \ldots$$

The intrinsic magnetic moment μ_s associated with the spin of the electron is called the Bohr magneton.

$$\mu_B = 9.27 \times 10^{-24}\ \text{J/T} \quad (22.28)$$

SUGGESTIONS, SKILLS, AND STRATEGIES

To remember the symbols for vectors that point away from you and towards you, think of a three-dimensional archery arrow as it turns to point first away from you (shown on the left), and then towards you, (shown on the right). The point on the arrow is represented by a dot; the feathers are represented by an 'x'.

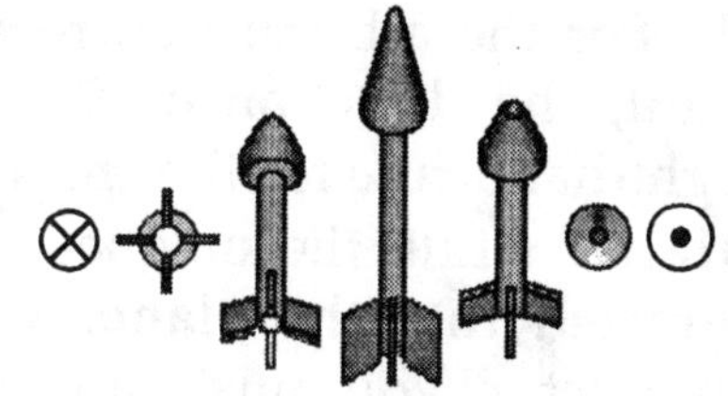

Equation 22.1, $\mathbf{F} = q\mathbf{v} \times \mathbf{B}$, serves as the definition of the magnetic field vector **B**. The direction of the magnetic force **F** is determined by the right-hand rule for the cross product as illustrated in the figure to the right. (This assumes that the charge q is a positive charge.)

The right-hand rule and the vector cross product are also used to determine the direction of the torque and the direction of the resulting rotation for a closed current loop in a magnetic field. When the four fingers on your right hand circle the current loop in the direction of the current, the thumb will point in the direction of the area vector **A**. Applying this rule to the situation shown in the figure to the right, the vector **A** is directed out of the plane of the rectangular loop facing you as shown.

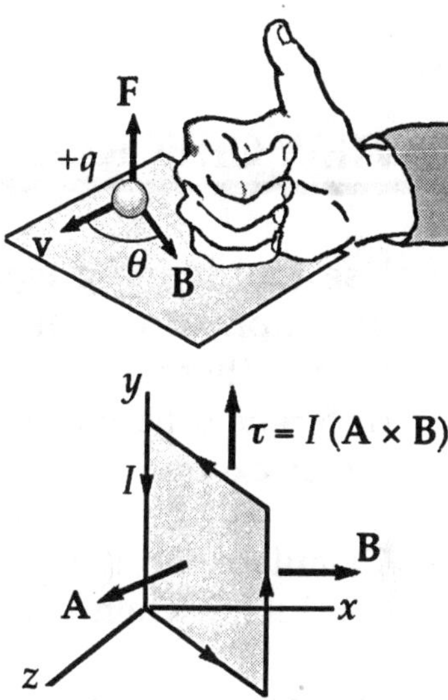

The resulting torque is parallel to the direction of $\mathbf{A} \times \mathbf{B}$ and is along the positive y axis as shown in the figure. This is consistent with the general rule for determining the direction of the cross product of two vectors. If the loop is considered to be hinged along the edge joining the y axis, then the rotation will be such that the angle θ between **A** and **B** decreases. The loop rotates until its perpendicular vector **A** is parallel to the magnetic field vector **B**. This is shown in the figure as a counterclockwise rotation about the y axis as seen from above.

It is important to remember that the **Biot-Savart law**, given by Equation 22.13,

$$d\mathbf{B} = \frac{\mu_0 I}{4\pi} \frac{d\mathbf{s} \times \hat{\mathbf{r}}}{r^2}$$

is a **vector** expression. The unit vector $\hat{\mathbf{r}}$ is directed from the element of conductor $d\mathbf{s}$ to the point P where the magnetic field is to be calculated, and r is the distance from $d\mathbf{s}$ to point P. For the arbitrary current element shown in the figure at right, the direction of **B** at point P, as determined by the right-hand rule for the cross product, is directed **out of the plane**; while the magnetic field at point P' due to the current in the element $d\mathbf{s}$ is directed **into the plane**. In order to find the **total** magnetic field at any point due to a conductor, you must sum up the contributions from all current elements making up the conductor. This means that the total **B** field is expressed as an integral over the entire length of the conductor:

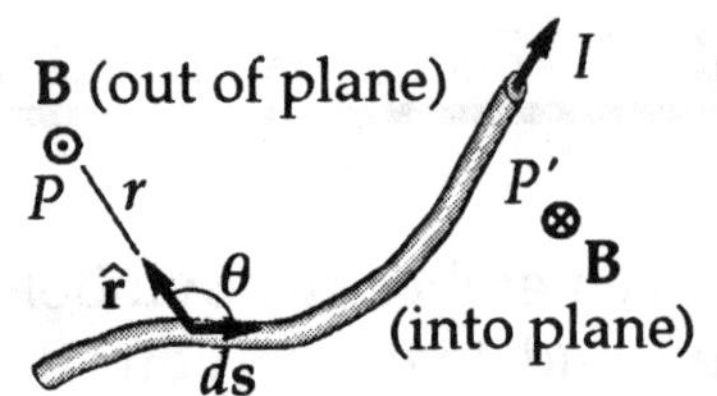

$$\mathbf{B} = \frac{\mu_0 I}{4\pi}\int \frac{d\mathbf{s}\times\hat{\mathbf{r}}}{r^2}$$

REVIEW CHECKLIST

▷ Use the defining equation for a magnetic field **B** to determine the magnitude and direction of the magnetic force exerted on an electric charge moving in a region where there is a magnetic field. You should understand clearly the important differences between the forces exerted on electric charges by electric fields and those forces exerted on moving electric charges by magnetic fields.

▷ Calculate the magnitude and direction of the magnetic force on a current-carrying conductor when placed in an external magnetic field.

▷ Determine the magnitude and direction of the torque exerted on a closed current loop in an external magnetic field. You should understand how to designate the direction of the area vector corresponding to a given current loop, and to incorporate the magnetic moment of the loop into the calculation of the torque on the loop.

▷ Use the Biot-Savart law to calculate the magnetic field at a specified point in the vicinity of a current element, and by integration find the total magnetic field due to a number of important geometric arrangements. Your use of the Biot-Savart law must include a clear understanding of the **direction** of the magnetic field contribution relative to the direction of the current element which produces it and the direction of the vector which locates the point at which the field is to be calculated.

▷ Use Ampère's law to calculate the magnetic field due to steady current configurations which have a sufficiently high degree of symmetry such as a long straight conductor, a long solenoid, and a toroidal coil.

ANSWERS TO SELECTED CONCEPTUAL QUESTIONS

1. Two charged particles are projected into a region where there is a magnetic field perpendicular to their velocities. If the charges are deflected in opposite directions, what can you say about them?

Answer We know the magnetic field is constant, and the velocity vectors are the same, but one force is the negative of the other. From $\mathbf{F} = q(\mathbf{v} \times \mathbf{B})$, we can conclude that the only thing which could cause the force to be of opposite sign is if the charges were of opposite sign.

3. Is it possible to orient a current loop in a uniform magnetic field so that the loop does not tend to rotate?

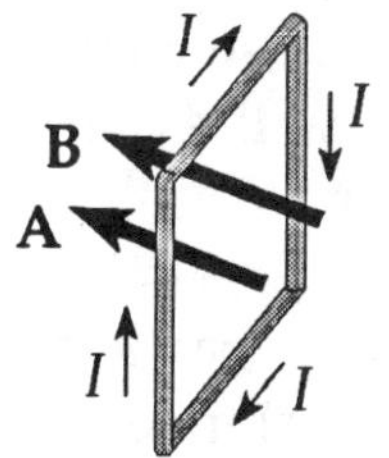

Answer Yes. If the magnetic field is perpendicular to the plane of the loop, the forces on opposite sides will be equal and opposite, but will produce no net torque.

□ □ □ □

9. Parallel wires exert magnetic forces on each other. What about perpendicular wires? Imagine two wires oriented perpendicular to each other, and almost touching. Each wire carries a current. Is there a force between the wires?

Answer There will be no net force on the wires, but there will be a torque. To understand this, imagine a fixed vertical wire, and a free horizontal wire, as shown to the right. The vertical wire carries a vertical current, and creates a magnetic field that circles the wire, as shown by the upper hand. Each segment of the horizontal wire (of length dL) also carries a current that interacts with the magnetic field according to the equation $d\mathbf{F} = I(d\mathbf{L} \times \mathbf{B})$. Applying the right-hand rule, we see that the horizontal wire experiences an upward force on one side, and an equal downward force on the other. The forces cancel, creating a torque around the point at which the wires cross.

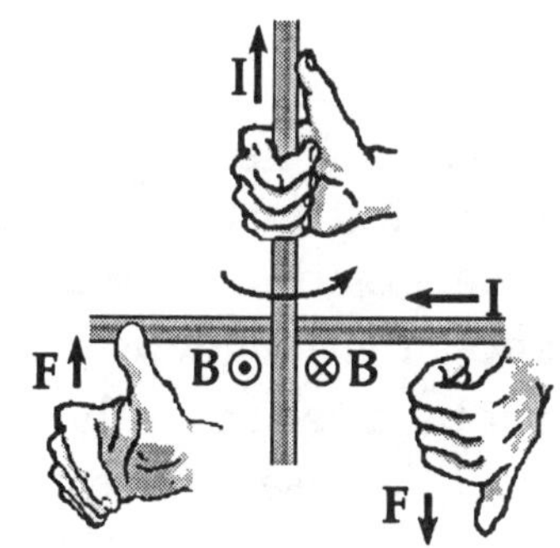

□ □ □ □

11. How can a current loop be used to determine the presence of a magnetic field in a given region of space?

Answer The loop can be mounted free to rotate around an axis. The loop will rotate about this axis when placed in an external magnetic field for some arbitrary orientation. As the current through the loop is increased, the torque on it will increase.

□ □ □ □

12. It is found that charged particles from outer space, called cosmic rays, strike the Earth more frequently at the poles than at the Equator. Why?

Answer You can imagine the magnetic field of the earth as being similar to that of a bar magnet, as shown to the right. A charged particle that approaches the earth's equator will experience a forces that deflect it away from the earth.

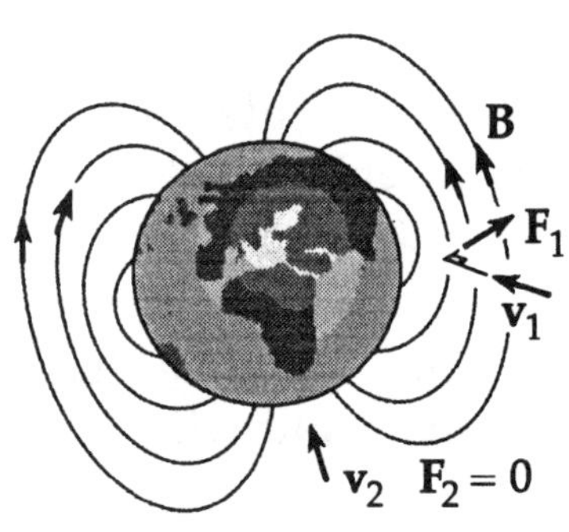

A charged particle that approaches the earth at the poles will experience a magnetic field that is parallel to its motion. Since the cross product of two parallel vectors is zero, it will experience no deflecting force, and will strike the earth directly.

□ □ □ □

13. Explain why two parallel wires carrying currents in opposite directions repel each other.

Answer The figure at the right will help you understand this result. The magnetic field due to wire 2 at the position of wire 1 is directed out of the paper. Hence, the magnetic force on wire 1, given by $I_1\mathbf{L}_1 \times \mathbf{B}_2$, must be directed to the left since $\mathbf{L}_1 \times \mathbf{B}_2$ is directed to the left. Likewise, you can show that the magnetic force on wire 2 due to the field of wire 1 is directed towards the right.

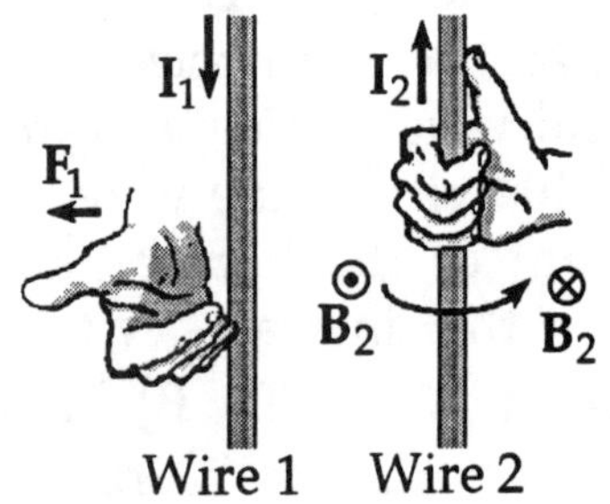

□ □ □ □

14. A hollow copper tube carries a current. Why is **B** = 0 inside the tube? Is **B** nonzero outside the tube?

Answer Let us apply Ampere's circuital law to the closed path labeled 1 in this figure. Since there is no current through this path, and because of the symmetry of the configuration, we see that the magnetic field inside the tube must be zero. On the other hand, the net current through the path labeled 2 is I, the current carried by the conductor. Therefore, the field outside the tube is nonzero.

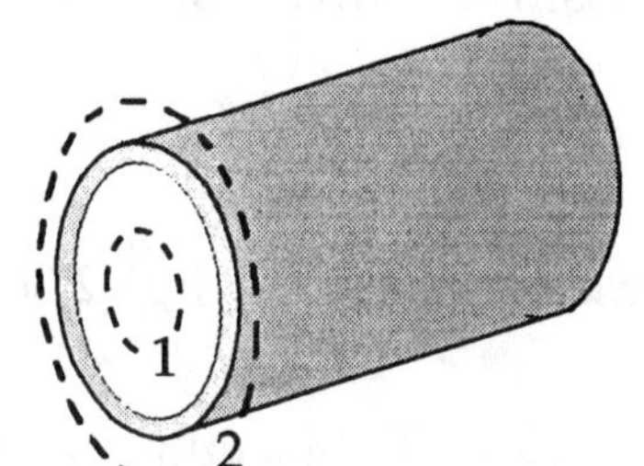

□ □ □ □

SOLUTIONS TO SELECTED END-OF-CHAPTER PROBLEMS

1. Determine the initial direction of the deflection of charged particles as they enter the magnetic fields shown in Figure P22.1.

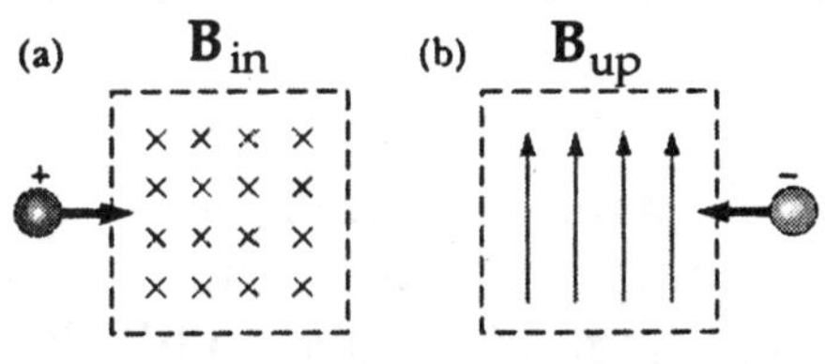

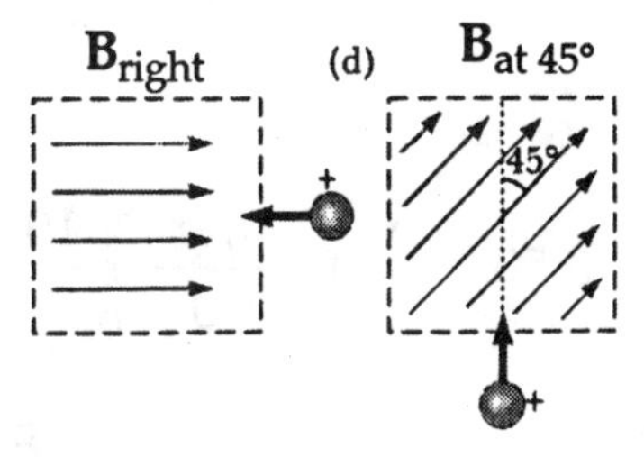

Figure P22.1

Solution

(a) By solution figure (a), $\mathbf{v} \times \mathbf{B}$ is $(\text{right}) \times (\text{away}) = \text{up}$ ◊

(b) By solution figure (b), $\mathbf{v} \times \mathbf{B}$ is $(\text{left}) \times (\text{up}) = \text{away}$. Since the charge is negative, $q\mathbf{v} \times \mathbf{B}$ is toward you ◊

(c) $\mathbf{v} \times \mathbf{B}$ is zero since the angle between **v** and **B** is 180° and sin 180° = 0. There is no deflection. ◊

(d) $\mathbf{v} \times \mathbf{B}$ is $(\text{up}) \times (\text{up and right})$, or away from you ◊

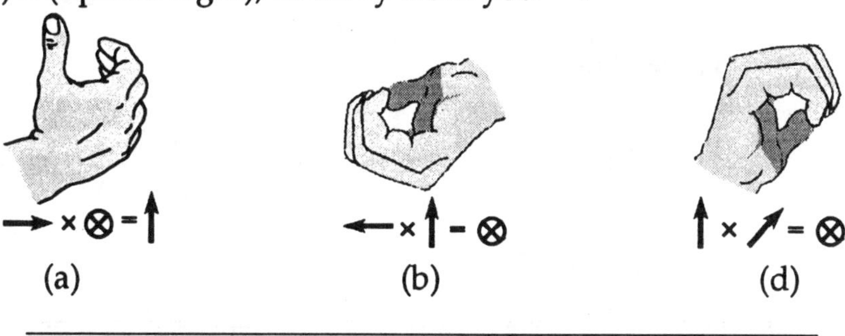

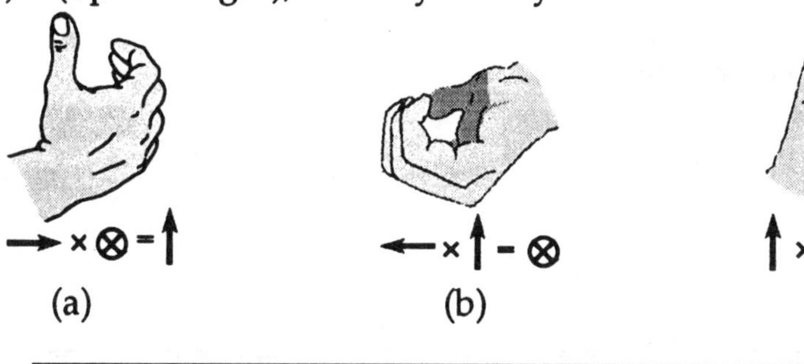

3. A proton moves perpendicular to a uniform magnetic field **B** at 1.00×10^7 m/s and experiences an acceleration of 2.00×10^{13} m/s^2 in the $+x$ direction when its velocity is in the positive z direction. Determine the magnitude and direction of the field.

Solution

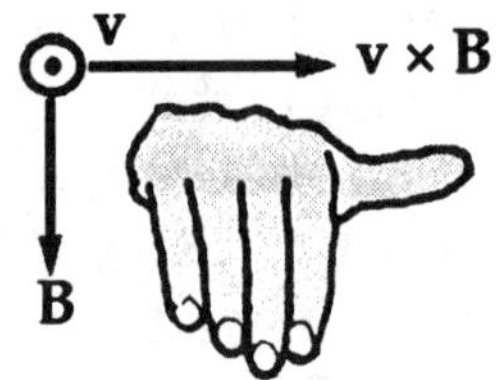

$$F = ma = (1.67\times10^{-27}\text{ kg})(2.00\times10^{13}\text{ m/s}^2)$$

F= ma
F=qvBsin90

$$F = 3.34\times10^{-14}\text{ N} = qvB\sin 90°$$

$$B = \frac{F}{qv} = \frac{3.34\times10^{-14}\text{ N}}{(1.60\times10^{-19}\text{ C})(1.00\times10^{7}\text{ m/s})}$$

$B = 2.09\times10^{-2}$ T ◊

The right-hand rule shows that **B** must be in the $-y$ direction ◊
This yields a force on the proton in the $+x$ direction when v is in the $+z$ direction.

---

9. A wire having a mass per unit length of 0.500 g/cm carries a 2.00-A current horizontally to the south. What are the direction and magnitude of the minimum magnetic field needed to lift this wire vertically upward?

Solution

$$\frac{m}{\ell} = \left(0.500\ \frac{\text{g}}{\text{cm}}\right)\left(\frac{1\text{ kg}}{1000\text{ g}}\right)\left(\frac{100\text{ cm}}{\text{m}}\right) = 5.00\times10^{-2}\text{ kg/m}$$

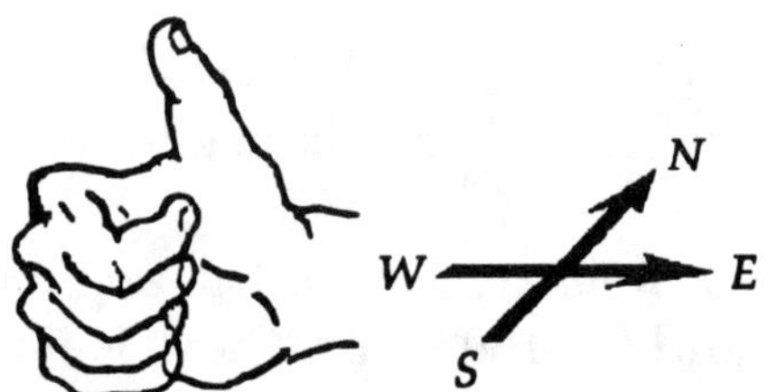

$I = 2.00$ A south. To make **F** upward, **B** must be in a direction given by right-hand rule: eastward ◊

$F = I\ell B\sin\theta$: must counter balance $w = mg$

$$mg = I\ell B\sin\theta$$

$$B = \left(\frac{m}{\ell}\right)\frac{g}{I\sin\theta} = \left(5.00\times10^{-2}\text{ kg/m}\right)\frac{9.80\text{ m/s}^2}{(2.00\text{ A})\sin 90°}$$

$B = 0.245$ T ◊

---

11. A strong magnet is placed under a horizontal conducting ring of radius r that carries current I, as in Figure P22.11. If the magnetic field lines make an angle θ with the vertical at the ring's location, what are the magnitude and direction of the resultant force on the ring?

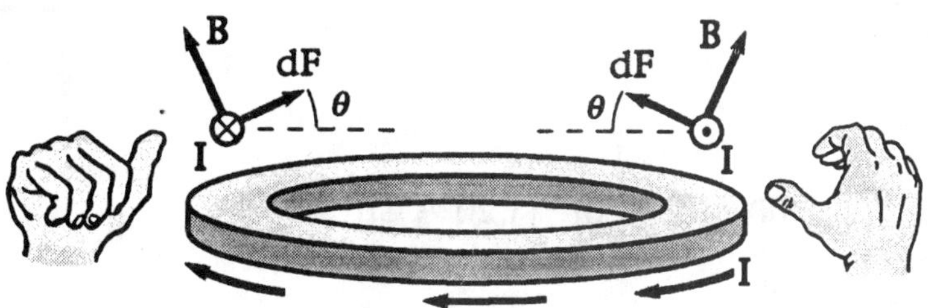

Figure P22.11

Solution

The magnetic force on each bit of ring is radial inward and upward, at an angle θ above the radial line, according to:

$$|d\mathbf{F}| = I|d\mathbf{s} \times \mathbf{B}| = I\, ds\, B$$

The radially inward components tend to squeeze the ring, but cancel out as forces. The upward components $I\, ds\, B \sin\theta$ all add to

F = I(2πr) B sinθ ↑

$$\mathbf{F} = I(2\pi r)B \sin\theta \text{ up } \Diamond.$$

The magnetic moment of the ring is down. This problem is a model for the force on a dipole in a nonuniform magnetic field, or for the force that one magnet exerts on another magnet.

---

13. A rectangular loop consists of $N = 100$ closely wrapped turns and has dimensions $a = 0.400$ m and $b = 0.300$ m. The loop is hinged along the y axis, and its plane makes an angle $\theta = 30.0°$ with the x axis (Fig. P22.13). What is the magnitude of the torque exerted on the loop by a uniform magnetic field $B = 0.800$ T directed along the x axis when the current is $I = 1.20$ A in the direction shown? What is the expected direction of rotation of the loop?

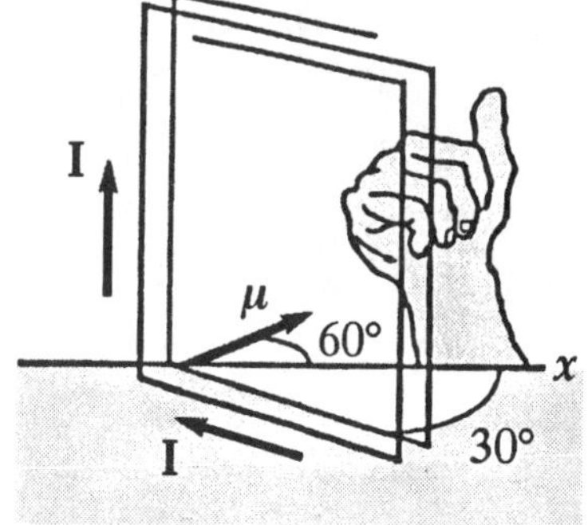

Figure P22.13

Solution The magnetic moment of the coil is $\mu = NIA$, perpendicular to its plane and making a 60° angle with the x axis as shown to the right. The torque on the dipole is then $\tau = \mu \times \mathbf{B} = NAIB\sin\theta$ down, having a magnitude of

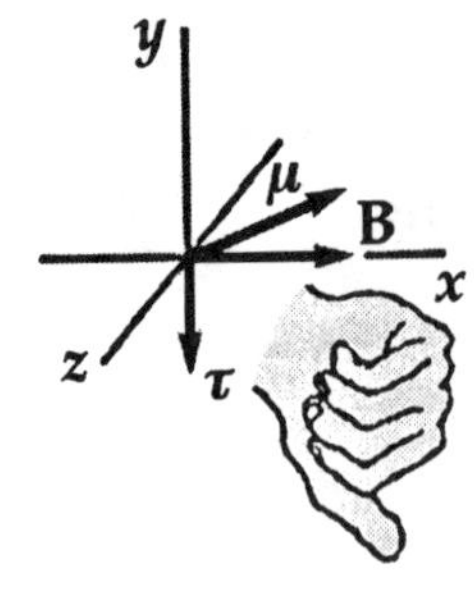

$$\tau = NBAI\ \sin\ \theta$$

$$\tau = (100)(0.800\ \text{T})(0.400 \times 0.300\ \text{m}^2)(1.20\ \text{A})\sin\ 60.0°$$

$$\tau = 9.98\ \text{N·m} \quad \Diamond$$

Note θ is the angle between the magnetic moment and the **B** field. Loop will rotate such to align the magnetic moment with the **B** field. Looking down along the y axis, loop will rotate in the clockwise direction.

---

19. Determine the magnetic field at a point P located a distance x from the corner of an infinitely long wire bent at a right angle, as in Figure P22.19. The wire carries a steady current I.

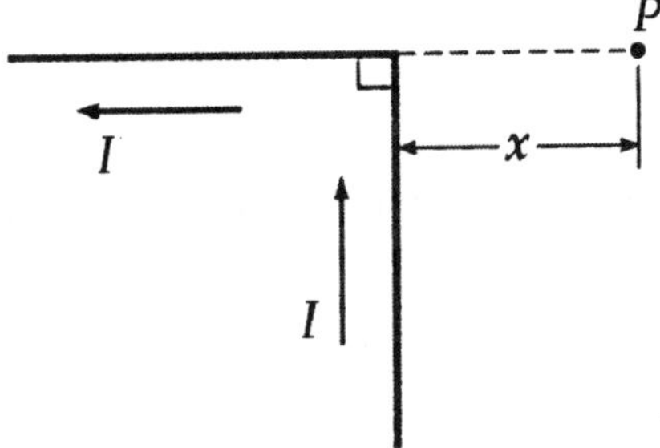

Figure P22.19

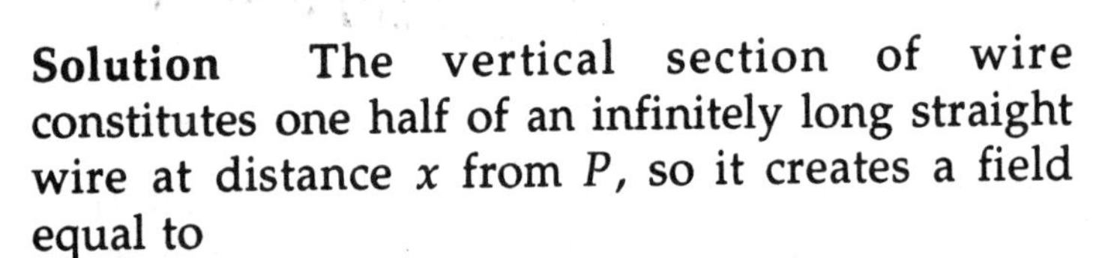

Solution The vertical section of wire constitutes one half of an infinitely long straight wire at distance x from P, so it creates a field equal to

$$B = \frac{1}{2}\left(\frac{\mu_0 I}{2\pi x}\right)$$

Hold your right hand with extended thumb in the direction of the current; the field is away from you, into the paper. For each bit of the horizontal section of wire $d\mathbf{s}$ is to the left and $\hat{\mathbf{r}}$ is to the right, so $d\mathbf{s} \times \hat{\mathbf{r}} = 0$. The horizontal current produces zero field at P. Thus,

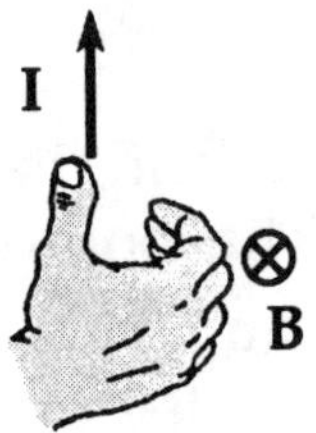

$$\mathbf{B} = \frac{\mu_0 I}{4\pi x}\ \text{into the paper} \quad \Diamond$$

---

21. In Figure P22.21, the current in the long, straight wire is $I_1 = 5.00$ A and the wire lies in the plane of the rectangular loop, which carries 10.0 A. The dimensions are $c = 0.100$ m, $a = 0.150$ m, and $\ell = 0.450$ m. Find the magnitude and direction of the net force exerted on the loop by the magnetic field created by the wire.

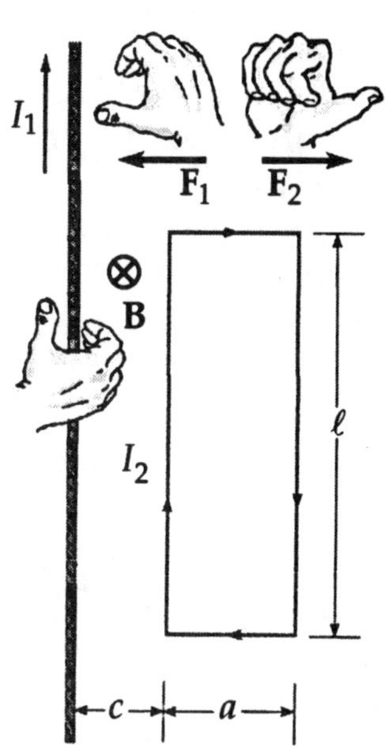

Figure P22.21 (modified)

Solution

By symmetry, the forces exerted on the segments of length a are equal and opposite and cancel. The magnetic field in the plane of I_2 to the right of I_1 is directed away from you into the plane. By the right-hand rule, $\mathbf{F} = I\boldsymbol{\ell} \times \mathbf{B}$ is directed toward the **left** for the near side of the loop and directed toward the **right** for the side at $c + a$.

Thus,

$$\mathbf{F} = \mathbf{F}_1 + \mathbf{F}_2 = \frac{\mu_0 I_1 I_2 \ell}{2\pi}\left(\frac{1}{c+a} - \frac{1}{c}\right)\mathbf{i}$$

$$\mathbf{F} = \frac{\mu_0 I_1 I_2 \ell}{2\pi}\left(\frac{-a}{c(c+a)}\right)\mathbf{i}$$

$$\mathbf{F} = \frac{(4\pi \times 10^{-7}\ \mathrm{N/A^2})(5.00\ \mathrm{A})(10.0\ \mathrm{A})(0.450\ \mathrm{m})}{2\pi}\left(\frac{-0.150\ \mathrm{m}}{(0.100\ \mathrm{m})(0.250\ \mathrm{m})}\right)\mathbf{i}$$

$\mathbf{F} = (-2.70 \times 10^{-5}\mathbf{i})\ \mathrm{N}$ or $F = 2.70 \times 10^{-5}\ \mathrm{N}$ toward the left ◊

23. Four long, parallel conductors carry equal currents of $I = 5.00$ A. Figure P22.23 is an end view of the conductors. The current direction is into the page at points A and B (indicated by the crosses) and out of the page at C and D (indicated by the dots). Calculate the magnitude and direction of the magnetic field at point P, located at the center of the square of edge length 0.200 m.

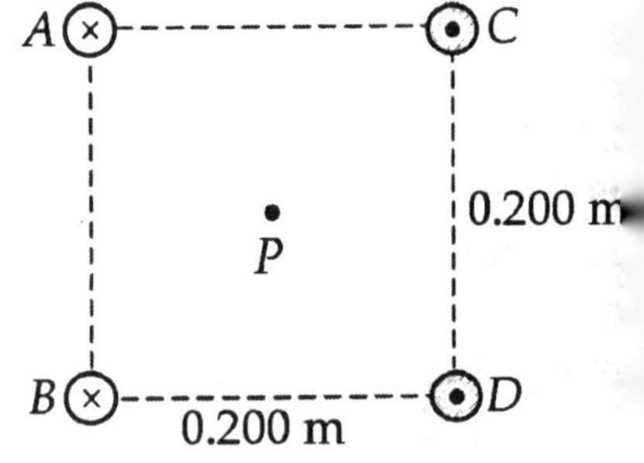

Figure P22.23

Solution

Each wire is distant from P by $(0.200 \text{ m}) \cos 45° = 0.141 \text{ m}$.

Each wire produces a field at P of equal magnitude:

$$B_A = \frac{\mu_0 I}{2\pi a} = \frac{(2.00 \times 10^{-7} \text{ T} \cdot \text{m / A})(5.00 \text{ A})}{(0.141 \text{ m})} = 7.07\ \mu\text{T}$$

Carrying currents away from you, the left-hand wires produce fields at P of 7.07 μT, in the following directions:

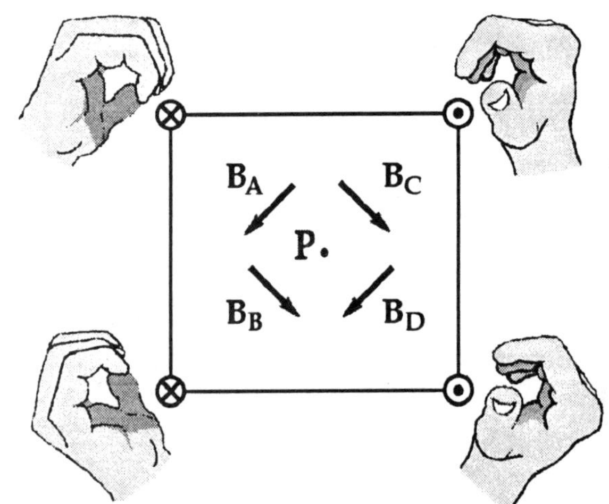

A: to the bottom and left, at 225°

B: to the bottom and right, at 315°;

Carrying currents toward you, the wires to the right also produce fields at P of 7.07 μT, in the following directions:

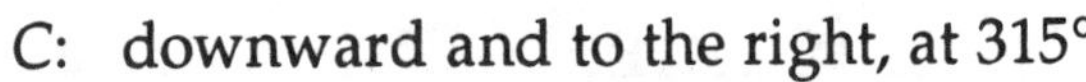

C: downward and to the right, at 315°

D: downward and to the left, at 225°.

The total field is then

$4(7.07\ \mu\text{T}) \sin 45° = 20.0\ \mu\text{T}$ toward the bottom of the page. ◊

27. A packed bundle of 100 long, straight, insulated wires forms a cylinder of radius $R = 0.500$ cm. (a) If each wire carries 2.00 A, what are the magnitude and direction of the magnetic force per unit length acting on a wire located 0.200 cm from the center of the bundle? (b) Would a wire on the outer edge of the bundle experience a force greater or smaller than the value calculated in part (a)?

Solution

The force **on** one wire is exerted **by** the other ninety-nine, through the magnetic field they create.

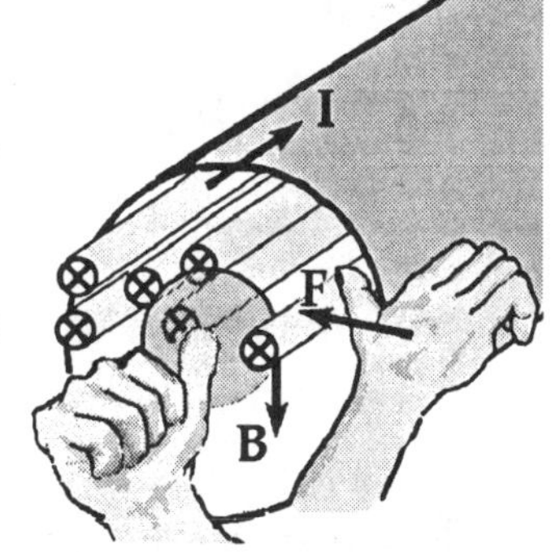

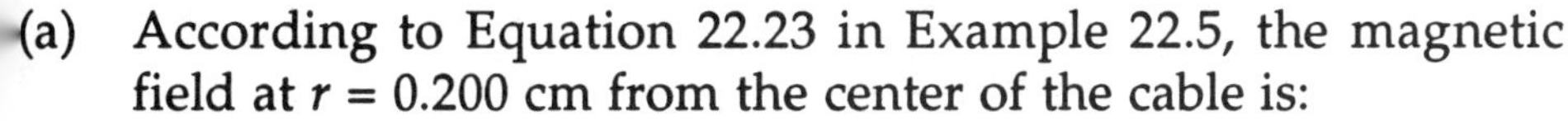

(a) According to Equation 22.23 in Example 22.5, the magnetic field at $r = 0.200$ cm from the center of the cable is:

$$B = \frac{\mu_0 I_0 r}{2\pi R^2} = \frac{(4\pi \times 10^{-7}\ \text{T}\cdot\text{m/A})(99)(2.00\ \text{A})(0.200 \times 10^{-2}\ \text{m})}{2\pi\,(0.500 \times 10^{-2}\ \text{m})^2}$$

$$B = 3.17 \times 10^{-3}\ \text{T}$$

This field points tangent to a circle of radius 0.2 mm. It will exert a force $\mathbf{F} = I\boldsymbol{\ell} \times \mathbf{B}$ toward the center of the bundle, on the hundredth wire:

$$\frac{F}{\ell} = IB\sin\theta = (2.00\ \text{A})(3.17 \times 10^{-3}\ \text{T})(\sin 90°) = 6.34\ \text{mN/m} \quad \text{toward the center} \quad \Diamond$$

(b) As is shown in Figure 22.20 of the text, the field is strongest at the outer surface of the cable, so the force on one strand is greater here, by a factor of 5/2. ◊

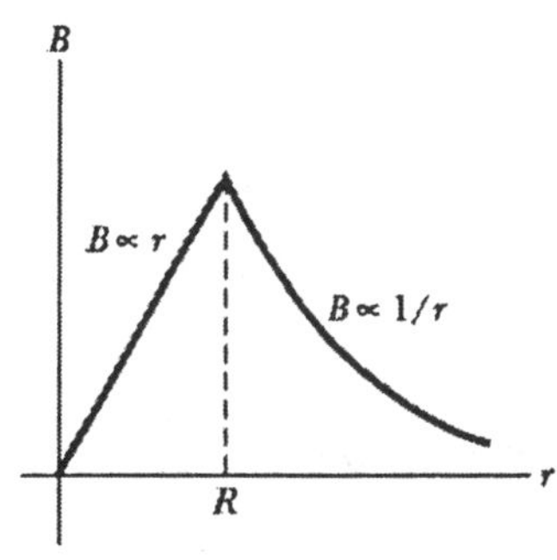

29. What current is required in the windings of a long solenoid that has 1000 turns uniformly distributed over a length of 0.400 m in order to produce at the center of the solenoid a magnetic field of magnitude 1.00×10^{-4} T ?

Solution $B = \mu_o \frac{N}{\ell} I$, so $I = \frac{B\ell}{\mu_o N} = \frac{(1.00 \times 10^{-4}\ \text{T} \cdot \text{A})(0.400\ \text{m})}{(4\pi \times 10^{-7}\ \text{T} \cdot \text{m})(1000)}$

Caution! If you use your calculator, it may not understand the keystrokes:

[4] [×] [π] [EXP] [+/−] [7]

You may need to use

[4] [EXP] [+/−] [7] [×] [π]

Make sure you get $I = 31.8$ mA ◊

---

33. The magnetic moment of the Earth is approximately 8.70×10^{22} A·m². (a) If this were caused by the complete magnetization of a huge iron deposit, how many unpaired electrons would this correspond to? (b) At 2 unpaired electrons per iron atom, how many kilograms of iron would this correspond to? (The density of iron is 7900 kg/m³, and there are approximately 8.50×10^{28} iron atoms/m³.)

Solution The magnetic moment of each unpaired electron is the Bohr magneton,

$$9.27 \times 10^{-24}\ \text{J/T} = \frac{9.27 \times 10^{-24}\ \text{J}}{\text{T}} \left(\frac{\text{N} \cdot \text{m}}{\text{J}}\right)\left(\frac{\text{T} \cdot \text{C} \cdot \text{m}}{\text{N} \cdot \text{s}}\right)\left(\frac{\text{A} \cdot \text{s}}{\text{C}}\right)$$

$$\mu_B = 9.27 \times 10^{-24}\ \text{A} \cdot \text{m}^2$$

(a) The number of unpaired electrons is $N = \frac{8.70 \times 10^{22}\ \text{A} \cdot \text{m}^2}{9.27 \times 10^{-24}\ \text{A} \cdot \text{m}^2} = 9.39 \times 10^{45}$ ◊

Each iron atom has two unpaired electrons, so the number of iron atoms required is

$$\tfrac{1}{2}N = \tfrac{1}{2}(9.39 \times 10^{45}) = 4.69 \times 10^{45}\ \text{atoms.}$$

(b) Mass $= \frac{(4.69 \times 10^{45}\ \text{atoms})(7900\ \text{kg/m}^3)}{8.50 \times 10^{28}\ \text{atoms/m}^3} = 4.36 \times 10^{20}$ kg ◊

---

43. A nonconducting ring of radius 10.0 cm is uniformly charged with a total positive charge 10.0 μC. The ring rotates at a constant angular speed 20.0 rad/s about an axis through its center, perpendicular to the plane of the ring. What is the magnitude of the magnetic field on the axis of the ring 5.00 cm from its center?

Solution The period of one revolution of the ring is $T = 2\pi/\omega$; because the ring is nonconducting, it carries a current just as a loop of wire would:

$$I = q/T = q\omega/2\pi = \left(10.0\times10^{-6}\ \text{C}\right)(20.0\ \text{rad/s})/2\pi = 31.8\times10^{-6}\ \text{A}$$

The magnetic field on its axis is then:

$$B = \frac{\mu_0 I R^2}{2\left(x^2+R^2\right)^{3/2}} = \frac{\left(4\pi\times10^{-7}\ \text{T}\cdot\text{m/A}\right)\left(31.8\times10^{-6}\ \text{A}\right)(0.100\ \text{m})^2}{2\left((0.0500\ \text{m})^2+(0.100\ \text{m})^2\right)^{3/2}} = 1.43\times10^{-10}\ \text{T} \quad \Diamond$$

---

44. A nonconducting ring of radius R is uniformly charged with a total positive charge q. The ring rotates at a constant angular speed ω about an axis through its center, perpendicular to the plane of the ring. What is the magnitude of the magnetic field on the axis of the ring a distance $R/2$ from its center?

Solution The time required for one revolution is given by $\omega = \dfrac{2\pi}{T}$ or $T = \dfrac{2\pi}{\omega}$

The spinning charged ring constitutes a loop carrying current $I = \dfrac{q}{T} = \dfrac{q\omega}{2\pi}$

so it creates magnetic field on its axis

$$B = \frac{\mu_0 I R^2}{2(x^2+R^2)^{3/2}} = \frac{\mu_0 q\omega R^2}{4\pi\left(\dfrac{R^2}{4}+R^2\right)^{3/2}}$$

Simplifying, $B = \dfrac{2\mu_0}{5\pi\sqrt{5}}\dfrac{q\omega}{R} \quad \Diamond$

Chapter

23

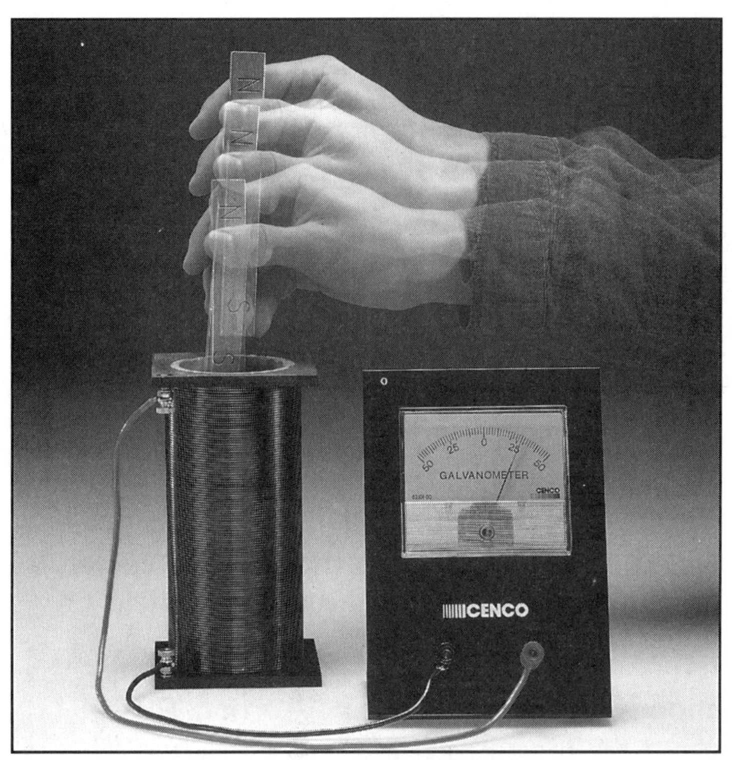

When a strong magnet is moved toward or away from the coil attached to a galvanometer, an electric current is induced, indicated by the momentary deflection of the galvanometer during the movement of the magnet. (Richard Megna, Fundamental Photographs)

FARADAY'S LAW AND INDUCTANCE

Chapter 23

FARADAY'S LAW AND INDUCTANCE

INTRODUCTION

Our studies so far have been concerned with electric fields due to charges and magnetic fields produced by moving charges. This chapter deals with electric fields that originate from changing magnetic fields.

Experiments conducted by Michael Faraday in England in 1831 and independently by Joseph Henry in the United States that same year showed that an electric current could be induced in a circuit by a changing magnetic field. The results of these experiments led to a basic and important law of electromagnetism known as Faraday's law of induction. This law says that the magnitude of the emf induced in a circuit equals the time rate of change of the magnetic flux through the circuit.

As we shall see, an induced emf can be produced in several ways. For instance, an induced emf and an induced current can be produced in a closed loop of wire when the wire moves into a magnetic field. We shall describe such experiments along with a number of important applications that make use of the phenomenon of electromagnetic induction.

We also describe an effect known as **self-induction,** in which a time-varying current in a conductor induces in the conductor an emf that opposes the change in the external emf that set up the current. Self-induction is the basis of the **inductor**, an electrical element that plays an important role in circuits that use time-varying currents. We discuss the energy stored in the magnetic field of an inductor and the energy density associated with a magnetic field.

NOTES FROM SELECTED CHAPTER SECTIONS

23.1 Faraday's Law of Induction

The emf induced in a circuit is proportional to the time rate of change of magnetic flux through the circuit.

An emf can be induced in the circuit in several ways:

- The magnitude of the magnetic field can change as a function of time.
- The area of the circuit can change with time.
- The direction of the magnetic field relative to the circuit can change with time.
- Any combination of the above can change.

In particular, it is important to note that the magnitude of the induced emf depends on the **rate at which the magnetic flux is changing.**

23.2 Motional emf

A potential difference will **appear** across a conductor moving in a magnetic field as long as the direction of motion through the field is not parallel to the field direction. If the motion is reversed, the polarity of the potential difference will also be reversed.

23.3 Lenz's Law

The polarity of the induced emf is such that it tends to produce a current that will create a magnetic flux to oppose the **change in flux** through the circuit.

23.5 Self-Inductance

The self-induced emf is always proportional to the **time rate of change** of current in the circuit.

The **inductance** of a device (an inductor) depends on its **geometry**.

23.6 *RL* Circuits

If a resistor and an inductor are connected in series to a battery, the current in the circuit will reach an **equilibrium** value ($\mathcal{E}/R$) after a time which is long compared to the **time constant** of the circuit, $\tau = L / R$.

23.7 Energy in a Magnetic Field

In an *RL* circuit, the rate at which energy is supplied by the battery equals the sum of the rate at which heat is dissipated in the resistor and the rate at which energy is stored in the inductor. The **energy density** is proportional to the **square of the magnetic field**.

EQUATIONS AND CONCEPTS

The total magnetic flux through a plane area, A, placed in a uniform magnetic field depends on the angle between the direction of the magnetic field and the direction perpendicular to the surface area.

$$\Phi \equiv B_{\perp}A = BA\cos\theta$$

$$\Phi_{\text{max}} = BA$$

The maximum flux through the area occurs when the magnetic field is perpendicular to the plane of the surface area. When the magnetic field is parallel to the plane of the surface area, the flux through the area is zero. The unit of magnetic flux is the weber, Wb.

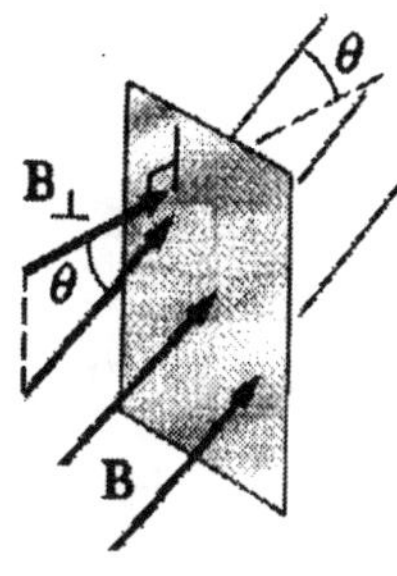

The magnetic flux threading a circuit is the integral of the normal component of the magnetic field over the area bounded by the circuit.

$$\Phi_B = \int \mathbf{B} \cdot d\mathbf{A} \tag{23.1}$$

Faraday's law of induction states that the emf induced in a circuit is the rate of change of magnetic flux through the circuit. The minus sign is included to indicate the polarity of the induced emf, which can be found by use of Lenz's law.

$$\mathcal{E} = -\frac{d\Phi_B}{dt} \tag{23.2}$$

Lenz's law states that the polarity of the induced emf (and the direction of the associated current in a closed circuit) produces a current whose magnetic field opposes the change in the flux through the loop. That is, the induced current tends to maintain the original flux through the circuit.

A "motional" emf is induced in a conductor of length ℓ, moving with speed v, perpendicular to a magnetic field.

$$\mathcal{E} = -B\ell v \qquad (23.5)$$

If the moving conductor is part of a complete circuit of resistance, R, a current will be induced in the circuit.

$$I = \frac{B\ell v}{R} \qquad (23.6)$$

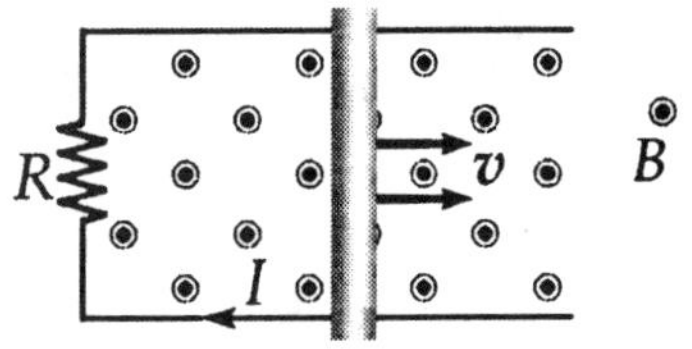

When a conducting coil of N turns and cross-sectional area A, rotates with a constant angular velocity in a magnetic field, the emf induced in the coil will vary sinusoidally in time. For a given coil, the maximum value of the induced emf will be proportional to the angular velocity of the loop.

$$\mathcal{E} = NAB\omega \sin \omega t \qquad (23.8)$$

$$\mathcal{E}_{max} = NAB\omega$$

When the current in a coil changes in time, a self-induced emf is present in the coil. The inductance, L, is a measure of the opposition of the coil to a change in the current.

$$\mathcal{E}_L = -L\frac{dI}{dt} \qquad (23.10)$$

he inductance of a given device, for xample a coil, depends on its physical ıakeup—diameter, number of turns, type f material on which the wire is wound, nd other geometric parameters. A circuit lement which has a large inductance is alled an inductor. The SI unit of nductance is the henry, H. A rate of hange of current of 1 ampere per second n an inductor of 1 henry will produce a elf-induced emf of 1 volt.

$$1\,\mathrm{H} = 1\frac{\mathrm{V\cdot s}}{\mathrm{A}} = 1\,\Omega\cdot\mathrm{s}$$

A coil, solenoid, toroid, coaxial cable, or other conducting device is characterized by a parameter called its **inductance,** L. The inductance can be calculated, knowing the current and magnetic flux.

$$L = \frac{N\Phi_B}{I} \quad (23.11)$$

The **inductance** of a particular circuit element can also be expressed as the ratio of the induced emf to the time rate of change of current in the circuit.

$$L = -\frac{\mathcal{E}_L}{dI/dt} \quad (23.12)$$

If the switch in the series circuit shown (which contains a battery, resistor, and inductor) is closed in position 1 at time $t = 0$, **current in the circuit** will increase in a characteristic fashion toward a maximum value of $(\mathcal{E}/R)$. This is shown in the graph to the right.

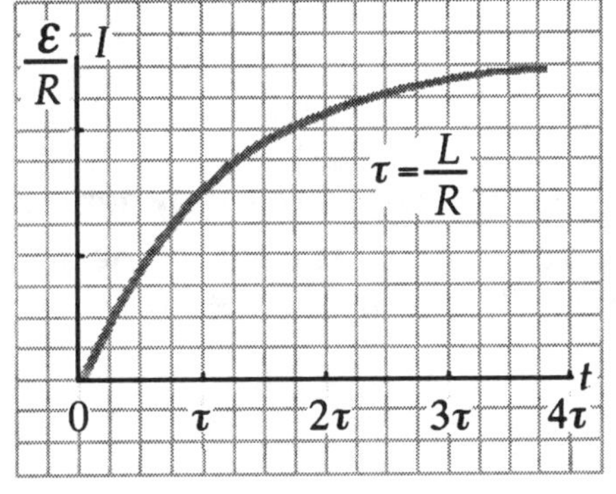

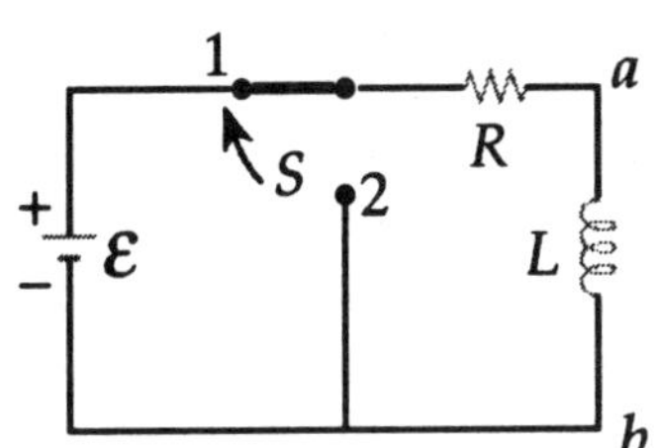

$$I(t) = \frac{\mathcal{E}}{R}(1 - e^{-t/\tau}) \quad (23.14)$$

Let the switch in the circuit shown be at position 1 with the current at its maximum value $I_0 = \mathcal{E}/R$. If the switch is thrown to position 2 at $t = 0$, the current **will decay** exponentially with time. The curve of the decay is shown in the second graph, to the right.

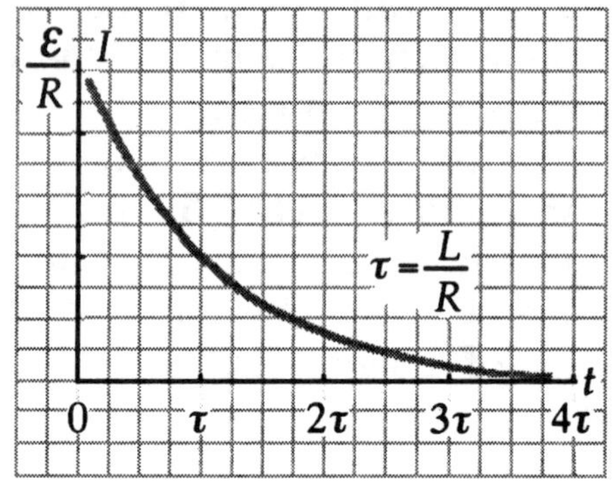

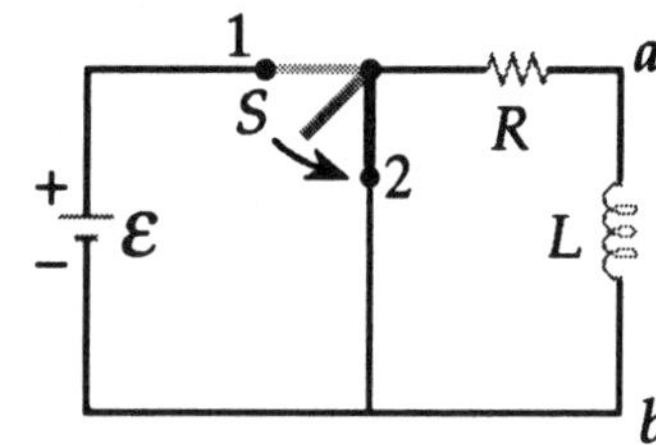

$$I(t) = \frac{\mathcal{E}}{R}e^{-t/\tau} = I_0 e^{-t/\tau} \quad (23.18)$$

where $\dfrac{\mathcal{E}}{R} = I_0$

In Equation 23.14 and Equation 23.18, the constant in the exponent is called the **time constant** of the circuit, τ. Physically, the time constant is the time required for the current to reach 63% of its final value.

$$\tau = L/R \quad (23.15)$$

The **stored energy** U_B in the magnetic field of an inductor is proportional to the square of the current in the inductor.

$$U_B = \tfrac{1}{2}LI^2 \quad (23.20)$$

It is often useful to express the energy in a magnetic field as **energy density** u_B; that is, energy per unit volume.

$$u_B = \frac{B^2}{2\mu_0} \quad (23.22)$$

UGGESTIONS, SKILLS, AND STRATEGIES

nstantaneous and Average Induced Emf:

t is important to distinguish clearly between the **instantaneous value** of emf induced in circuit and the **average value** of the emf induced in the circuit over a finite time nterval.

To calculate the **average induced emf**, it is often useful to write Equation 23.3 as

$$\mathcal{E}_{av} = -N\left(\frac{d\Phi_B}{dt}\right)_{av} = -N\frac{\Delta\Phi_B}{\Delta t}$$

or

$$\mathcal{E}_{av} = -N\left(\frac{\Phi_{B,f} - \Phi_{B,i}}{\Delta t}\right)$$

where the subscripts i and f refer to the magnetic flux through the circuit at the initial and final moments of the time interval Δt. For a circuit (or loop) in a single plane, $\Phi_B = BA\cos\theta$, where θ is the angle between the direction of the normal to plane of the circuit (conducting loop) and the direction of the magnetic field.

Equation 23.4 can be used to calculate the **instantaneous value of an induced emf**. For a multiple turn loop, the induced emf is

$$\mathcal{E} = -N\frac{d}{dt}(BA\cos\theta)$$

where in a particular case B, A, θ, or any combination of those parameters can be time dependent while the others remain constant. The expression resulting from the differentiation is then evaluated using the values of B, A, and θ corresponding to the specified value.

REVIEW CHECKLIST

- ▷ Calculate the emf (or current) induced in a circuit when the magnetic flux throug the circuit is changing in time. The variation in flux might be due to a change i (a) the area of the circuit, (b) the magnitude of the magnetic field, (c) the direction c the magnetic field, or (d) the orientation/location of the circuit in the magnetic field.

- ▷ Calculate the emf induced between the ends of a conducting bar as it moves throug a region where there is a constant magnetic field (motional emf). Apply Lenz's lav to determine the direction of an induced emf or current. You should als understand that Lenz's law is a consequence of the law of conservation of energy.

- ▷ Calculate the inductance of a device of suitable geometry.

- ▷ Calculate the magnitude and direction of the self-induced emf in a circuit containin one or more inductive elements when the current changes with time.

- ▷ Determine instantaneous values of the current in an *RL* circuit while the current is either increasing or decreasing with time.

- ▷ Calculate the total magnetic energy stored in a magnetic field. You should be able to perform this calculation if (1) you are given the values of the inductance of the device with which the field is associated and the current in the circuit, or (2) given the value of the magnetic field intensity throughout the region of space in which the magnetic field exists. In the latter case, you must integrate the expression for the energy density u_B over an appropriate volume.

ANSWERS TO SELECTED CONCEPTUAL QUESTIONS

1. A loop of wire is placed in a uniform magnetic field. For what orientation of the loop is the magnetic flux a maximum? For what orientation is the flux zero?

Answer The flux is calculated as $\Phi_B \equiv B_\perp A = BA\cos\theta$. The flux is therefore maximum when the entire magnetic field vector is perpendicular to the loop of wire. We may also deduce that the flux is zero when there is no component of the magnetic field that is perpendicular to the loop.

□ □ □ □

As the conducting bar in Figure Q23.2 moves to the right, an electric field is set up directed downward in the conductor. If the bar were moving to the left, explain why the electric field would be upward.

Answer If the bar were moving to the left, the magnetic force on the negative charge carriers in the bar would be upward, causing an accumulation of negative charge on the top, and an accumulation of positive charge on the bottom. Hence, the electric field in the bar would be directed upwards.

Figure Q23.2

□ □ □ □

3. As the bar in Figure Q23.2 moves perpendicular to the field, is an external force required to keep it moving with constant speed?

Answer **No.** The motion of the bar within the magnetic field exerts a force on the charges in the bar, such that some negative charges migrate to one side of the bar, while some positive charges migrate to the other side of the bar.

This will continue, creating an electric field **E** (as shown), until the force caused by the magnetic field is balanced by the force caused by the electric field, and $\mathbf{E} = \mathbf{v} \times \mathbf{B}$.

Once this occurs, a steady state is reached, and all the internal forces will be balanced by equal and opposite internal forces. Therefore, by Newton's First Law, the speed will remain constant without an applied external force.

□ □ □ □

6. How is electrical energy produced in dams? (That is, how is the energy of motion of the water converted to ac electricity)?

Answer As the water falls, it gains kinetic energy. It is then forced to pass through a water wheel, transferring some of its energy to the rotor of an large AC electric generator.

The rotor of the generator is supplied with a small amount of DC current, which powers electromagnets in the rotor. Because the rotor is spinning, the electromagnets then create a magnetic flux that changes with time, according to the equation $\Phi_B = BA\cos\omega t$.

Coils of wire that are placed near the magnet then experience an induced electromagnetic force according to the equation $\mathcal{E} = -N\,d\Phi_B/dt$.

Finally, a small amount of this electricity is used to supply the rotor with its DC current; the rest is sent out over power lines to supply customers with electricity.

□ □ □ □

8. The bar in Figure Q23.8 moves on rails to the right with a velocity **v**, and the uniform, constant magnetic field is out of the page. Why is the induced current clockwise? If the bar were moving to the left, what would be the direction of the induced current?

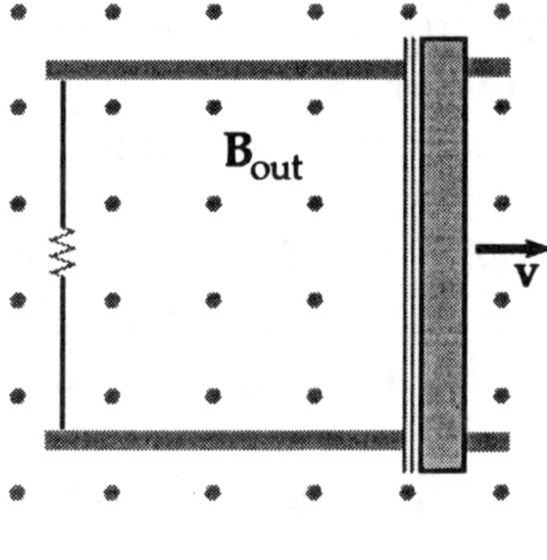

Figure Q23.8

Answer

The external flux is out of the paper; as the area A enclosed by the loop increases, the external flux increases according to

$$\Phi_B = BA\cos\theta = BA$$

Under the influence of this flux, free electric charges will move in a current to create a magnetic flux inside the loop to oppose the change in flux.

In this case, the flux of the current must point into the paper in order oppose the increasing magnetic flux. By the right-hand rule (with your thumb pointing in the direction of the current along each wire), the current must be clockwise.

If the bar were moving toward the left, the area would decrease, and the flux would decrease. The **change** in the flux pointing out of the paper, then, would be negative. In order to cancel this change, the current would have to create a flux inside the loop that was positive, pointing out of the paper. This time, by the right-hand rule, we see that the current must be counter-clockwise.

□ □ □ □

16. If the current in an inductor is doubled, by what factor does the stored energy change?

Answer The energy stored in an inductor carrying a current I is given by $U = \frac{1}{2}LI^2$. Therefore, doubling the current will quadruple the energy stored in the inductor.

□ □ □ □

7. Suppose the switch in Figure 23.17 has been closed for a long time and is suddenly opened. Does the current instantaneously drop to zero? Why does a spark appear at the switch contacts at the instant the switch is thrown open?

Answer No. The inductor has stored energy (much like a flywheel), which causes current to continue flowing for a time, even though the switch is open.

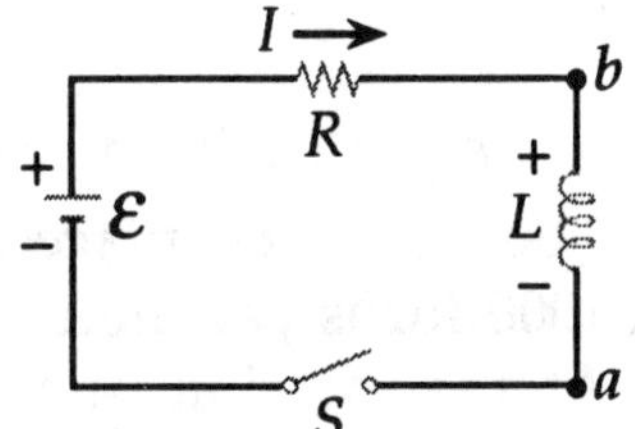

Figure 23.17

As a result, as the current flows, it creates an electric field across the inductor, and (inherently) across the switch. This electric field induces an electromotive force that is strongest at the moment the switch is opened. Since the gap is also small at that initial moment, the electromotive force is sometimes strong enough to ionize the air, and cause a spark to jump across the gap in the switch.

The switch can be protected from sparking by a 100 kΩ resistor, connected in parallel with the inductor. During normal operation, very little power will flow through the resistor; but when the switch is opened, any induced emf will be discharged through the resistor, rather than causing a spark.

□ □ □ □

SOLUTIONS TO SELECTED END-OF-CHAPTER PROBLEMS

1. A strong electromagnet has a field of 1.60 T and a cross-sectional area of 0.200 m². If we place a coil having 200 turns and a total resistance of 20.0 Ω around the electromagnet, and then turn off the power to the electromagnet in 20.0 ms, what is the current induced in the coil?

Solution

The induced voltage is $$\mathcal{E} = -N\frac{d(B\cdot A)}{dt} = -N\left(\frac{0 - B_i A\cos\theta}{t}\right)$$

Solving, $$\mathcal{E} = \frac{+200(1.60\text{ T})(0.200\text{ m}^2)(\cos 0°)}{20.0\times 10^{-3}\text{ s}}\left(\frac{1\text{ N}\cdot\text{s}}{\text{T}\cdot\text{C}\cdot\text{m}}\right)\left(\frac{1\text{ V}\cdot\text{C}}{\text{N}\cdot\text{m}}\right) = 3200\text{ V}$$

$$I = \frac{\mathcal{E}}{R} = \frac{3200\text{ V}}{20.0\ \Omega} = 160\text{ A} \quad \Diamond$$

The positive sign means that the current in the coil flows in the same direction as the current in the electromagnet.

---

3. An aluminum ring of radius 5.00 cm and resistance $3.00 \times 10^{-4}\ \Omega$ is placed on top of a long air-core solenoid with 1000 turns per meter and radius 3.00 cm, as in Figure P23.3. At the location of the ring, the magnetic field due to the current in the solenoid is one half that at the center of the solenoid. If the current in the solenoid is increasing at a rate of 270 A/s, (a) what is the induced current in the ring? (b) At the center of the ring, what is the magnetic field produced by the induced current in the ring? (c) What is the direction of this field?

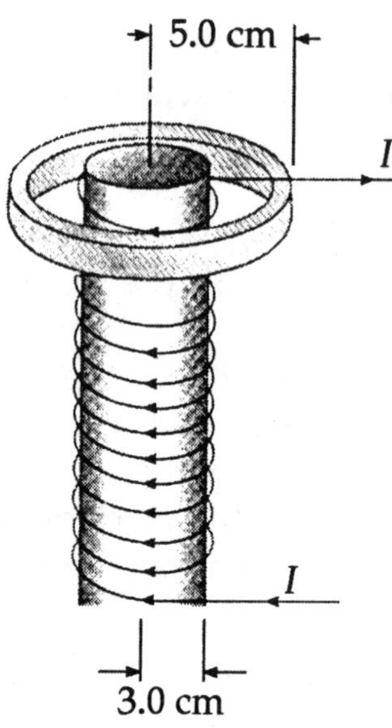

Figure P23.3

Solution

$$\mathcal{E} = -\frac{d}{dt}(BA\cos\theta) = -\frac{d}{dt}(0.500\ \mu_o nIA\cos 0°)$$

$$\mathcal{E} = -0.500\ \mu_o nA\frac{dI}{dt}$$

Note that A must be interpreted as the area of the solenoid, where the field is strong:

$$\mathcal{E} = -0.500(4\pi \times 10^{-7}\ \text{T}\cdot\text{m/A})(1000\ \text{turns/m})\left[\pi(0.0300\ \text{m})^2\right](270\ \text{A/s})$$

Applying conversion factors of $1\ \dfrac{\text{N}\cdot\text{s}}{\text{C}\cdot\text{m}\cdot\text{T}}$ and $1\ \dfrac{\text{V}\cdot\text{C}}{\text{N}\cdot\text{m}}$,

$$\mathcal{E} = -4.80 \times 10^{-4}\ \text{V}$$

(a) $I_{\text{ring}} = \dfrac{|\mathcal{E}|}{R} = \dfrac{0.000480}{0.000300} = 1.60\ \text{A}$ ◊

(b) $B_{\text{ring}} = \dfrac{\mu_o I}{2R} = 2.01 \times 10^{-5}\ \text{T}$ ◊

(c) The coil's field points downward, and is increasing, so B_{ring} points upward. ◊

5. A long, straight wire carries a current $I = I_o \sin(\omega t + \phi)$ and lies in the plane of a rectangular loop of N turns of wire, as shown in Figure P23.5. The quantities I_o, ω, and ϕ are all constants. Determine the emf induced in the loop by the magnetic field created by the current in the straight wire. Assume I_o = 50.0 A, $\omega = 200\pi\,\text{s}^{-1}$, N = 100, $a = b = 5.00$ cm, and ℓ = 20.0 cm.

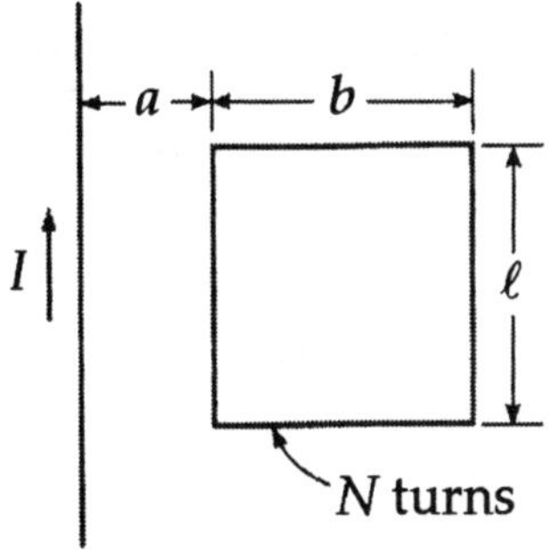

Figure P23.5

Solution

The loop is the boundary of a rectangular area. The magnetic field produced by the current in the straight wire is perpendicular to the plane of the area at all points. The magnitude of the field is

$$B = \frac{\mu_0 I}{2\pi r}$$

Thus the flux is

$$N\Phi_m = \frac{\mu_0 N\ell}{2\pi} I \int_a^{a+b} \frac{dr}{r} = \frac{\mu_0 N\ell}{2\pi} I_o \ln\left(\frac{a+b}{a}\right) \sin(\omega t + \phi)$$

Finally, the induced emf is in absolute value

$$|\mathcal{E}| = N\frac{d\,\Phi_m}{dt} = \frac{\mu_0 N\ell}{2\pi} I_o \omega \ln\left(\frac{a+b}{a}\right) \cos(\omega t + \phi)$$

$$|\mathcal{E}| = \left(4\pi \times 10^{-7}\,\frac{\text{T}\cdot\text{m}}{\text{A}}\right) \frac{(100)(0.200\text{ m})}{2\pi} (50.0\text{ A})(200\pi\,\text{s}^{-1}) \ln\left(\frac{10.0\text{ cm}}{5.00\text{ cm}}\right) \cos(\omega t + \phi)$$

$$|\mathcal{E}| = (87.1\text{ mV})\cos(200\pi t + \phi) \quad \lozenge$$

Related Comments

The term $\sin(\omega t + \phi)$ in the expression for the current in the straight wire does not change appreciably when ωt changes by 0.1 rad or less. Thus, the current does not change appreciably during a time interval

$$t < 0.100/\left(200\pi\ \mathrm{s}^{-1}\right) = 1.59 \times 10^{-4}\ \mathrm{s}.$$

We define a critical length,

$$c\,t = (3.00 \times 10^{8}\ \mathrm{m/s})(1.59 \times 10^{-4}\ \mathrm{s}) = 4.77 \times 10^{4}\ \mathrm{m},$$

equal to the distance to which field changes could be propagated during an interval of 1.59×10^{-4} s. This length is so much larger than any dimension of the loop or its distance from the wire that, although we consider the straight wire to be infinitely long, we can also safely ignore the field propagation effects in the vicinity of the loop. Moreover, the phase angle can be considered to be constant along the wire in the vicinity of the loop. If the frequency ω were much larger, say, $200\pi \times 10^{5}\ \mathrm{s}^{-1}$, the corresponding critical length would be only 48 cm. In this situation, propagation effects would be important and the above expression of $\mathcal{E}$ would require modification. As a "rule of thumb," we can consider field propagation effects for circuits of laboratory size to be negligible for frequencies, $f = \omega/2\pi$, that are less than about 10^{6} Hz.

15. A conducting rectangular loop of mass M, resistance R, and dimensions w by ℓ falls from rest into a magnetic field **B** as in Figure P23.15. The loop accelerates until it reaches a terminal speed v_t. (a) Show that $v_t = MgR/B^2w^2$. (b) Why is v_t proportional to R? (c) Why is it inversely proportional to B^2?

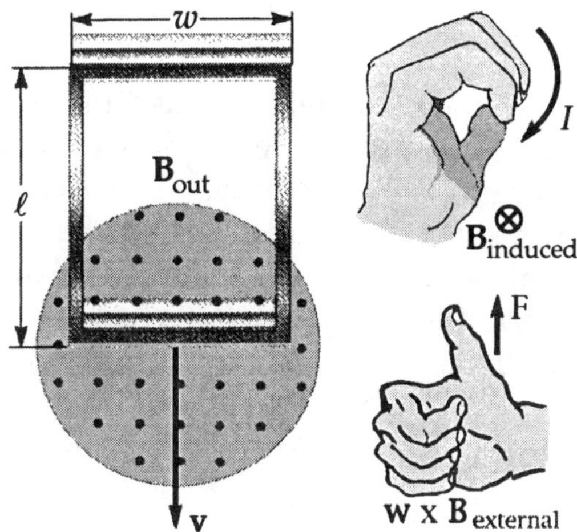

Figure P23.15

Solution Let y represent the vertical dimension of the lower part of the loop where it is inside the strong magnetic field. As the loop falls, y increases; the loop encloses increasing flux toward you and has induced in it an emf to produce a current to make its own magnetic field away from you. This current is to the left in the bottom side of the loop, and feels an upward force in the external field.

(a) Symbolically, the flux is $\Phi = BA\cos\theta = Bwy\cos 0°$

The emf is $\mathcal{E} = -N\frac{d}{dt}(Bwy) = -Bw\frac{dy}{dt} = -Bwv$

The magnitude of the current is $I = \frac{|\mathcal{E}|}{R} = \frac{Bwv}{R}$

and the force is $\mathbf{F} = I\,\mathbf{w}\times\mathbf{B} = \left(\frac{Bwv}{R}\right)wB\ \sin 90° = \frac{B^2w^2v}{R}$ up

At terminal speed, the loop is in equilibrium:

$\Sigma F_y = 0$ becomes $\frac{+B^2w^2v_t}{R} - Mg = 0.$ Thus, $v_t = \frac{MgR}{B^2w^2}$ ◊

(b) The emf is directly proportional to v, but the current is inversely proportional to R. A large R means a small current at a given speed, so the loop must travel faster to get F_{mag} = weight.

(c) At a given speed, the current is directly proportional to the magnetic field. But the force is proportional to the product of the current and the field. For a small B, the speed must increase to compensate for both the small B and also the current, so $v \propto B^{-2}$.

17. A magnetic field directed into the page changes with time according to $B = (0.0300t^2 + 1.40)$ T, where t is in seconds. The field has a circular cross-section of radius R = 2.50 cm (Fig. P23.17). What are the magnitude and direction of the electric field at point P_1 when t = 3.00 s and r_1 = 0.0200 m?

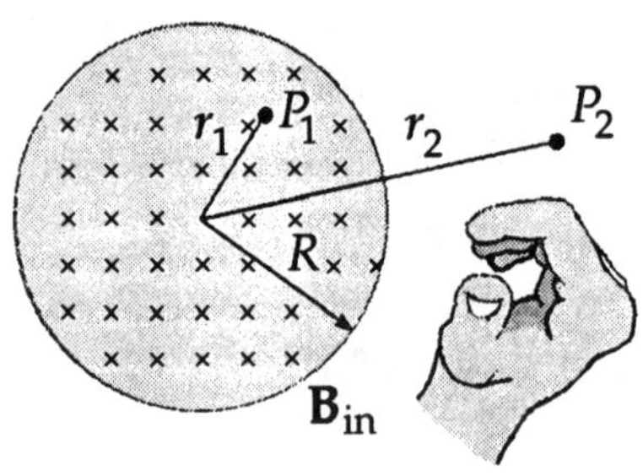

Figure P23.17

Solution $\oint \mathbf{E} \cdot ds = -\frac{d\Phi_B}{dt}$

Consider a circular integration path of radius r_1:

$$E(2\pi r_1) = -\frac{d}{dt}(BA) = -A\left(\frac{dB}{dt}\right)$$

$$|E| = \frac{A}{2\pi r_1}\frac{d}{dt}(0.0300t^2 + 1.40)\text{T} = \frac{\pi r_1^2}{2\pi r_1}(0.0600t) = \frac{r_1}{2}(0.0600t)$$

At $t = 3.00$ s, $$E = \left(\frac{0.0200 \text{ m}}{2}\right)(0.0600 \text{ T/sec})(3.00 \text{ sec}) = 1.80 \times 10^{-3} \text{ N/C} \quad \lozenge$$

If there were a circle of wire of radius r_1, it would enclose increasing magnetic flux away from you. It would carry counterclockwise current to make its own magnetic field toward you, to oppose the change. Even without the wire and current, the counterclockwise electric field that would cause the current is lurking. At point P_1, it is upward and to the left, perpendicular to r_1. ◊

---

23. A 10.0-mH inductor carries a current $I = I_{max} \sin \omega t$, with I_{max} = 5.00 A and $\omega/2\pi$ = 60.0 Hz. What is the back emf as a function of time?

Solution $$\mathcal{E}_{back} = -\mathcal{E}_L = L\frac{dI}{dt} = L\frac{d}{dt}(I_{max} \sin \omega t)$$

$$\mathcal{E}_{back} = L\omega I_{max} \cos \omega t = (0.0100 \text{ H})(120\pi \text{ s}^{-1})(5.00 \text{ A})\cos(120\pi t)$$

$$\mathcal{E}_{back} = (18.8 \text{ V})\cos(377t) \quad \lozenge$$

---

29. A 12.0-V battery is about to be connected to a series circuit containing a 10.0-Ω resistor and a 2.00-H inductor. How long will it take the current to reach (a) 50.0% and (b) 90.0% of its final value?

Solution The time constant is $\tau = \frac{L}{R} = 0.200\text{ s}$:

(a) In $I = \frac{\mathcal{E}(1 - e^{-t/\tau})}{R}$, the final value which the current approaches is

$$I = \frac{\mathcal{E}(1 - e^{-\infty})}{R} = \frac{\mathcal{E}}{R}$$

We have at 50%, $(0.500)\left(\frac{\mathcal{E}}{R}\right) = \frac{\mathcal{E}(1 - e^{-t/0.200})}{R}$

Solving for t, $0.500 = 1 - e^{-t/0.200\text{ s}}$

$$e^{-t/0.200} = 0.500$$

$$e^{t/0.200} = 2.00$$

$$\frac{t}{0.200} = \ln 2.00$$

Thus, $t = 0.139\text{ s}$ ◊

(b) At 90%, $0.900 = 1 - e^{-t/\tau}$

and $t = \tau \ln\left(\frac{1}{1 - 0.900}\right)$

$$t = 0.200 \ln 10.0 = 0.461\text{ sec}$$ ◊

---

37. The magnetic field inside a superconducting solenoid is 4.50 T. The solenoid has an inner diameter of 6.20 cm and a length of 26.0 cm. Determine (a) the magnetic energy density in the field and (b) the energy stored in the magnetic field within the solenoid.

Solution The magnetic energy density is given by Equation 23.22,

(a) $$u_B = \frac{B^2}{2\mu_o} = \frac{(4.50\ \text{T})^2}{2(1.26 \times 10^{-6}\ \text{T} \cdot \text{m} / \text{A})}$$

$$u_B = 8.06 \times 10^6\ \text{J} / \text{m}^3 = 8.06\ \text{MJ} / \text{m}^3 \quad \Diamond$$

(b) The magnetic energy stored in the field equals u_B times the volume of the solenoid (the volume in which B is non-zero).

$$U_B = u_B V = u_B \pi r^2 h = (8.06 \times 10^6\ \text{J} / \text{m}^3)\pi(0.0310\ \text{m})^2(0.260\ \text{m})$$

and $$U_B = 6.32\ \text{kJ} \quad \Diamond$$

---

43. A loop of area 0.100 m^2 is rotating at 60.0 rev/s with the axis of rotation perpendicular to a 0.200-T magnetic field. (a) If there are 1000 turns on the loop, what is the maximum voltage induced in it? (b) When the maximum induced voltage occurs, what is the orientation of the loop with respect to the magnetic field?

Solution For a coil rotating in a magnetic field,

(a) By Equation 23.8, $\mathcal{E} = NBA\omega \sin\theta$,

so $$\mathcal{E}_{\text{max}} = NBA\omega$$

Solving, $$\mathcal{E}_{\text{max}} = (1000)(0.200\ \text{T})\left(0.100\ \text{m}^2\right)\left(60.0\ \frac{\text{rev}}{\text{s}}\right)\left(2\pi\ \frac{\text{rad}}{\text{rev}}\right) = 7540\ \text{V} \quad \Diamond$$

(b) $\mathcal{E}$ approaches $\mathcal{E}_{\text{max}}$ when $\sin\theta$ approaches 1, or $\theta = \pm\frac{\pi}{2}$

Therefore, at maximum emf, the plane of the coil is parallel to the field. ◊

---

47. The magnetic flux threading a metal ring varies with time t according to $\Phi_B = 3(at^3 - bt^2)$ T·m², with $a = 2.00$ s^{-3} and $b = 6.00$ s^{-2}. The resistance of the ring is 3.00 Ω. Determine the maximum current induced in the ring during the interval from $t = 0$ to $t = 2.00$ s.

Solution Substituting the given values, $\Phi_B = \left(6.00t^3 - 18.0t^2\right)\,\text{T}\cdot\text{m}^2$

Therefore, the emf induced is $\mathcal{E} = -\frac{d\Phi_B}{dt} = -18.0t^2 + 36.0t$

The maximum $\mathcal{E}$ occurs when $\frac{d\mathcal{E}}{dt} = -36.0t + 36.0 = 0,$ which gives $t = 1.00$ s.

Thus, maximum current (at $t = 1.00$ s) is $I_{max} = \frac{\mathcal{E}}{R} = \frac{(-18.0 + 36.0)\text{V}}{3.00\ \Omega} = 6.00\ \text{A}$ ◊

---

49. Assume that the switch in Figure P23.49 is initially in position 1. Show that if the switch is thrown from position 1 to position 2, all the energy stored in the magnetic field of the inductor is dissipated as thermal energy in the resistor.

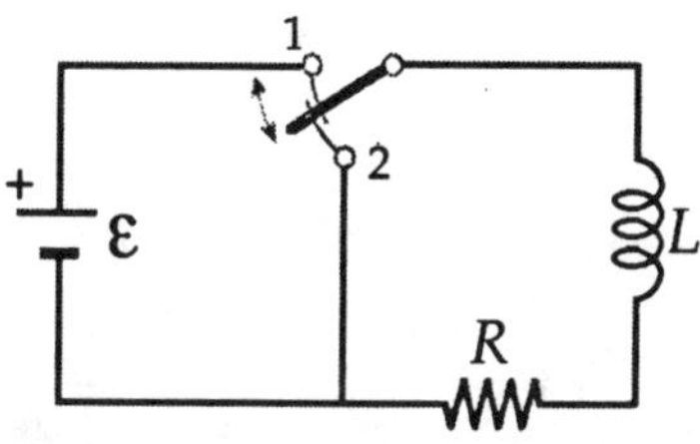

Figure P23.49

Solution The initial current in the inductor will be $\mathcal{E}/R$; therefore, the energy initially stored in the inductor will be

$$U_L = \tfrac{1}{2}LI^2 = \tfrac{1}{2}L\mathcal{E}^2/R^2.$$

After the switch is thrown, $I = \frac{\mathcal{E}}{R}e^{-Rt/L}$

The heat power of the resistor is $P = \frac{dU}{dt} = I^2R = \frac{\mathcal{E}^2}{R}e^{-2Rt/L}$

and the total heat generated is $U = \int_{\text{all } t} dU = \int_0^\infty P\,dt = \int_0^\infty \frac{\mathcal{E}^2}{R}e^{-2Rt/L}dt$

Substituting, $U = \frac{\mathcal{E}^2}{R}\left(-\frac{L}{2R}\right)\int_0^\infty e^{-2Rt/L}\left(-\frac{2Rdt}{L}\right) = -\frac{L\mathcal{E}^2}{2R^2}e^{-2Rt/L}\Big]_0^\infty$

and $U = -\frac{L\mathcal{E}^2}{2R^2}(0-1) = +\frac{L\mathcal{E}^2}{2R^2}$ ◊

Chapter 24

This is one of 27 radio telescopes of the Very Large Array. The telescopes are arranged in a Y-shaped pattern; any section of the sky can be examined by simply aiming several of the telescopes, and correlating the data from several telescopes. (Danny Lehman)

ELECTROMAGNETIC WAVES

Chapter 24

ELECTROMAGNETIC WAVES

INTRODUCTION

By definition, mechanical disturbances, such as sound waves, water waves, and waves on a string, require the presence of a medium. This chapter is concerned with the properties of electromagnetic waves that (unlike mechanical waves) can propagate through empty space.

The consequences of Maxwell's equations are far-reaching and very dramatic for the history of physics. One of them, the Ampere-Maxwell law, predicts that a time-varying electric field produces a magnetic field just as a time-varying magnetic field produces an electric field (Faraday's law). From this generalization, Maxwell introduced the concept of displacement current, a new source of a magnetic field. Thus, Maxwell's theory provided the final important link between electric and magnetic fields.

NOTES FROM SELECTED CHAPTER SECTIONS

24.2 Maxwell's Wonderful Equations

Electromagnetic waves are generated by accelerating electric charges. The radiated waves consist of oscillating electric and magnetic fields, which are **at right angles to each other** and also **at right angles to the direction of wave propagation.**

The fundamental laws describing the behavior of electric and magnetic fields are Maxwell's equations. In this unified theory of electromagnetism, Maxwell showed that electromagnetic waves are a natural consequence of these fundamental laws.

The theory he developed is based upon the following pieces of information:

- A charge creates an electric field. Electric fields originate on positive charges and terminate on negative charges. The electric field due to a point charge is described by Gauss's law.
- Magnetic field lines always form closed loops; that is, they do not begin or end anywhere.
- A varying magnetic field induces an emf and hence an electric field. This is a statement of Faraday's law.

- A moving charge (constituting a current) creates a magnetic field, as summarized in Ampère's law.

- A varying electric field creates a magnetic field. This is Maxwell's addition to Ampère's law.

24.3 Electromagnetic Waves

Following is a summary of the properties of electromagnetic waves:

- The solutions of Maxwell's third and fourth equations are wavelike, where both **E** and **B** satisfy the same wave equation.

- Electromagnetic waves travel through empty space with the speed of light,

$$c = \frac{1}{\sqrt{\mu_0 \epsilon_0}}$$

- The electric and magnetic field components of plane electromagnetic waves are perpendicular to each other and also perpendicular to the direction of wave propagation. The latter property can be summarized by saying that electromagnetic waves are transverse waves.

- The magnitudes of **E** and **B** in empty space are related by $E/B = c$.

- Electromagnetic waves obey the principle of superposition.

24.5 Production of Electromagnetic Waves by an Antenna

The fundamental mechanism responsible for radiation by an antenna is the acceleration of a charged particle. Whenever a charged particle undergoes an acceleration, it must radiate energy. An alternating voltage applied to the wires of an antenna forces electric charges in the antenna to oscillate.

24.6 Energy Carried by Electromagnetic Waves

The magnitude of the Poynting vector represents the rate at which energy flows through a unit surface area perpendicular to the flow.

For an electromagnetic wave, the instantaneous energy density associated with the magnetic field equals the instantaneous energy density associated with the electric field. Hence, in a given volume, the energy is equally shared by the two fields.

24.7 Momentum and Radiation Pressure

Electromagnetic waves have momentum and exert pressure on surfaces on which they are incident. The pressure exerted by a normally incident wave on a **totally reflecting** surface is **double** that exerted on a surface which **completely absorbs** the incident wave.

EQUATIONS AND CONCEPTS

Maxwell's equations are the fundamental laws governing the behavior of electric and magnetic fields. Electromagnetic waves are a natural consequence of these laws.

$$\oint \mathbf{E} \cdot d\mathbf{A} = \frac{Q}{\epsilon_0} \quad (24.4)$$

$$\oint \mathbf{B} \cdot d\mathbf{A} = 0 \quad (24.5)$$

You should notice that the integrals in Equations 24.4 and 24.5 are **surface integrals** in which the normal components of electric and magnetic fields are integrated over a **closed surface** while Equations 24.6 and 24.7 involve line integrals in which the tangential components of electric and magnetic fields are integrated around a **closed path**.

$$\oint \mathbf{E} \cdot d\mathbf{s} = -\frac{d\Phi_B}{dt} \quad (24.6)$$

where $\Phi_B = \mathbf{B} \cdot \mathbf{A}$

$$\oint \mathbf{B} \cdot d\mathbf{s} = \mu_0 I + \epsilon_0 \mu_0 \frac{d\Phi_E}{dt} \quad (24.7)$$

where $\Phi_E = \mathbf{E} \cdot \mathbf{A}$

Both **E** and **B** satisfy a differential equation which has the form of the general wave equation. These are the wave equations for electromagnetic waves in free space (where $Q = 0$ and $I = 0$). As stated here, they represent linearly polarized waves traveling with a speed c.

$$\frac{\partial^2 E}{\partial x^2} = \mu_0 \epsilon_0 \frac{\partial^2 E}{\partial t^2} \quad (24.12)$$

$$\frac{\partial^2 B}{\partial x^2} = \mu_0 \epsilon_0 \frac{\partial^2 B}{\partial t^2} \quad (24.13)$$

$$c = \frac{1}{\sqrt{\mu_0 \epsilon_0}} \quad (24.14)$$

The electric and magnetic fields vary in position and time as **sinusoidal transverse waves**. Their planes of vibration are perpendicular to each other and perpendicular to the direction of propagation.

$$E = E_{max} \cos(kx - \omega t) \quad (24.15)$$

$$B = B_{max} \cos(kx - \omega t) \quad (24.16)$$

The ratio of the magnitude of the electric field to the magnitude of the magnetic field is constant and equal to the speed of light c.

$$\frac{E}{B} = c \quad (24.17)$$

The **Poynting vector S** describes the energy flow associated with an electromagnetic wave. The direction of **S** is along the direction of propagation and the magnitude of **S** is the rate at which electromagnetic energy crosses a unit surface area perpendicular to the direction of **S**.

$$\mathbf{S} \equiv \frac{1}{\mu_o} \mathbf{E} \times \mathbf{B} \quad (24.19)$$

The **wave intensity** is the time average of the magnitude of the Poynting vector. E_{max} and B_{max} are the **maximum values** of the field magnitudes.

$$I = S_{av} = \frac{E_{max}^2}{2\mu_o c} = \frac{c}{2\mu_o} B_{max}^2 \quad (24.21)$$

The electric and magnetic fields have **equal instantaneous energy densities.**

$$u_B = u_E = \tfrac{1}{2}\epsilon_o E^2 = \frac{B^2}{2\mu_o}$$

The total instantaneous energy density u is proportional to E^2 and B^2 while the **total average energy density** is proportional to E_{max}^2 and B_{max}^2. The average energy density is also proportional to the wave intensity.

$$u = \epsilon_o E^2 = \frac{B^2}{\mu_o}$$

$$u_{av} = \tfrac{1}{2}\epsilon_o E_{max}^2 = \frac{B_{max}^2}{2\mu_o} \qquad (24.24)$$

$$I = S_{av} = cu_{av} \qquad (24.25)$$

The **linear momentum** p **delivered to an absorbing surface** by an electromagnetic wave at normal incidence depends on the fraction of the total energy absorbed.

$$p = \frac{U}{c} \quad \text{(complete absorption)} \qquad (24.26)$$

$$p = \frac{2U}{c} \quad \text{(complete reflection)}$$

An absorbing surface (at normal incidence) will experience a **radiation pressure** P which depends on the magnitude of the Poynting vector and the degree of absorption.

$$P = \frac{S}{c} \quad \text{(complete absorption)} \qquad (24.27)$$

$$P = \frac{2S}{c} \quad \text{(complete reflection)}$$

REVIEW CHECKLIST

▷ Describe the essential features of the apparatus and procedure used by Hertz in his experiments leading to the discovery and understanding of the source and nature of electromagnetic waves.

▷ For a properly described plane electromagnetic wave, calculate the values for the Poynting vector (magnitude), wave intensity, and instantaneous and average energy densities.

▷ Calculate the radiation pressure on a surface and the linear momentum delivered to a surface by an electromagnetic wave.

▷ Understand the production of electromagnetic waves and radiation of energy by an oscillating dipole. Use a diagram to show the relative directions for **E**, **B**, and **S**.

ANSWERS TO SELECTED CONCEPTUAL QUESTIONS

10. Suppose a creature from another planet had eyes that were sensitive to infrared radiation. Describe what he would see if he looked around the room you are now in. That is, what would be bright, and what would be dim?

Answer Light bulbs and the toaster glow brightly in the infrared. Somewhat fainter are the back of the refrigerator and the back of the television set, while the TV screen is dark. The pipes under the sink show the same weak glow as the walls until you turn on the faucets. Then the pipe on the right gets darker while that on the left develops a rich gleam that quickly runs up along its length. The food on your plate shines; so does human skin, the same color for all races. Clothing is dark as a rule, but your seat glows like a monkey's rump when you get up from a chair, and you leave a patch of the same glow on your chair. Your face appears lit from within, like a jack-o-lantern; your nostrils and openings of your ear canals are bright; brighter still are the pupils of your eyes.

□ □ □ □

13. Radio stations often advertise "instant news." If what they mean is that you hear the news at the instant they speak it, is their claim true? About how long would it take for a message to travel across this country by radio waves, assuming that these waves could travel this great distance and still be detected?

Answer Radio waves move at the speed of light. They can travel around the curved surface of the Earth, bouncing between the ground and the ionosphere, which has an altitude that is small when compared to the radius of the earth. The distance across the lower forty-eight states is approximately 5000 km, requiring a time of $(5\times10^6\text{ m})/(3\times10^8\text{ m/s})\sim10^{-2}\text{ s}$. To go halfway around the Earth takes only 0.07 s. In other words, a speech can be heard on the other side of the world before it is heard at the back of the room.

□ □ □ □

14. Light from the Sun takes approximately $8\frac{1}{3}$ minutes to reach the Earth. During this time the Earth has continued to move in its orbit around the Sun. How far is the actual location of the Sun from its image in the sky?

Answer The Sun's angular speed in our sky is our daily rate of rotation, 360° every 24 hours, or 15° every hour. Using this ratio, we may calculate that in $8\frac{1}{3}$ minutes the image of the sun will move to the west by $\theta=[15°/60\text{ min}](8.3\text{ min})=2.1°$. This is about four times the angular diameter of the Sun.

□ □ □ □

SOLUTIONS TO SELECTED END-OF-CHAPTER PROBLEMS

5. Figure 24.4 shows a plane electromagnetic sinusoidal wave propagating in the x direction. The wavelength is 50.0 m, and the electric field vibrates in the xy plane with an amplitude of 22.0 V/m. Calculate (a) the sinusoidal frequency and (b) the magnitude and direction of **B** when the electric field has its maximum value in the negative y direction. (c) Write an expression for B in the form

$$B = B_{\text{max}} \cos(kx - \omega t)$$

with numerical values for B_{max}, k, and ω.

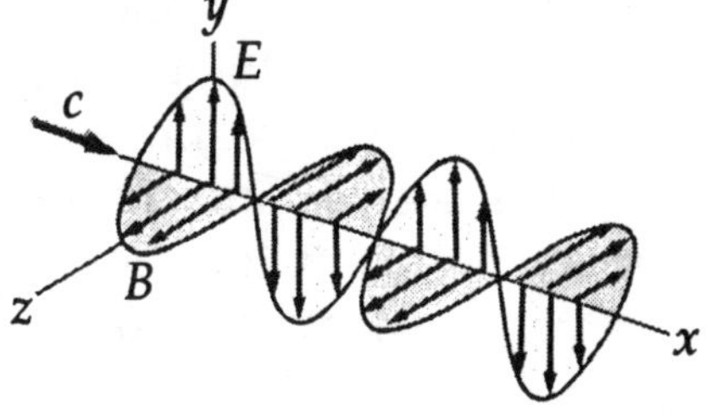

Figure 24.4

Solution

(a) $c = f\lambda$

$$f = \frac{c}{\lambda} = \frac{3.00 \times 10^8 \text{ m/s}}{50.0 \text{ m}} = 6.00 \times 10^6 \text{ Hz} \quad \Diamond$$

(b) $c = \dfrac{E}{B}$

$$B = \frac{E}{c} = \frac{22.0 \text{ V/m}}{3.00 \times 10^8 \text{ m/s}} = 7.33 \times 10^{-8} \text{ T} = 73.3 \text{ nT} \quad \Diamond$$

B is directed along **negative z direction** when **E** is in the negative y direction; therefore, $\mathbf{S} = \mathbf{E} \times \mathbf{B}/\mu_0$ will propagate in the direction of $(-\mathbf{j}) \times (-\mathbf{k}) = +\mathbf{i}$.

(c) $B = B_{\text{max}} \cos(kx - \omega t)$

$$k = \frac{2\pi}{\lambda} = \frac{2\pi}{50.0 \text{ m}} = 0.126 \text{ m}^{-1}$$

$$\omega = 2\pi f = (2\pi \text{ rad})\left(6.00 \times 10^6 \text{ Hz}\right) = 3.77 \text{ rad/s}$$

Thus, $$B = (73.3 \text{ nT}) \cos\left((0.126 \text{ rad/m})x - \left(3.77 \times 10^7 \text{ rad/s}\right)t\right) \quad \Diamond$$

11. A television set uses a dipole-receiving antenna for VHF channels and a loop antenna for UHF channels. The UHF antenna produces a voltage from the changing magnetic flux through the loop. (a) Using Faraday's law, derive an expression for the amplitude of the voltage that appears in a single-turn circular loop antenna with a radius r, small compared to the wavelength of the wave. The TV station broadcasts a signal with a frequency f, and the signal has an electric field amplitude E_{max} and magnetic field amplitude B_{max} at the receiving antenna's location. (b) If the electric field in the signal points vertically, what should be the orientation of the loop for best reception?

Solution We can approximate the magnetic field as uniform over the area of the loop while it oscillates in time as $B = B_{max} \cos \omega t$. The induced voltage is

$$\mathcal{E} = -\frac{d\,\Phi_B}{dt} = -\frac{d}{dt}(BA\cos\theta) = -A\,\frac{d}{dt}(B_{max}\cos\omega t\cos\theta)$$

$$\mathcal{E} = AB_{max}\,\omega(\sin\omega t\cos\theta)$$

$$\mathcal{E}(t) = 2\pi f B_{max} A\sin 2\pi f t\cos\theta = 2\pi^2 r^2 f B_{max}\cos\theta\sin 2\pi f t$$

(a) The amplitude of this emf is $\mathcal{E}_{max} = 2\pi^2 r^2 f B_{max}\cos\theta$, where θ is the angle between the magnetic field and the normal to the loop. ◊

(b) If **E** is vertical, then **B** is horizontal, so the plane of the loop should be vertical, and the normal of the plane should point toward the transmitter. This will make $\theta = 0°$, so $\cos\theta$ takes on its maximum value.

---

13. Two radio-transmitting antennas are separated by half the broadcast wavelength and are driven in phase with each other. In which directions are (a) the strongest and (b) the weakest signals radiated?

Solution

(a) Along the perpendicular bisector of the line joining the antennas, you are equally distant from both. They oscillate in phase, so along this line you receive the two signals in phase. They interfere constructively to produce the strongest signal.

(b) Along the extended line joining the sources, the wave from the farther antenna must travel one-half wavelength farther to reach you, so you receive the waves 180° out of phase. They interfere destructively to produce the weakest signal.

---

15. What is the average magnitude of the Poynting vector 5.00 miles from a radio transmitter broadcasting isotropically with an average power of 250 kW?

Solution To avoid confusion in this chapter, we use p to represent momentum, P to represent pressure, and $\mathcal{P}$ to denote power. The Poynting vector is the rate at which electromagnetic energy crosses a unit area; alternatively, we can write this in terms of the power of the broadcast station:

$$S_{av} = \frac{\mathcal{P}}{A} = \frac{\mathcal{P}}{4\pi r^2}$$

In meters, $r = (5.00 \text{ mi})(1609 \text{ m/mi}) = 8045 \text{ m}$

and the magnitude is $S = \dfrac{250 \times 10^3 \text{ W}}{(4\pi)(8.045 \times 10^3 \text{ m})^2} = 3.07 \times 10^{-4} \text{ W/m}^2$ ◊

---

17. A community plans to build a facility to convert solar radiation to electrical power. They require 1.00 MW of power, and the system to be installed has an efficiency of 30.0% (that is, 30.0% of the solar energy incident on the surface is converted to electrical energy). What must be the effective area of a perfectly absorbing surface used in such an installation, assuming a constant energy flux of 1000 W/m^2?

Solution

At 30.0% efficiency, the power $\mathcal{P} = 0.300SA$

$$A = \frac{\mathcal{P}}{0.300S} = \frac{1.00 \times 10^6 \text{ W}}{(0.300)(1000 \text{ W/m}^2)} = 3330 \text{ m}^2 \quad ◊$$

This area is approximately 0.75 acres. ◊

---

19. The filament of an incandescent lamp has a 150-Ω resistance and carries a direct current of 1.00 A. The filament is 8.00 cm long and 0.900 mm in radius. (a) Calculate the magnitude of the Poynting vector at the surface of the filament. (b) Find the magnitude of the electric and magnetic fields at the surface of the filament.

Solution In this problem, the Poynting vector does not represent the light radiated or the heat convected away from the filament. Rather, it represents the energy flow in the static electric and magnetic fields created in the surrounding empty space, by the power supply pushing current through the filament.

The rate at which the resistor converts electromagnetic energy into heat is

$$\mathcal{P} = I^2R = 150 \text{ W},$$

and the surface area is $A = 2\pi rL = 2\pi(0.900 \times 10^{-3} \text{ m})(0.0800 \text{ m}) = 4.52 \times 10^{-4} \text{ m}^2.$

(a) The Poynting vector is $S = \dfrac{\mathcal{P}}{A} = 3.32 \times 10^5 \text{ W/m}^2$ ◊ (points radially inward)

(b) $$B = \mu_0 \frac{I}{2\pi r} = \frac{\mu_0(1)}{2\pi(0.900 \times 10^{-3})} = 2.22 \times 10^{-4} \text{ T} \quad ◊$$

$$E = \frac{\Delta V}{\Delta x} = \frac{IR}{L} = \frac{150 \text{ V}}{0.0800 \text{ m}} = 1880 \text{ V/m} \quad ◊$$

Note: We could also calculate the Poynting vector from $S = \dfrac{EB}{\mu_0} = 3.32 \times 10^5 \text{ W/m}^2$ ◊

---

25. A radio wave transmits 25.0 W/m² of power per unit area. A flat surface of area A is perpendicular to the direction of propagation of the wave. Calculate the radiation pressure on it if the surface is a perfect absorber.

Solution For complete absorption, $P = \dfrac{S}{c}$

$$P = \frac{25.0 \text{ W/m}^2}{3.00 \times 10^8 \text{ m/s}} = 8.33 \times 10^{-8} \text{ N/m}^2 \quad ◊$$

---

27. A 15.0-mW helium-neon laser (λ = 632.8 nm) emits a beam of circular cross section the diameter of which is 2.00 mm. (a) Find the maximum electric field in the beam. (b) What total energy is contained in a 1.00-m length of the beam? (c) Find the momentum carried by a 1.00-m length of the beam.

Solution

The intensity of the light is the average magnitude of the Poynting vector:

$$I = \frac{\mathscr{P}}{\pi r^2} = \frac{E_{max}^2}{2\mu_o c}$$

(a) Therefore, the maximum electric field is

$$E_{max} = \sqrt{\frac{\mathscr{P}(2\mu_o c)}{\pi r^2}} = 1.90 \times 10^3 \text{ N/C} \quad \Diamond$$

(b) The power being 15.0 mW means that 15.0 mJ passes through a cross section of the beam in one second. This energy is uniformly spread through a beam length of 3.00×10^8 m, since that is how far the front end of the energy travels in one second. Thus, the energy in just a one-meter length is

$$\frac{15.0 \times 10^{-3} \text{ J/s}}{3.00 \times 10^8 \text{ m/s}}(1.00 \text{ m}) = 5.00 \times 10^{-11} \text{ J} \quad \Diamond$$

(c) The linear momentum carried by a 1.00 m length of the beam is the momentum that would be received by an absorbing surface, under complete absorption:

$$p = \frac{U}{c} = \frac{5.00 \times 10^{-11} \text{ J}}{3.00 \times 10^8 \text{ m/s}} = 1.67 \times 10^{-19} \text{ kg} \cdot \text{m/s} \quad \Diamond$$

35. Plane-polarized light is incident on a single polarizing disk with the direction of $\mathbf{E}_0$ parallel to the direction of the transmission axis. Through what angle should the disk be rotated so that the intensity in the transmitted beam is reduced by a factor of (a) 3.00, (b) 5.00, (c) 10.0?

Solution We define the initial angle, at which all the light is transmitted to be $\theta = 0$. Turning the disk to another angle will then reduce the transmitted light by an intensity factor of $I = I_0 \cos^2 \theta$.

(a) For $I = I_0 / 3.00$, $\cos\theta = \dfrac{1}{\sqrt{3.00}}$ and $\theta = 54.7°$ ◊

(b) For $I = I_0 / 5.00$, $\cos\theta = \dfrac{1}{\sqrt{5.00}}$ and $\theta = 63.4°$ ◊

(c) For $I = I_0 / 10.0$, $\cos\theta = \dfrac{1}{\sqrt{10.0}}$ and $\theta = 71.6°$ ◊

---

39. A dish antenna having a diameter of 20.0 m receives (at normal incidence) a radio signal from a distant source, as shown in Figure P24.39. The radio signal is a continuous sinusoidal wave with amplitude E_{max} = 0.200 μV/m. Assume the antenna absorbs all the radiation that falls on the dish. (a) What is the amplitude of the magnetic field in this wave? (b) What is the intensity of the radiation received by this antenna? (c) What is the power received by the antenna? (d) What force is exerted on the antenna by the radio waves?

Figure P24.39

Solution (a) $B_{max} = \dfrac{E_{max}}{c} = 6.67 \times 10^{-16}$ T ◊

(b) $S_{av} = \dfrac{E_{max}^2}{2\mu_0 c} = 5.31 \times 10^{-17}$ W/m^2 ◊

(c) $\mathcal{P}_{av} = S_{av} A = 1.67 \times 10^{-14}$ W ◊ (Do not mix up this power with the pressure $P = S/c$, which we use to find the force.)

(d) $F = PA = \left(\dfrac{S_{av}}{c}\right) A = 5.56 \times 10^{-23}$ N ◊ (the weight of ~3000 Hydrogen atoms!)

---

41. In 1965, Penzias and Wilson discovered the cosmic microwave radiation left over from the Big Bang expansion of the Universe. Suppose the energy density of this radiation is 4.00×10^{-14} J/m^3. Determine the corresponding electric field amplitude.

Solution $u = \frac{1}{2}\epsilon_o E_{\max}^2$ (Eq. 24.24)

$$E_{\max} = \sqrt{\frac{2u}{\epsilon_o}} = \sqrt{\frac{2\left(4.00\times10^{-14}\ \mathrm{N/m^2}\right)}{8.85\times10^{-12}\ \mathrm{C^2/N\cdot m^2}}} = 95.1\ \mathrm{mV/m} \quad \lozenge$$

---

43. A linearly polarized microwave of wavelength 1.50 cm is directed along the positive x axis. The electric field vector has a maximum value of 175 V/m and vibrates in the xy plane. (a) Assume that the magnetic field component of the wave can be written in the form $B = B_{\max}\sin(kx - \omega t)$ and give values for $B_{\max}$, k, and ω. Also, determine in which plane the magnetic field vector vibrates. (b) Calculate the magnitude of the Poynting vector for this wave. (c) What maximum radiation pressure would this wave exert if directed at normal incidence onto a perfectly reflecting sheet? (d) What maximum acceleration would be imparted to a 500-g sheet (perfectly reflecting and at normal incidence) of dimensions 1.00 m × 0.750 m?

Solution $B = B_{\max}\sin(kx - \omega t)$

(a) $$B_{\max} = \frac{E_{\max}}{c} = \frac{175\ \mathrm{V/m}}{3.00\times10^{8}\ \mathrm{m/s}} = 5.83\times10^{-7}\ \mathrm{T} \quad \lozenge$$

$$k = \frac{2\pi}{\lambda} = \frac{2\pi}{0.015\ \mathrm{m}} = 419\ \mathrm{m^{-1}} \quad \lozenge$$

$$\omega = kc = (419\ \mathrm{m^{-1}})(3.00\times10^{8}\ \mathrm{m/s}) = 1.26\times10^{11}\ \mathrm{rad/s} \quad \lozenge$$

(b) $$S_{av} = \frac{E_{\max}B_{\max}}{2\mu_o} = \frac{(175\ \mathrm{V/m})(5.83\times10^{-7}\ \mathrm{T})}{2\times4\pi\times10^{-7}\ \mathrm{N/A^2}} = 40.6\ \mathrm{W/m^2} \quad \lozenge$$

(c) For perfect reflection, $$P_r = \frac{2S}{c} = \frac{(2)(40.6\ \mathrm{W/m^2})}{3.00\times10^{8}\ \mathrm{m/s}} = 2.71\times10^{-7}\ \mathrm{N/m^2} \quad \lozenge$$

(d) $$a = \frac{F}{m} = \frac{P_r A}{m} = \frac{(2.71\times10^{-7}\ \mathrm{N/m^2})(0.750\ \mathrm{m^2})}{0.500\ \mathrm{kg}} = 4.06\times10^{-7}\ \mathrm{m/s^2} \quad \lozenge$$

Chapter 25

This is a photograph of a rainbow over Niagra Falls in Ontario, Canada. The colors of the rainbow are produced when white light reflects off the interior surfaces of water droplets. Upon exiting the water droplet, different colors are refracted by different amounts, and color separation occurs. (John Edwards / Tony Stoine Images)

REFLECTION AND REFRACTION OF LIGHT

Chapter 25

REFLECTION AND REFRACTION OF LIGHT

NTRODUCTION

The chief architect of the particle theory of light was Newton. With this theory he provided simple explanations of some known experimental facts concerning the nature of light, namely the laws of reflection and refraction.

In 1678 a Dutch physicist and astronomer, Christian Huygens (1629 - 1695), showed that a wave theory of light could also explain the laws of reflection and refraction. The wave theory did not receive immediate acceptance for several reasons. All the waves known at the time (sound, water, and so on) traveled through some sort of medium, but light from the Sun could travel to Earth through empty space. Furthermore, it was argued that if light were some form of wave, it would bend around obstacles; hence, we should be able to see around corners. It is now known that light does indeed bend around the edges of objects. This phenomenon, known as **diffraction**, is not easy to observe because light waves have such short wavelengths. For more than a century most scientists rejected the wave theory and adhered to Newton's particle theory. This was, for the most part, due to Newton's great reputation as a scientist.

NOTES FROM SELECTED CHAPTER SECTIONS

25.2 The Ray Approximation in Geometric Optics

In the ray approximation, it is assumed that a wave travels through a medium along a straight line in the direction of its rays. A ray for a given wave is a straight line perpendicular to the wavefront. This approximation also neglects diffraction effects.

25.3 Reflection and Refraction

A line drawn perpendicular to a surface at the point where an incident ray strikes the surface is called the **normal line**. Angles of reflection and refraction are measured relative to the normal.

When an incident ray undergoes partial reflection and partial refraction, the incident, reflected and refracted rays are all **in the same plane**.

The path of a light ray through a refracting surface is reversible.

As light travels from one medium into another, the **frequency does not change**.

25.4 Dispersion and Prisms

For a given material, the index of refraction is a function of the wavelength of light passing through the material. This effect is called dispersion. In particular, when light passes through a prism, a given ray is refracted at two surfaces and emerges bent away from its original direction by an angle of deviation, δ. Due to dispersion, δ is different for different wavelengths.

25.5 Huygens's Principle

Every point on a given wavefront can be considered as a point source for a **secondary wavelet**. At some later time, the new position of the wavefront is determined by the surface tangent to the set of secondary wavelets.

25.6 Total Internal Reflection

Total internal reflection, illustrated in Figure 25.1, is possible only when light rays traveling in one medium are incident on a boundary between the first medium and a second medium of **lesser** index of refraction.

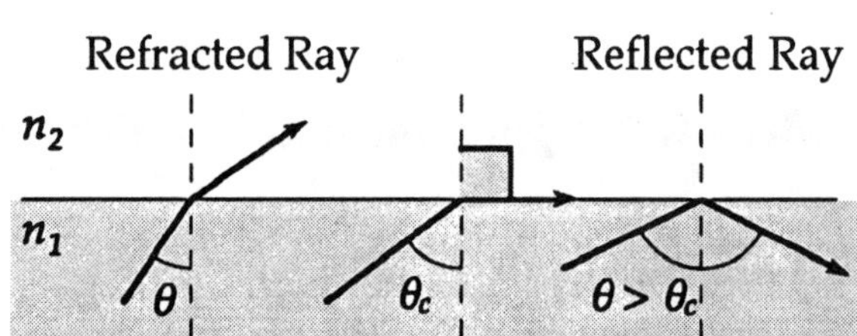

Figure 25.1 Total internal reflection of light occurs at angles of incidence $\theta_1 \geq \theta_c$ where $n_1 > n_2$

EQUATIONS AND CONCEPTS

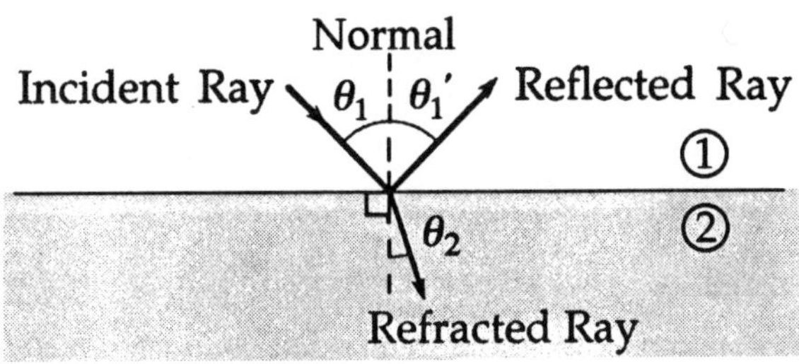

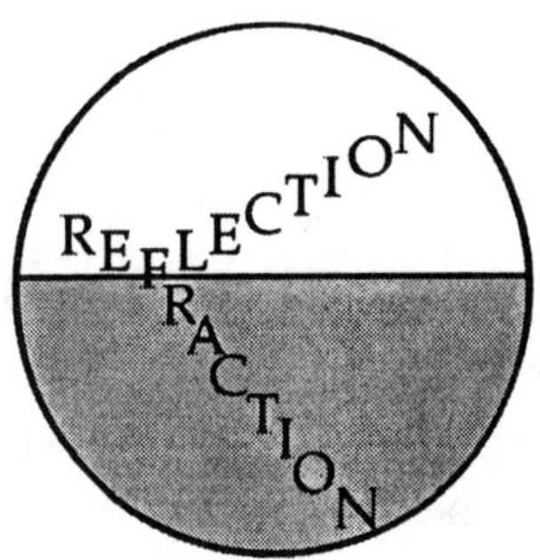

Figure 25.2 The reflection and refraction of light waves at an interface. The direction of the waves are perpendicular to the wavefronts.

Consider the situation in Figure 25.2, in which a light ray is incident obliquely on a smooth, planar surface which forms the boundary between two transparent media of different indices of refraction. A portion of the ray will be reflected back into the original medium, while the remaining fraction will be transmitted into the second medium.

The energy of a photon is proportional to the frequency of the associated electromagnetic wave.

$$E = hf \tag{25.1}$$

The ratio between the energy and the frequency is known as Planck's constant.

$$h = 6.63 \times 10^{-34}\ \text{J}\cdot\text{s}$$

The law of reflection states that the angle of incidence (the angle measured between the incident ray and the normal line) equals the angle of reflection (the angle measured between the reflected ray and the normal line).

$$\theta_1' = \theta_1 \quad (25.2$$

This is one form of the statement of Snell's law. The angle of refraction (measured relative to the normal line) depends on the angle of incidence, and also on the ratio of the speeds of light in the two media on either side of the refracting surface.

$$\frac{\sin\theta_2}{\sin\theta_1} = \frac{v_2}{v_1} = \text{constant} \quad (25.3$$

This is the most widely used and most practical form of Snell's law. This equation involves a parameter called the index of refraction, the value of which is characteristic of a particular medium. The index of refraction is defined in Equation 25.4 and Equation 25.7.

$$n_1 \sin\theta_1 = n_2 \sin\theta_2 \quad (25.8)$$

Each transparent medium is characterized by a dimensionless number n, the index of refraction, which equals the ratio of the speed of light in vacuum to the speed of light in the medium.

$$n = \frac{c}{v} \quad (25.4)$$

The frequency of a wave is characteristic of the source. Therefore, as light travels from one medium into another of different index of refraction, the frequency remains constant but the wavelength changes. The index of refraction of a given medium can be expressed as the ratio of the wavelength of light in vacuum to the wavelength in that medium.

$$n = \frac{\lambda_o}{\lambda_n} \quad (25.7)$$

For angles of incidence equal to or greater than the critical angle, the incident ray will be totally internally reflected back into the first medium.

$$\sin\theta_c = \frac{n_2}{n_1} \quad (\text{for } n_1 > n_2) \qquad (25.9)$$

Total internal reflection is possible only when a light ray is directed from a medium of high index of refraction into a medium of lower index of refraction.

SUGGESTIONS, SKILLS, AND STRATEGIES

It is helpful in this chapter to review the laws of trigonometry. Most important, of course, is the definition of the trigonometric functions:

$$\tan\theta = \frac{\text{Opposite length}}{\text{Adjacent length}}$$

$$\sin\theta = \frac{\text{Opposite length}}{\text{Hypotenuse}} \qquad \sin(-\theta) = -\sin\theta$$

$$\cos\theta = \frac{\text{Adjacent length}}{\text{Hypotenuse}} \qquad \cos(-\theta) = \cos\theta$$

Hypotenuse
Opposite
θ
90°
Adjacent

(Valid for right triangles only)

It will be useful to remember that the interior angles of a triangle add to 180°. It may also be useful to convert trigonometric functions from one form to another, using the trigonometric versions of the Pythagorean theorem:

$$\sin^2\theta = 1 - \cos^2\theta \qquad \tan^2\theta = \frac{\sin^2\theta}{1-\sin^2\theta}$$

Finally, you should review the angle-sum relations:

$$\sin(\theta+\phi) = \sin\theta\cos\phi + \cos\theta\sin\phi \qquad \cos(\theta+\phi) = \cos\theta\cos\phi - \sin\theta\sin\phi$$

REVIEW CHECKLIST

▷ Understand Huygens' principle and the use of this technique to construct th subsequent position and shape of a given wavefront.

▷ Determine the directions of the reflected and refracted rays when a light ray i. incident obliquely on the interface between two optical media.

▷ Understand the conditions under which total internal reflection can occur in a medium and determine the critical angle for a given pair of adjacent media.

ANSWERS TO SELECTED CONCEPTUAL QUESTIONS

4. As light travels from vacuum (n = 1) to a medium such as glass (n > 1), does the wavelength of the light change? Does the frequency change? Does the speed change? Explain.

Answer You can think about this based upon the fact that the light wave must be continuous. Therefore, the frequency of the light must remain constant, since any one crest coming up to the interface must create one crest in the new medium. However, the speed of light does vary in different media, as described by Snell's Law. In addition, since the speed of the light wave is equal to the light wave's frequency times its wavelength, and the frequency does not change, we can be sure that the wavelength must change. The speed decreases upon entering the new medium; therefore the wavelength decreases as well.

□ □ □ □

Chapter 25

Explain why a diamond loses most of its sparkle when submerged in carbon disulfide and why an imitation diamond of cubic zirconia loses all of its sparkle in corn syrup.

Answer A ball covered with mirrors sparkles by reflecting light from its surface. On the other hand, a faceted diamond lets in light at the top, reflects it by total internal reflection in the bottom half, and sends the light out through the top again. Its high index of refraction means that in air, the critical angle for total internal reflection, $\theta_c = \mathrm{Arcsin}(n_{\mathrm{air}}/n_{\mathrm{diamond}})$, is small. Thus, light rays enter through a large area, and exit through a very small area with a much higher intensity. When a diamond is immersed in carbon disulfide, the critical angle is increased to $\theta_c = \mathrm{Arcsin}(n_{\mathrm{CS}_2}/n_{\mathrm{diamond}})$. As a result, the light is emitted from the diamond over a larger area, and appears less intense.

An imitation diamond of cubic zirconia loses all of its sparkle in corn syrup because its index of refraction is the same as that of corn syrup. This stone will therefore reflect almost no light, either internally or externally. If the absorption spectrum (color) of the stone is the same as that of the corn syrup, we can expect the stone to be invisible.

□ □ □ □

10. The level of water in a clear, colorless glass is easily observed with the naked eye. The level of liquid helium in a clear glass vessel is extremely difficult to see with the naked eye. Explain.

Answer The index of refraction of water is 1.333, which is different from that of air, which is about 1. On the other hand, the index of refraction of liquid helium is much closer to that of air. Consequently, light undergoes less refraction in helium than it does in water.

□ □ □ □

12. Why does a diamond show flashes of color when observed under white light?

Answer The diamond acts like a prism in dispersing the light into its spectra components. Different colors are observed as a consequence of the manner in which the index of refraction varies with the wavelength.

□ □ □ □

13. Explain why a diamond sparkles more than a glass crystal of the same shape and size.

Answer Diamond has a larger index of refraction than glass, and consequently has a smaller critical angle for internal reflection. This results in a greater amount of light being internally reflected, and more light being emitted within the critical angles.

□ □ □ □

14. Why do astronomers looking at distant galaxies talk about looking backward in time?

Answer Light travels through a vacuum at a speed of 300 000 km per second. Thus, an image we see from a distant star or galaxy must have been generated some time ago.

For example, the star Altair is 16 light-years away; if we look at an image of Altair today, we know only what was happening 16 years ago.

This may not initially seem significant; however, astronomers who look at other galaxies can gain an idea of what galaxies looked like when they were significantly younger. Thus, it actually makes sense to speak of "looking backward in time."

□ □ □ □

5. A solar eclipse occurs when the Moon gets between the Earth and the Sun. Use a diagram to show why some areas of the Earth see a total eclipse, other areas see a partial eclipse, and most areas see no eclipse.

Answer

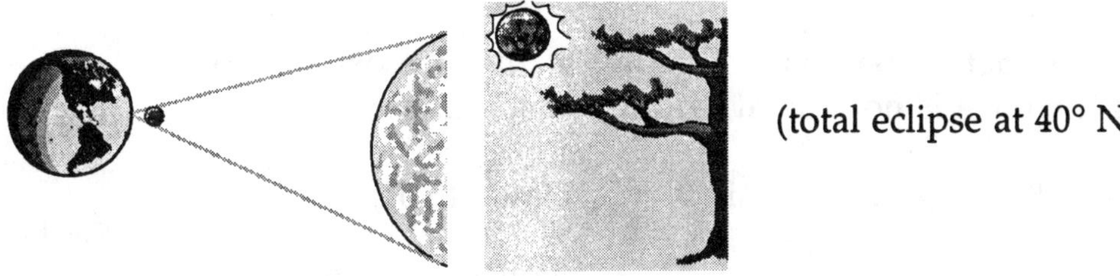

(total eclipse at 40° N)

These diagrams, though not drawn to scale, demonstrate the way the moon blocks the sun when viewed from different locations on the Earth.

(Partial eclipse at 10° N)

Since the moon and the sun appear approximately the same size in the sky, a slight change in latitude can make the difference between seeing a total eclipse, no eclipse, or a partial eclipse.

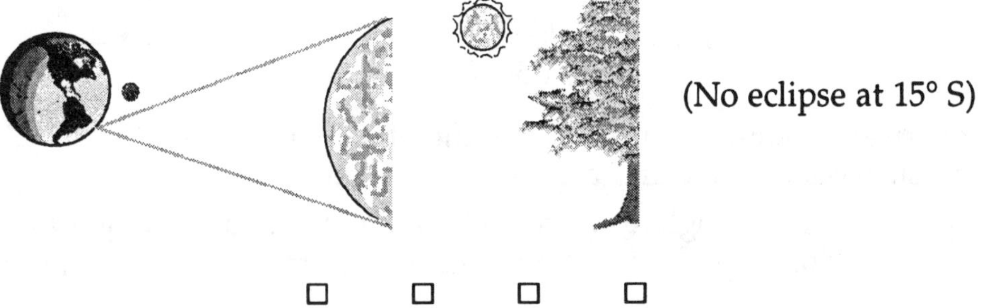

(No eclipse at 15° S)

☐ ☐ ☐ ☐

20. Suppose you are told that only two colors of light (X and Y) are sent through a glass prism and that X is bent more than Y. Which color travels more slowly in the prism?

Answer If the light slows down upon entering the prism, a light ray that is bent more suffers a greater loss of speed as it enters the new medium. Therefore, the X light rays travel more slowly.

☐ ☐ ☐ ☐

SOLUTIONS TO SELECTED END-OF-CHAPTER PROBLEMS

3. An underwater scuba diver sees the Sun at an apparent angle of 45.0° from the vertical. Where is the Sun?

Solution Refraction happens as sunlight in air crosses into water. The interface is horizontal, so the normal is vertical:

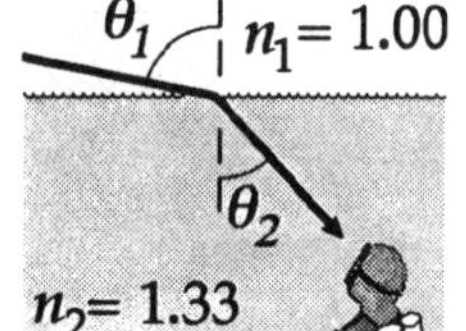

$n_1 \sin\theta_1 = n_2 \sin\theta_2$ gives $\sin\theta_1 = 1.333 \sin(45.0°)$

$$\sin\theta_1 = (1.333)(0.707) = 0.943$$

The sunlight is at θ_1 = 70.5° to the vertical, so the Sun is 19.5° above the horizon. ◊

---

5. A ray of light strikes a flat block of glass (n = 1.50) of thickness 2.00 cm at an angle of 30.0° with the normal. Trace the light beam through the glass and find the angles of incidence and refraction at each surface.

Solution At entry: $n_1 \sin\theta_1 = n_2 \sin\theta_2$

$$1.00 \sin 30° = 1.50 \sin\theta_2$$

and $\theta_2 = \sin^{-1}\left(\frac{0.500}{1.50}\right) = 19.5°$ ◊

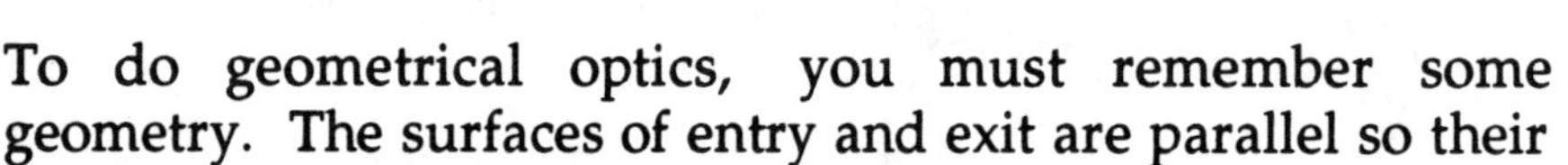

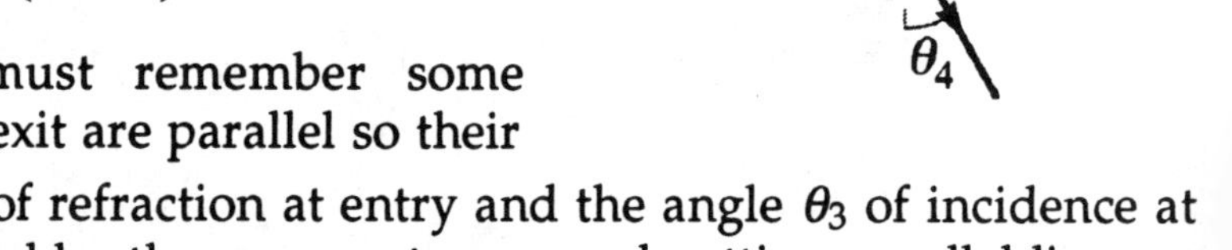

To do geometrical optics, you must remember some geometry. The surfaces of entry and exit are parallel so their normals are parallel. Then angle θ_2 of refraction at entry and the angle θ_3 of incidence at exit are alternate interior angles formed by the ray as a transversal cutting parallel lines.

So $\theta_3 = \theta_2 = 19.5°$ ◊

At the exit, $n_2 \sin\theta_3 = n_1 \sin\theta_4$

$1.5 \sin 19.5° = 1 \sin\theta_4$ and $\theta_4 = 30.0°$ ◊

The exiting ray in air is parallel to the original ray in air. Thus, a car windshield of uniform thickness will not distort, but shows the driver the actual direction to every object outside.

.9. The index of refraction for violet light in silica flint glass is 1.66, and that for red light is 1.62. What is the angular dispersion of visible light passing through a prism of apex angle 60.0° if the angle of incidence is 50.0° (Fig. P25.19)?

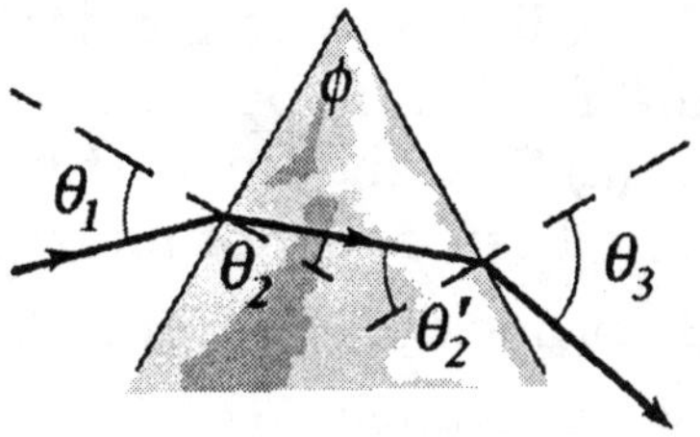

Figure P25.19

Solution Call the angles of incidence and refraction, at the surfaces of entry and exit, θ_1, θ_2, θ_2', and θ_3, in the order as shown. The apex angle (ϕ = 60.0°) is the angle between the surfaces of entry and exit. The ray in the glass forms a triangle with these surfaces, in which the interior angles must add to 180°. Thus,

$$(90° - \theta_2) + \phi + (90 - \theta_2') = 180°$$

and $$\theta_2' = \phi - \theta_2$$

This is a general rule for light going through prisms.

For the incoming ray, $$\sin\theta_2 = \frac{\sin\theta_1}{n}$$

$$(\theta_2)_{\text{violet}} = \sin^{-1}\left(\frac{\sin 50.0°}{1.66}\right) = 27.48°$$

$$(\theta_2)_{\text{red}} = \sin^{-1}\left(\frac{\sin 50.0°}{1.62}\right) = 28.22°$$

For the outgoing ray, $$\theta_2' = 60° - \theta_2 \quad \text{and} \quad \sin\theta_3 = n\sin\theta_2'$$

$$(\theta_3)_{\text{violet}} = \sin^{-1}[1.66 \sin 32.52°] = 63.17°$$

$$(\theta_3)_{\text{red}} = \sin^{-1}[1.62 \sin 31.78°] = 58.56°$$

The dispersion is the difference between these two angles:

$$\Delta\theta_3 = 63.17° - 58.56° = 4.61° \quad \Diamond$$

21. A light ray is incident on a prism and refracted at the first surface, as shown in Figure P25.21. Suppose the apex angle of the prism is 60.0° and its index of refraction is 1.50. (a) Find the smallest value of the angle of incidence at the first surface for which the refracted ray does not undergo total internal reflection at the second surface. (b) For what angle of incidence θ_1 does the light ray pass symmetrically through the prism and leave at the same angle θ_1?

Solution (a) The figures for this problem are included with problem 22. The limiting condition for total internal reflection at the second surface is

$$n \sin\theta_3 = 1 \qquad \text{or} \qquad \theta_3 = \sin^{-1}(1/n) = \sin^{-1}(1/1.50) = 41.8°$$

From the solution to problem 19, $\theta_2 + \theta_3 = \Phi$

so $\theta_2 = 60.0° - 41.8° = 18.2°$

Therefore, at entry, $(1.00)\sin\theta_1 = (1.50)\sin 18.2°$

$$\theta_1 = 27.9° \quad \lozenge$$

The photograph on page 744 of the text shows this effect nicely.

(b) The requirements of symmetry in this case also require that the light ray **inside** the prism must be horizontal (or specifically, perpendicular to the bisector of the apex angle.) In such an instance, the light will bend by the same amount entering the prism as it does exiting the prism.

From geometry, we may calculate that the base angles ϕ' are both 60.0°; in addition, since ϕ' and θ_2 together make up a right (90°) angle,

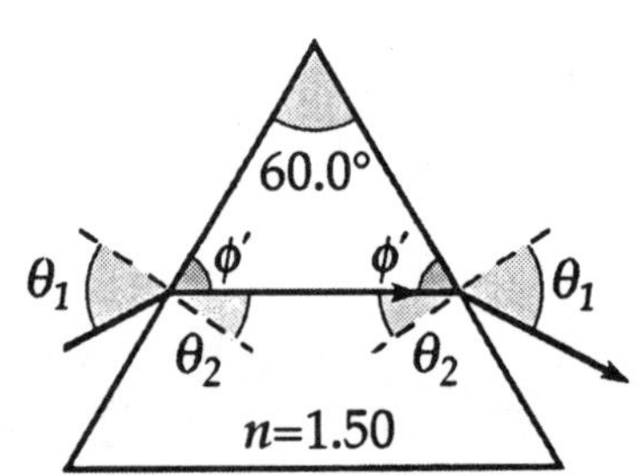

$$\theta_2 = 90.0° - \phi' = 30.0°$$

Applying Snell's law,

$$(1.00)\sin\theta_1 = (1.50)\sin 30.0°$$

and $\theta_1 = 48.6° \quad \lozenge$

This symmetric passage is easy to identify experimentally, since it happens to result in minimum deviation of the light.

22. A light ray is incident on a prism and refracted at the first surface, as shown in Figure P25.21. Let Φ represent the apex angle of the prism and n its index of refraction. Find in terms of n and Φ the smallest allowed value of the angle of incidence at the first surface for which the refracted ray does not undergo total internal reflection at the second surface.

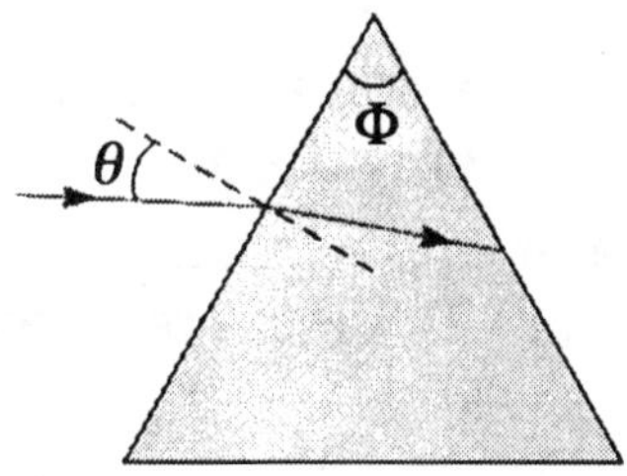

Figure P25.21

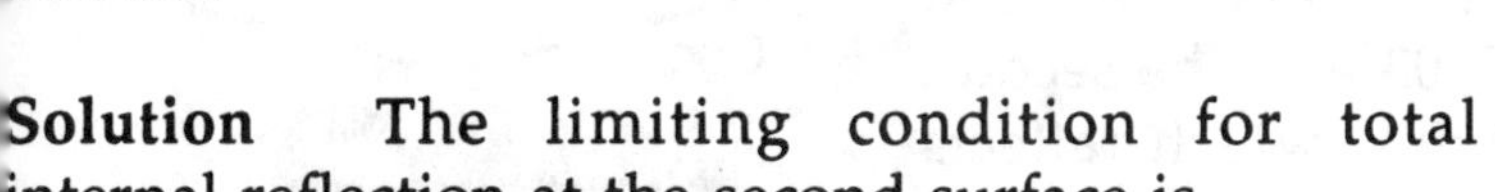

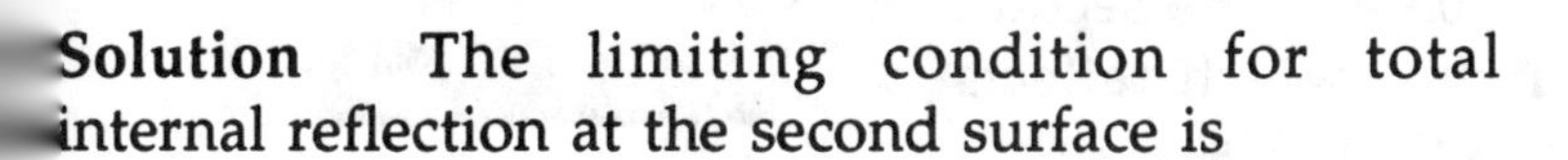

Solution The limiting condition for total internal reflection at the second surface is

$$n\sin\theta_3 = 1 \qquad \text{or} \qquad \sin\theta_3 = 1/n$$

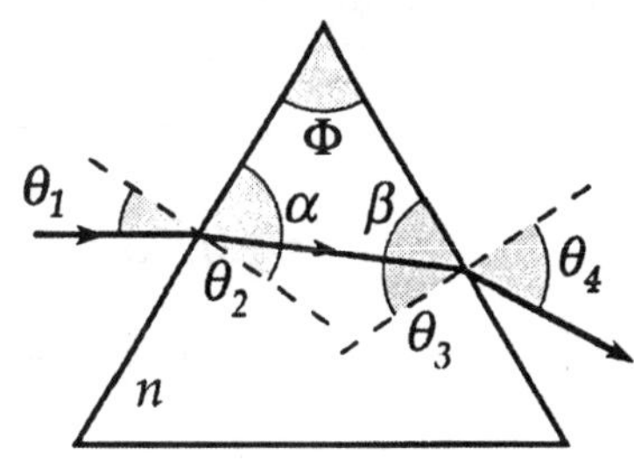

From the trigonometric version of the Pythagorean theorem (given in **Suggestions, Skills and Strategies**),

$$\cos\theta_3 = \sqrt{1-\sin^2\theta_3} = \sqrt{1-\frac{1}{n^2}}$$

Next, we can relate θ_3 to θ_2 by remembering that the interior angles of any triangle add up to 180°:

$$\Phi + \alpha + \beta = 180°$$

Since $\alpha + \theta_2 = 90°$ and $\beta + \theta_3 = 90°$,

$$\theta_2 = \Phi - \theta_3$$

Applying Snell's law at the first refracting surface,

$$(1.00)\sin\theta_1 = n\sin(\Phi - \theta_3)$$

From the angle-sum relation for sines,

$$\sin\theta_1 = n(\sin\Phi\cos\theta_3 - \cos\Phi\sin\theta_3)$$

Substituting for the θ_3 terms,

$$\sin\theta_1 = n\left(\sqrt{1-\frac{1}{n^2}}\sin\Phi - \frac{1}{n}\cos\Phi\right)$$

Therefore,

$$\theta_1 = \sin^{-1}\left(\sqrt{n^2-1}\sin\Phi - \cos\Phi\right) \quad \Diamond$$

25. Consider a common mirage formed by superheated air just above the roadway. If an observer viewing from 2.00 m above the road (where n = 1.0003) has the illusion of seeing water up the road at θ_1 = 88.8°, find the index of refraction of the air just above the road surface. (**Hint:** Treat this as a problem in total internal reflection.)

Solution

Think of the air as in two discrete layers, the first medium being cooler air with n_1 = 1.0003 and the second medium being hot air with a lower index, which reflects light from the sky by total internal reflection. Use:

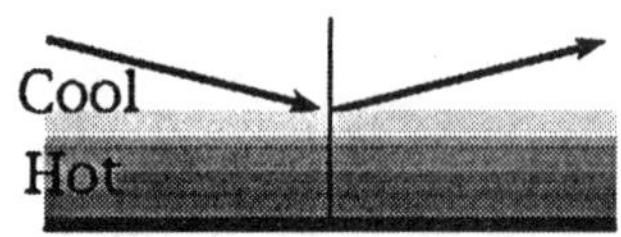

$$n_1 \sin\theta_1 \geq n_2 \sin 90°.$$

$$1.0003 \sin 88.8° \geq n_2$$

$$n_2 \leq 1.00008 \quad \lozenge$$

---

33. A small light source is on the bottom of a swimming pool 1.00 m deep. The light escaping from the water forms a circle on the water surface. What is the diameter of this circle?

Solution At the edge of the circle, the light is totally internally reflected:

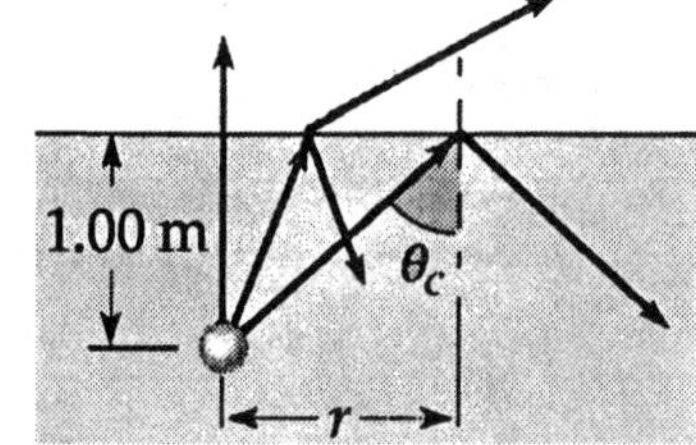

$$n_1 \sin\theta_c = n_2 \sin 90°$$

$$(1.333)\sin\theta_c = 1$$

$$\theta_c = \sin^{-1}(0.750) = 48.6°$$

The radius then satisfies $\tan\theta_c = \dfrac{r}{(1.00 \text{ m})}$

So the diameter is $d = 2\ \tan\theta_c = 2\ \tan\ 48.6° = 2.27 \text{ m} \quad \lozenge$

---

41. A hiker stands on a mountain peak near sunset and observes a rainbow caused by water droplets in the air about 8.00 km away. The valley is 2.00 km below the mountain peak and entirely flat. What fraction of the complete circular arc of the rainbow is visible to the hiker?

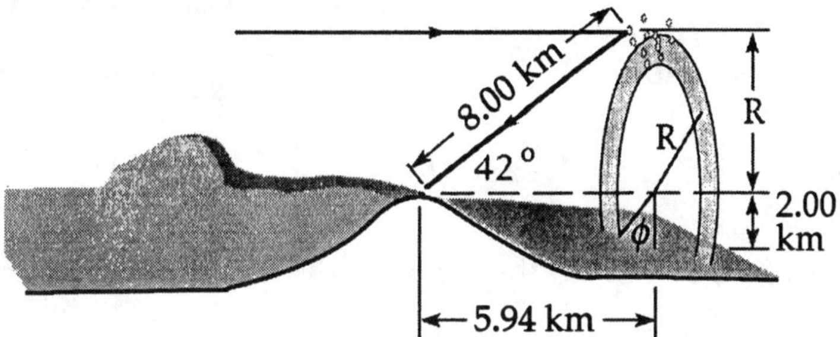

Solution Horizontal light rays from the setting Sun pass above the hiker. The light rays are twice refracted and once reflected, as in Figure 25.15 of the text, by just the certain special raindrops at 40.0° to 42.0° from the hiker's shadow, and reach the hiker as the rainbow. The hiker sees a greater percentage of the violet inner edge, so we consider the red outer edge. The radius R of the circle of droplets is

$$R = (8.00\text{ km})(\sin 42.0°) = 5.35\text{ km}$$

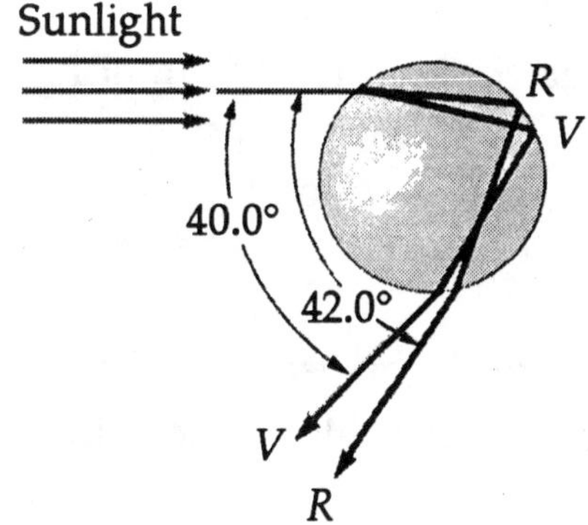

Figure 25.15 (text)

Then the angle ϕ, between the vertical and the radius where the bow touches the ground, is given by

$$\cos\phi = \frac{2.00\text{ km}}{R} = \frac{2.00\text{ km}}{5.35\text{ km}} = 0.374 \quad \text{or} \quad \phi = 68.1°$$

The angle filled by the visible bow is $360° - (2 \times 68.1°) = 224°$, so the visible bow is

$$\frac{224°}{360°} = 62.2\% \text{ of a circle} \quad \lozenge$$

This striking view motivated Charles Wilson's 1906 invention of the cloud chamber, a standard tool of nuclear physics. Look for a full-circle rainbow around your shadow when you fly in an airplane.

43. A laser beam strikes one end of a slab of material, as shown in Figure P25.43. The index of refraction of the slab is 1.48. Determine the number of internal reflections of the beam before it emerges from the opposite end of the slab.

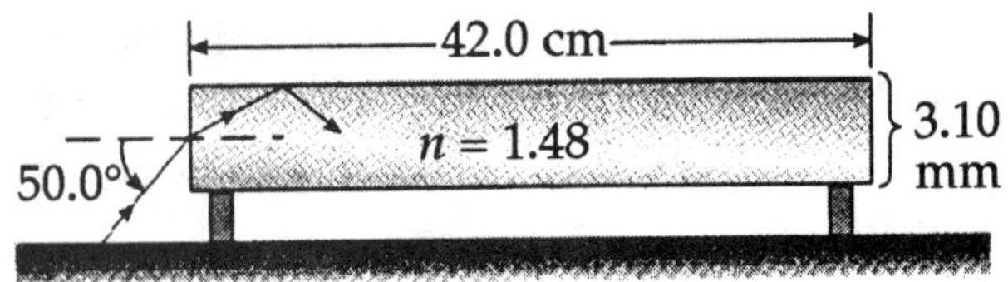

Figure P25.43

Solution On entrance, $\sin 50.0° = 1.48 \sin \theta_2$ so $\theta_2 = 31.17°$

The beam strikes the top face at horizontal coordinate $x_1 = \dfrac{1.55 \text{ mm}}{\tan 31.17°} = 2.562 \text{ mm}$

Thereafter, the beam strikes a face every $2x_1 = 5.124$ mm.

Since the slab is 420 mm long, the beam makes another $\dfrac{420 - 2.562}{5.124} = 81$ reflections;

Therefore, the total is 82 reflections. ◊

---

45. The light beam in Figure P25.45 strikes surface 2 at the critical angle. Determine the angle of incidence θ_1.

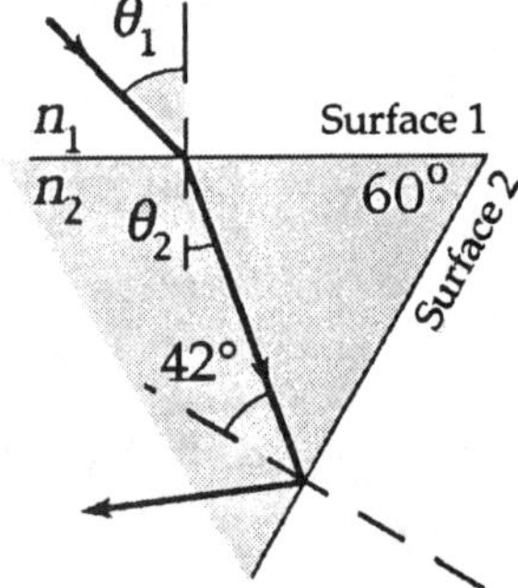

Figure P25.45

Solution Call the index of refraction of the prism material n_2. Use n_1 to represent the surrounding medium. At surface 2,

$$n_2 \sin 42° = n_1 \sin 90°.$$

So $$\frac{n_2}{n_1} = \frac{1}{\sin 42°} = 1.49$$

Call the angle of refraction θ_2 at the surface 1. The ray inside the prism forms with surfaces 1 and 2 a triangle whose interior angles add to 180°. Thus,

$$(90° - \theta_2) + 60° + (90° - 42°) = 180° \quad \text{and} \quad \theta_2 = 18.0°$$

At surface 1, $n_1 \sin\theta_1 = n_2 \sin 18.0°$

$$\sin\theta_1 = (n_2/n_1) \sin 18.0° \qquad \sin\theta_1 = 1.49(\sin 18.0°)$$

$$\theta_1 = 27.5° \quad ◊$$

---

49. A light ray of wavelength 589 nm is incident at an angle θ on the top surface of a block of polystyrene, as shown in Figure P25.49. (a) Find the maximum value of θ for which the refracted ray undergoes total internal reflection at the left vertical face of the block. Repeat the calculation for the case in which the polystyrene block is immersed in (b) water and (c) carbon disulfide.

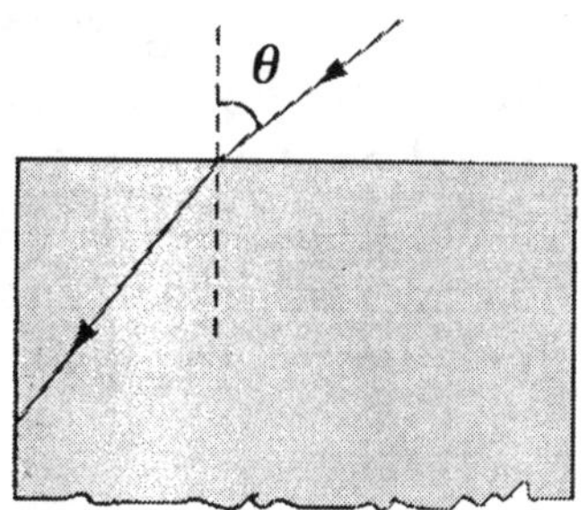

Figure P25.49

Solution The index of refraction (for 589 nm light) of each material is listed below:

Air:	1.00
Water:	1.33
Polystyrene:	1.49
Carbon disulfide:	1.63

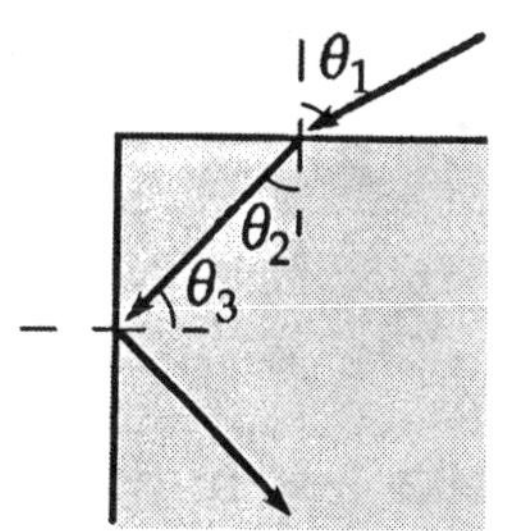

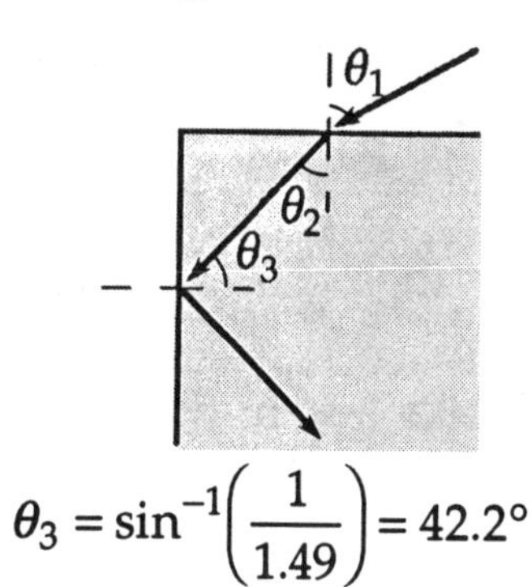

(a) For polystyrene **surrounded by air,**

internal reflection requires $\theta_3 = \sin^{-1}\left(\frac{1}{1.49}\right) = 42.2°$

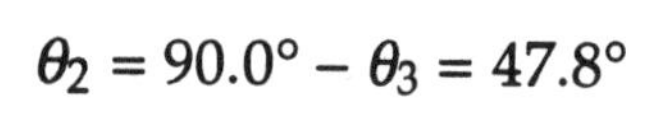

and then from the geometry, $\theta_2 = 90.0° - \theta_3 = 47.8°$

From Snell's law, $\sin\theta_1 = (1.49)\sin 47.8°$

This has no solution; thus, the real maximum value for θ_1 is 90.0°: total internal reflection always occurs. ◊

(b) For polystyrene **surrounded by water,** we have $\theta_3 = \sin^{-1}\left(\frac{1.33}{1.49}\right) = 63.2°$

and $\theta_2 = 26.8°$

From Snell's law, $n_1 \sin\theta_1 = n_2 \sin\theta_2$: $1.33\sin\theta_1 = 1.49\sin 26.8°$

and $\theta_1 = 30.3°$ ◊

(c) **This is not possible** since the beam is initially traveling in a medium of lower index of refraction.

51. A drinking glass is 4.00 cm wide at the bottom, as shown in Figure P25.51. When an observer's eye is placed as shown, the observer sees the edge of the bottom of the glass. When this glass is filled with water, the observer sees the center of the bottom of the glass. Find the height of the glass.

h

4.00 cm

Figure P25.51

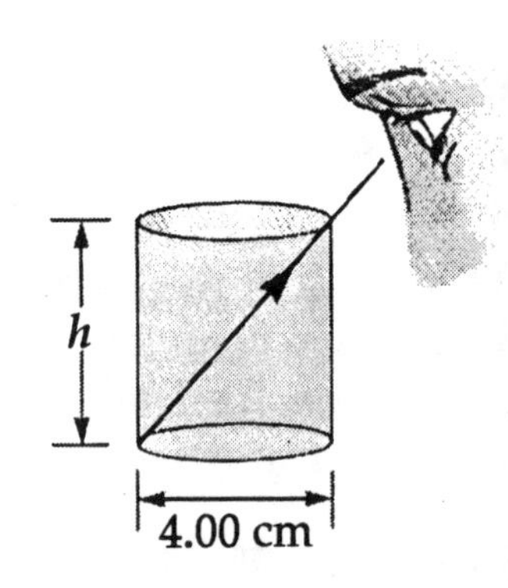

Solution $\tan\theta_1 = \dfrac{4.00 \text{ cm}}{h}$ and $\tan\theta_2 = \dfrac{2.00 \text{ cm}}{h}$

$$\tan^2\theta_1 = (2.00\tan\theta_2)^2 = 4.00\tan^2\theta_2$$

$$\frac{\sin^2\theta_1}{(1-\sin^2\theta_1)} = 4.00\frac{\sin^2\theta_2}{(1-\sin^2\theta_2)} \qquad (1)$$

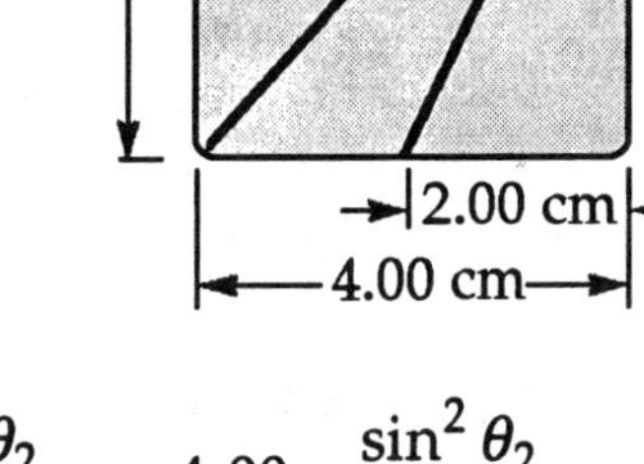

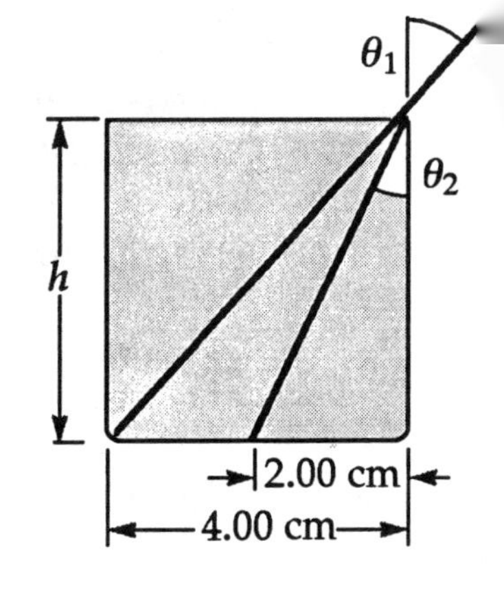

Snell's law in this case is: $n_1\sin\theta_1 = n_2\sin\theta_2$

$$\sin\theta_1 = 1.333\ \sin\theta_2$$

Squaring both sides, $\sin^2\theta_1 = 1.777\ \sin^2\theta_2 \qquad (2)$

Substituting (2) into (1) yields $\dfrac{1.777\ \sin^2\theta_2}{(1-1.777\ \sin^2\theta_2)} = 4.00\dfrac{\sin^2\theta_2}{(1-\sin^2\theta_2)}$

Defining $x = \sin^2\theta_2$ and solving for x, $\dfrac{0.444}{(1-1.778x)} = \dfrac{1}{(1-x)}$

$$0.444 - 0.444x = 1 - 1.777x$$

$$x = 0.417$$

From x we can solve for θ_2: $\theta_2 = \sin^{-1}\sqrt{0.417} = 40.2°$

and $h = \dfrac{(2.00 \text{ cm})}{\tan\theta_2} = \dfrac{(2.00 \text{ cm})}{\tan(40.2°)} = 2.37 \text{ cm}$ ◊

Chapter 26

These are some common navigational and orienteering instruments. A magnifying glass is used to view fine print and small objects. A compass is used to determine directions. Binoculars are used to view distant objects. (Murray Alcosser / The Image Bank)

MIRRORS AND LENSES

Chapter 26

MIRRORS AND LENSES

INTRODUCTION

This chapter is concerned with the images formed when spherical waves fall on flat and spherical surfaces. We find that images can be formed by reflection or by refraction and that mirrors and lenses work because of this reflection and refraction. Such devices, commonly used in optical instruments and systems, are described in detail. In this chapter, we continue to use the ray approximation and to assume that light travels in straight lines. These assumptions are valid because here we are studying the field called **geometric optics**. In subsequent chapters, we shall concern ourselves with interference and diffraction effects, or the field of **wave optics.**

NOTES FROM SELECTED CHAPTER SECTIONS

26.1 Images Formed by Flat Mirrors

The image formed by a plane mirror has the following properties:

- The image is as far behind the mirror as the object is in front.
- The image is unmagnified, virtual, and upright. (By upright, we mean that, if the object arrow points upward, so does the image arrow.)
- The image has an apparent right-left reversal.

26.2 Images Formed by Spherical Mirrors

A spherical mirror is a reflecting surface which has the shape of a segment of a sphere.

Real images are formed at a point when **reflected light actually passes through the image point. Virtual images** are formed at a point when light rays **appear to diverge from the image point.**

The point of intersection of any two of the following rays in a ray diagram for mirrors locates the image:

- The first ray is drawn from the top of the object parallel to the principal axis and is reflected back through the focal point, *F*.

- The second ray is drawn from the top of the object through the focal point and is reflected parallel to the axis.

- The third ray is drawn from the top of the object through the center of curvature, *C*, and is reflected back on itself.

- The fourth ray is drawn from the top of the object to the center of the mirror's surface and reflects on the other side of the principal axis, with an angle of reflection equal to its angle of incidence.

26.4 Thin Lenses

The following three rays form the ray diagram for a thin lens:

- The first ray is drawn parallel to the principal axis. After being refracted by the lens, this ray passes through (or appears to come from) one of the focal points.

- The second ray is drawn through the center of the lens. This ray continues in a straight line.

- The third ray is drawn through the other focal point *F*, and emerges from the lens parallel to the principal axis.

26.5 Lens Aberrations

Aberrations are responsible for the formation of imperfect images by lenses and mirrors. Spherical aberration is due to the variation in focal points for parallel incident rays that strike the lens at various distances from the optical axis. Chromatic aberration arises from the fact that light of different wavelengths focuses at different points when refracted by a lens.

EQUATIONS AND CONCEPTS

The **mirror equation** is used to locate the position of an image formed by reflection of paraxial rays. The focal point of a spherical mirror is located midway between the center of curvature and the center of the mirror.

$$\frac{1}{p}+\frac{1}{q}=\frac{1}{f} \qquad (26.6)$$

$$f=\frac{R}{2} \qquad (26.5)$$

The **lateral magnification** of a spherical mirror can be stated either as a ratio of image size to object size or in terms of the ratio of image distance to object distance.

$$M=\frac{h'}{h}=-\frac{q}{p} \qquad (26.2)$$

A magnified image of an object can be formed by a single spherical refracting surface of radius R which separates two media whose indices of refraction are n_1 and n_2.

$$\frac{n_1}{p}+\frac{n_2}{q}=\frac{n_2-n_1}{R} \qquad (26.8)$$

$$M=\frac{h'}{h}=-\frac{n_1 q}{n_2 p} \qquad (26.9)$$

A special case is that of the virtual image formed by a **planar refracting surface** ($R=\infty$).

$$\frac{n_1}{p}=-\frac{n_2}{q}$$

$$q=-\frac{n_2}{n_1}p \qquad (26.10)$$

The **thin lens** is an important component in many optical instruments. The location of the image formed by a given object is determined by the characteristic properties of the lens (index of refraction n and radii of curvature R_1 and R_2). The equation for the lateral magnification of a thin lens has the same form as that for a spherical mirror.

$$\frac{1}{f}=(n-1)\left(\frac{1}{R_1}-\frac{1}{R_2}\right) \qquad (26.13)$$

$$\frac{1}{p}+\frac{1}{q}=\frac{1}{f} \qquad (26.12)$$

$$M=\frac{h'}{h}=-\frac{q}{p} \qquad (26.11)$$

In using the equations of this chapter, you must be very careful to use the correct algebraic sign for each quantity. The sign conventions appropriate for the form of the equations stated here are summarized in SUGGESTIONS, SKILLS AND STRATEGIES.

SUGGESTIONS, SKILLS, AND STRATEGIES

A major portion of this chapter is devoted to the development and presentation of equations which can be used to determine the location and nature of images formed by various optical components acting either singly or in combination. It is essential that these equations be used with the correct algebraic sign associated with each quantity involved. You must understand clearly the sign conventions for mirrors, refracting surfaces, and lenses. The following discussion represents a review of these sign conventions.

SIGN CONVENTIONS FOR MIRRORS

Equations: $\frac{1}{p}+\frac{1}{q}=\frac{1}{f}=\frac{2}{R}$ $\qquad M=\frac{h'}{h}=-\frac{q}{p}$

The front side of the mirror is the region on which light rays are incident and reflected.

p is + if the object is in front of the mirror (real object).
p is – if the object is in back of the mirror (virtual object).

q is + if the image is in front of the mirror (real image).
q is – if the image is in back of the mirror (virtual image).

Both f and R are + if the center of curvature is in front of the mirror (concave mirror).

Both f and R are – if the center of curvature is in back of the mirror (convex mirror).

If M is positive, the image is upright.
If M is negative, the image is inverted.

You should check the sign conventions as stated against the situations described in Figure 26.1. You may also wish to check the image location in each diagram by drawing in ray number two, as described in Section 26.2.

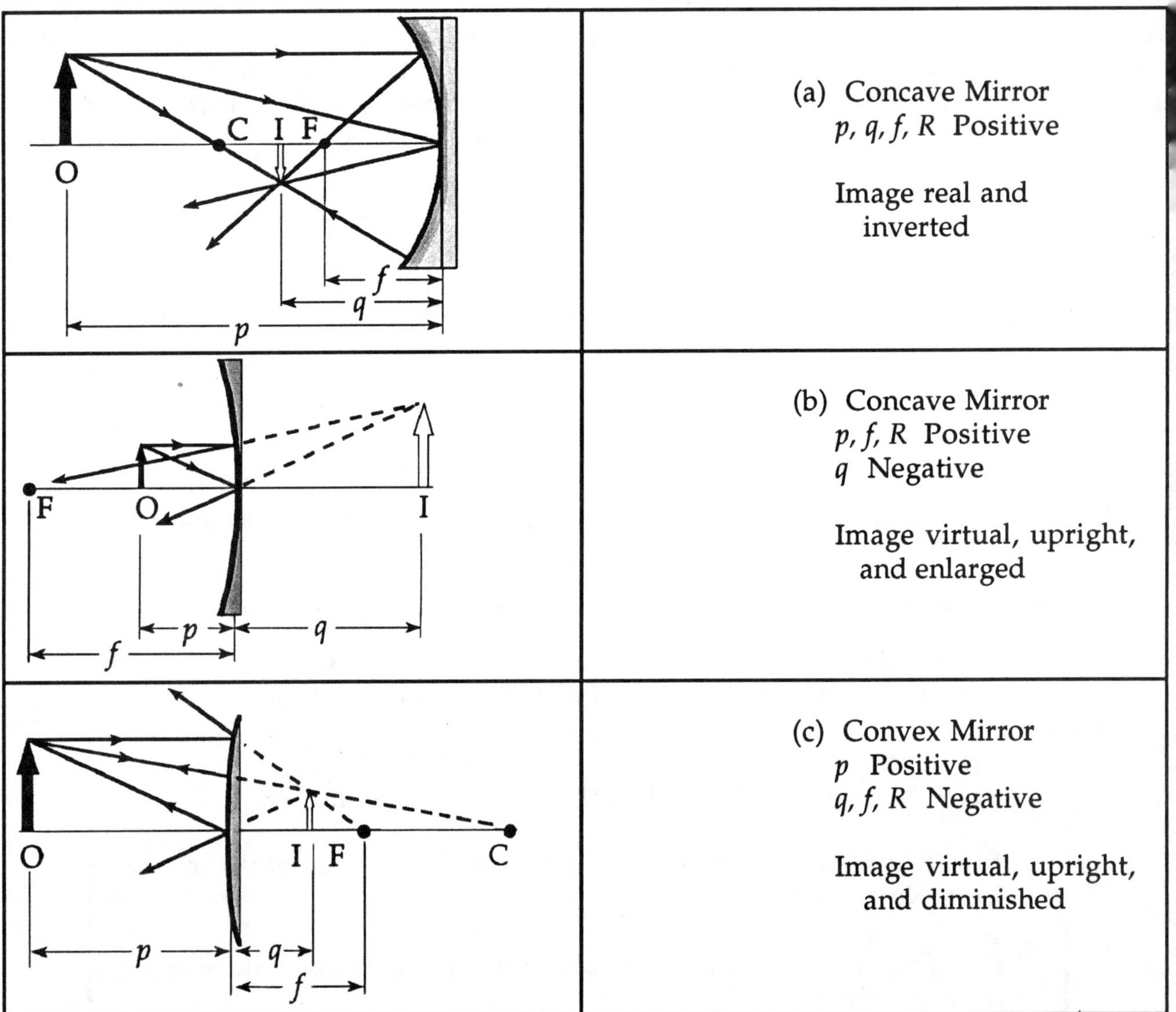

Figure 26.1 Figures describing sign conventions for mirrors.

SIGN CONVENTIONS FOR REFRACTING SURFACES

Equations: $\frac{n_1}{p} + \frac{n_2}{q} = \frac{n_2 - n_1}{R}$ $\quad\quad M = \frac{h'}{h} = -\frac{n_1 q}{n_2 p}$

n the following table, the **front** side of the surface is the side **from which the light is ncident.**

p is + if the object is in front of the surface (real object). p is – if the object is in back of the surface (virtual object).
q is + if the image is in back of the surface (real image). q is – if the image is in front of the surface (virtual image).
R is + if the center of curvature is in back of the surface. R is – if the center of curvature is in front of the surface.
n_1 refers to the index of refraction of the medium on the side of the interface from which the light comes.
n_2 is the index of refraction of the medium into which the light is transmitted after refraction at the interface.

Review the above sign conventions for the situations shown in Figure 26.2.

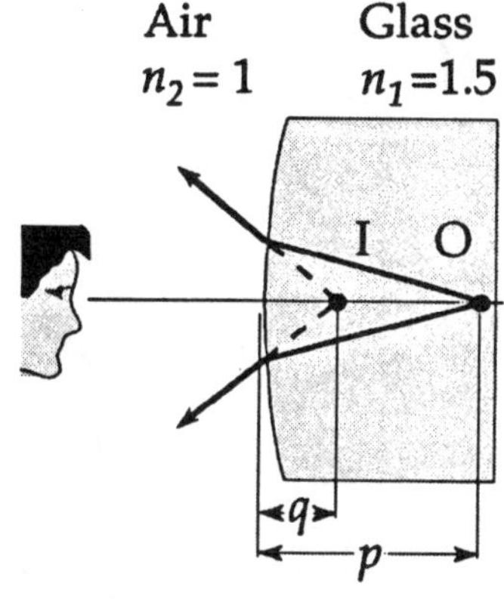

p + (real object)
q – (virtual image)
R – (concave to incident light)
n_1 and n_2 as shown

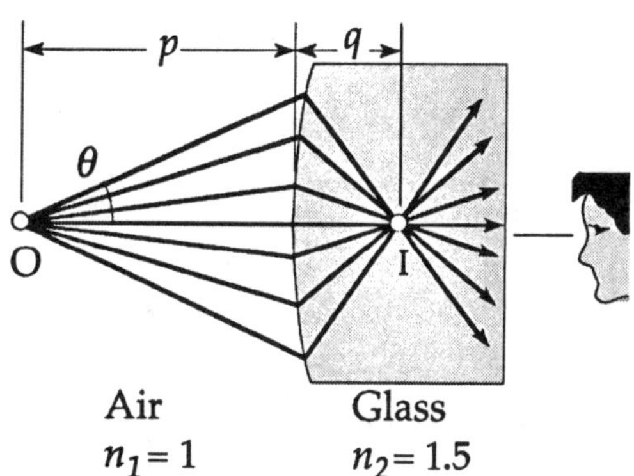

p + (real object)
q + (real image)
R + (convex to incident light)
n_1 and n_2 as shown

Figure 26.2 These figures describe sign conventions for refracting surfaces. In the first case, the object appears closer than it actually is. In the second case, an object beyond the glass appears to be in the glass.

SIGN CONVENTIONS FOR THIN LENSES

Equations: $\dfrac{1}{p}+\dfrac{1}{q}=\dfrac{1}{f}=(n-1)\left(\dfrac{1}{R_1}-\dfrac{1}{R_2}\right)$ $\qquad M=\dfrac{h'}{h}=-\dfrac{q}{p}$

In the following table, the **front** of the lens is the **side from which the light is incident.**

p is + if the object is in front of the lens. p is – if the object is in back of the lens.
q is + if the image is in back of the lens. q is – if the image is in front of the lens.
f is + if the lens is converging. f is – if the lens is diverging.
R_1 and R_2 are + if the center of curvature is in back of the lens.
R_1 and R_2 are – if the center of curvature is in front of the lens.

The sign conventions for thin lenses are illustrated by the examples shown in Figure 26.3.

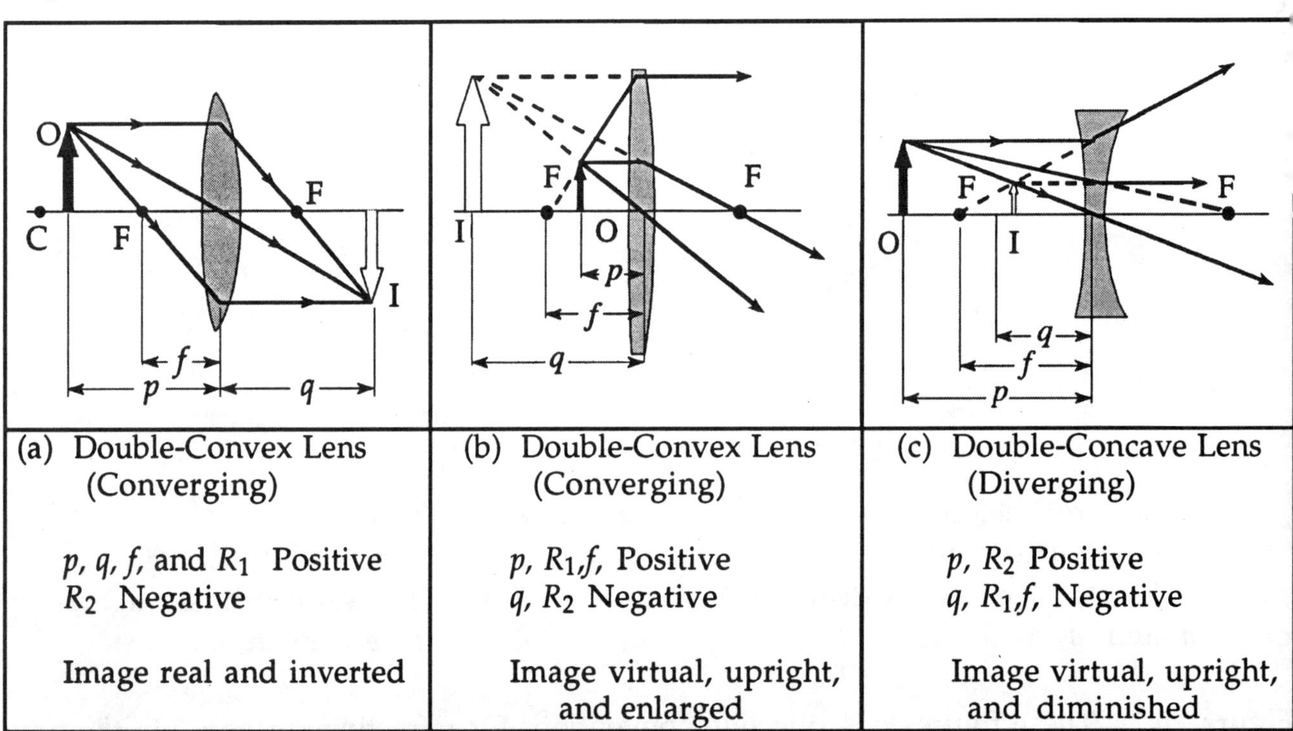

Figure 26.3 Figures describing the sign conventions for various thin lenses.

REVIEW CHECKLIST

> Identify the following properties which characterize an image formed by a lens or mirror system with respect to an object: position, magnification, orientation (i.e. inverted or upright) and whether real or virtual.

> Understand the relationship of the algebraic signs associated with calculated quantities to the nature of the image and object: real or virtual, upright or inverted.

> Calculate the location of the image of a specified object as formed by a plane mirror, spherical mirror, plane refracting surface, spherical refracting surface, thin lens, or a combination of two or more of these devices. Determine the magnification and character of the image in each case.

> Construct ray diagrams to determine the location and nature of the image of a given object when the geometrical characteristics of the optical device (lens or mirror) are known.

ANSWERS TO SELECTED CONCEPTUAL QUESTIONS

2. The side-view mirror on late-model cars warns the user that objects may be closer than they appear. What kind of mirror is being used and why was this type selected?

Answer The mirror would probably be a convex mirror. This type would have been selected because the designers wish to give the driver a wide field of view, and a virtual upright image for all distances.

□ □ □ □

4. Why does a clear stream always appear to be shallower than it actually is?

Answer The apparent depth is less than the true depth due to refraction of light originating from objects such as stones at the bottom of the stream.

□ □ □ □

5. A person spearfishing from a boat sees a fish that is 3 m from the boat at a depth of 1 m. In order to spear the fish, should the person aim at, above, or below the image of the fish?

Answer As the light rays exit the water, they refract away from their normals and towards the boat. As the diagram to the right shows, the fish appears to be at a depth which is less than its actual depth. Therefore, the person should aim below the image of the fish, in order to hit it.

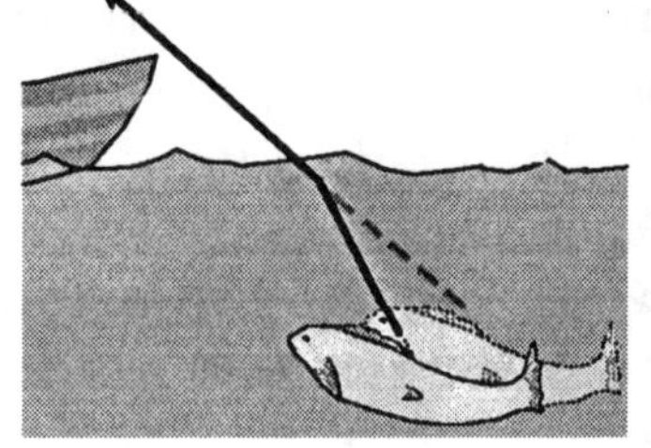

6. Explain why a fish in a spherical goldfish bowl appears larger than it really is.

Answer A spherical goldfish bowl acts as a single refracting surface; the magnification of any object in the bowl can be calculated from Equations 26.8 and 26.9 to be

$$M = 1 + \left(\frac{n_w - 1}{R}\right)q = 1 + \frac{0.33q}{R}$$

where R is negative.

Since the image distance q is also always negative, this gives a magnification that is always greater than 1. Therefore, the image of the goldfish will always be larger than the object.

7. If a cylinder of solid glass or clear plastic is placed above the words LEAD OXIDE and viewed from the side, as shown in Figure Q26.7, the LEAD appears inverted but the OXIDE does not. Explain.

Answer Both words are actually inverted. However, the word "OXIDE" does not appear to be inverted because of the symmetry of the characters. That is, when the word "OXIDE" is inverted, it still looks the same.

□ □ □ □

8. A mirage is formed when the air gets gradually cooler as the height above the ground increases. What might happen if the air grows gradually warmer as the height is increased? This often happens over bodies of water or snow-covered ground; The effect is called **looming**.

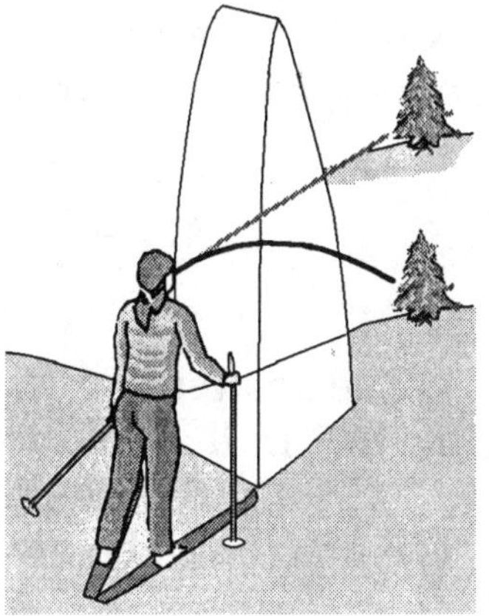

Answer In the situation described above, the air ends up acting like a prism, with its densest (thickest) segment near the ground. Light leaving the object and headed toward the sky is refracted downward, and an image can be seen hovering in the air above the real object.

□ □ □ □

9. Lenses used in eyeglasses, whether converging or diverging, are always designed such that the middle of the lens curves away from the eye, like the center lenses of Figure 26.17a and 26.17b. Why?

Answer With the meniscus design, when you direct your gaze near the outer circumference of the lens you receive a ray that has passed through glass with more nearly parallel surfaces of entry and exit. Thus, the lens minimally distorts the direction to the object you are looking at. If you wear glasses, you can demonstrate this by turning them around and looking through them the wrong way, maximizing the distortion.

□ □ □ □

10. Describe lenses that can be used to start a fire.

Answer Normally, solar energy is not intense enough to start a fire on its own, so rays from the sun must be converged and focused. A lens that could start such a fire would therefore have to be converging. It should also be noted that a larger lens will start a fire more quickly than a small one.

□ □ □ □

12. Consider the image formed by a thin converging lens. Under what conditions is the image (a) inverted, (b) upright, (c) real, (d) virtual, (e) larger than the object, and (f) smaller than the object?

Answer It is possible to calculate the answers to questions (a), (b), (e), and (f) by combining Equations 26.2 and 26.12:

$$M = -\frac{q}{p} \quad \text{and} \quad \frac{1}{p} + \frac{1}{q} = \frac{1}{f} \quad \text{become} \quad M = \frac{1}{1 - p/f}$$

The answer to questions (a), (b), (c), and (d) may also be determined by referring to Figure 26.3 in this solutions manual.

(a) The image will be inverted ($M < 0$) if the object is beyond the focal point of the lens.

(b) It will be upright ($M > 0$) when the object is closer to the lens than the focal point.

(c) The image will be real ($q > 0$) when the object is located beyond the focal point.

(d) The image will be virtual ($q < 0$) when the object is closer than the focal point.

(e) The image will be larger than the object $(|M| > 1)$ if the object is closer than twice the focal length of the lens.

(f) The image will be reduced $(|M| < 1)$ if the object is located beyond twice the focal length of the lens.

□ □ □ □

18. A pinhole camera can be constructed by punching a small hole in one side of a cardboard box. If the opposite side is cut out and replaced with tissue paper, the image of a distant object is formed by light passing through the pinhole onto the tissue paper (the screen). No lens is involved here. In effect, the pinhole replaces the lens of the camera. Explain why an image is formed, and describe the nature of the image.

Answer The pinhole camera depends upon the fact that light tends to travel in straight lines. As a result, photons that hit a particular spot on the screen come from one specific angular direction.

The image produced is real and inverted, and typically is reduced.

□ □ □ □

19. If you want to examine the fine detail of an object with a magnifying lens of focal length 15 cm, where should the object be placed in order to observe a magnified image of the object?

Answer The object must be placed closer than the focal length, 15 cm. Ideally, it should be placed so that a virtual image would be formed a comfortable distance from the eye (about 20 cm).

□ □ □ □

20. Large telescopes are usually reflecting rather than refracting. List some reasons for this choice.

Answer In contrast to a lens, a mirror is immune to chromatic aberration. The mirror has only one surface to shape, and can be made of glass with inhomogeneities inside of its volume. Thus your budget will buy a bigger mirror than a lens. Supported only at the edges, a lens larger than a meter in diameter tends to sag under its own weight, while a mirror can be supported all across its back side; motors placed there can actively correct the shape of the mirror. To maximize light throughput, an achromatic doublet lens needs antireflective coatings on several lens surfaces; a mirror needs a highly reflective coating on only one surface.

□ □ □ □

21. A classic science fiction novel, **The Invisible Man**, tells of a person who becomes invisible by changing the index of refraction of his body to that of air. This story has been criticized by students who know how the eye works; they claim the invisible man himself would be unable to see. On the basis of your knowledge of the eye, could he see or not?

Answer The cornea of the eye works by refracting light, and creating a real image at the retina. Therefore, unless the invisible man wore a pair of glasses that focused the light, everything he saw would be a blur, and an image would not be formed.

This is actually a very real problem for some former cataract patients. The cornea, which turns from clear to a milky-white in the case of cataracts, can be removed, but the patient must then either have it replaced, or must wear very thick glasses.

□ □ □ □

SOLUTIONS TO SELECTED END-OF-CHAPTER PROBLEMS

3. Determine the minimum height of a vertical flat mirror in which a person 5'10" in height can see his or her full image. (A ray diagram would be helpful.)

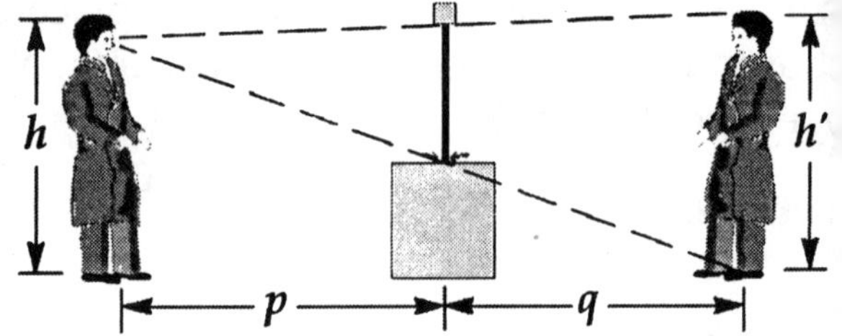

Solution The flatness of the mirror is described by $R = \infty$, $f = \infty$, and $1/f = 0$. By our general mirror equation,

$$\frac{1}{p} + \frac{1}{q} = \frac{1}{f}, \quad \text{or} \quad q = -p$$

Thus, the image is as far behind the mirror as the person is in front. The magnification is then

$$M = \frac{-q}{p} = 1 = \frac{h'}{h}, \quad \text{so} \quad h' = h = 70''$$

The required height of the mirror is defined by the triangle from the person's eyes to the top and bottom of his image, as shown. From the geometry of the triangle, we see that the mirror must be:

$$h'\left(\frac{p}{p-q}\right) = h'\left(\frac{p}{2p}\right) = \frac{h'}{2}$$

Thus, the mirror must be at least 35" high. ◊

---

9. A spherical convex mirror has a radius of curvature of 40.0 cm. Determine the position of the virtual image and magnification for object distances of (a) 30.0 cm and (b) 60.0 cm. (c) Are the images upright or inverted?

Solution The convex mirror is described by $f = R/2 = (-40.0 \text{ cm})/2 = -20.0$ cm.

(a) $\frac{1}{p}+\frac{1}{q}=\frac{1}{f}$, so $\frac{1}{30.0 \text{ cm}}+\frac{1}{q}=\frac{1}{-20.0 \text{ cm}}$ and $q = -12.0$ cm ◊

$$M = \frac{-q}{p} = -\frac{-12.0 \text{ cm}}{30.0 \text{ cm}} = +0.400$$ ◊

(c) The image is behind the mirror, upright, virtual, and diminished. ◊

(b) $\frac{1}{60.0 \text{ cm}}+\frac{1}{q}=-\frac{1}{20.0 \text{ cm}}$

$q = -15.0$ cm ◊

$$M = \frac{-q}{p} = -\frac{-15.0 \text{ cm}}{60.0 \text{ cm}} = +0.250$$ ◊

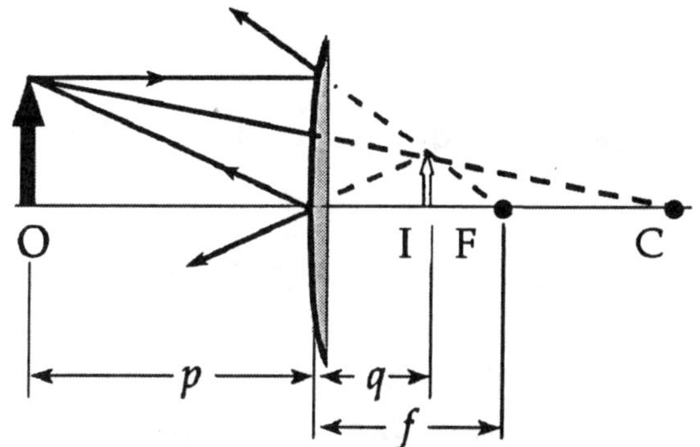

(c) The principal ray diagram is an essential complement to the numerical description of the image. Draw the rays into this diagram for a 60-cm object distance:

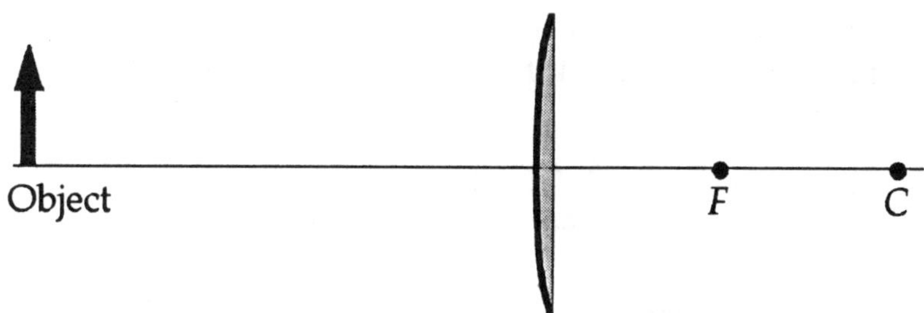

Use your diagram to figure out that the image is behind the mirror, upright, virtual and diminished. ◊

11. A spherical mirror is to be used to form, on a screen located 5.00 m from the object, an image five times the size of the object. (a) Describe the type of mirror required. (b) Where should the mirror be positioned relative to the object?

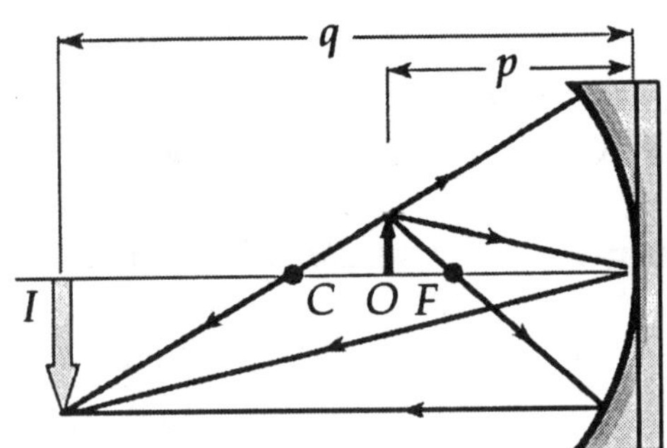

Solution We are given that $q = (p + 5.00\text{ m})$. Since the image must be real,

$$M = -5.00 = -\frac{q}{p}, \quad \text{or} \quad q = 5.00p$$

(b) Solving for the distance of the mirror from the object, $p + 5.00 = 5.00p$

$$p = 1.25\text{ m} \quad \lozenge$$

(a) Applying Equation 26.6, $\frac{1}{f} = \frac{1}{p} + \frac{1}{q} = \frac{1}{1.25\text{ m}} + \frac{1}{6.25\text{ m}}$

so that the focal length of the mirror is $f = 1.04\text{ m}$

Noting that the image is real, inverted and enlarged, we can say that the mirror must be concave, and must have a radius of curvature of $R = 2f = 2.08\text{ m}$ ◊

---

15. A cubical block of ice 50.0 cm on a side is placed on a level floor over a speck of dust. Find the location of the image of the speck if the index of refraction of ice is 1.309.

Solution The upper surface of the block is a single refracting surface with zero curvature, and with infinite radius of curvature.

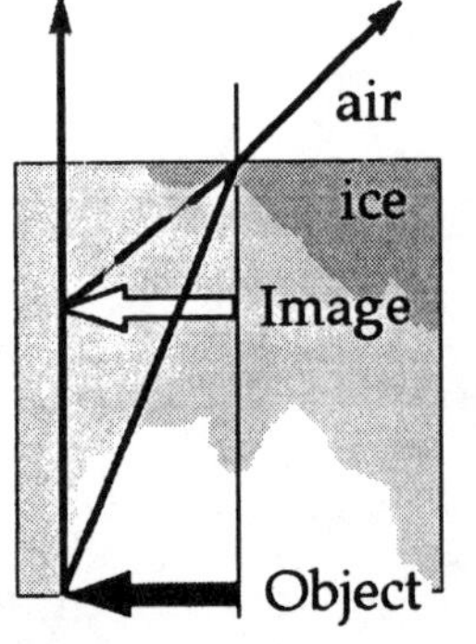

$$\frac{n_1}{p} + \frac{n_2}{q} = \frac{n_2 - n_1}{R} \quad \text{or} \quad \frac{1.309}{50.0\text{ cm}} + \frac{1}{q} = \frac{1.00 - 1.309}{\infty} = 0$$

$$q = \frac{-50.0\text{ cm}}{1.309} = -38.2\text{ cm}$$

The speck of dust on the floor appears to be 38.2 cm below the upper surface of the ice. The image is virtual, upright, and actual size. ◊

•. A glass hemisphere is used as a
ıperweight with its flat face resting on a stack
f papers. The radius of the circular cross-
ection is 4.00 cm, and the index of refraction
f the glass is 1.55. The center of the
emisphere is directly over a letter "O" that is
50 mm in diameter. What is the diameter of
ıe image of the letter as seen looking along a
ertical radius?

olution The image is formed by refraction
t a single surface:

$$\frac{n_1}{p}+\frac{n_2}{q}=\frac{n_2-n_1}{R}$$

or

$$\frac{1.55}{4.00\text{ cm}}+\frac{1.00}{q}=\frac{1.00-1.55}{-4.00\text{ cm}}$$

Thus,

$$q=\frac{1}{[0.550/4.00\text{ cm}-1.55/4.00\text{ cm}]}=-4.00\text{ cm}$$

The virtual image is at the same location as the object.

To find the image diameter h', consider the magnification equation 26.9:

$$M=\frac{h'}{h}=-\frac{n_1q}{n_2p}$$

Then,

$$h'=-\frac{n_1qh}{n_2p}=\frac{-1.55(-4.00\text{ cm})(2.50\text{ mm})}{1(4.00\text{ cm})}=3.88\text{ mm} \quad \Diamond$$

The image is virtual, upright, and enlarged.

21. The left face of a biconvex lens has a radius of curvature of 12.0 cm, and the rigl face has a radius of curvature of 18.0 cm. The index of refraction of the glass is 1.4 (a) Calculate the focal length of the lens. (b) Calculate the focal length if the radii c curvature of the two faces are interchanged.

Solution Convex outward on both sides, the lens has the centers of curvature of it surfaces on opposite sides. The second surface has negative radius:

(a) $$\frac{1}{f}=(n-1)\left(\frac{1}{R_1}-\frac{1}{R_2}\right)=(1.44-1.00)\left(\frac{1}{12.0\text{ cm}}-\frac{1}{-18.0\text{ cm}}\right)$$

$f=16.4$ cm ◊

(b) $$\frac{1}{f}=(0.440)\left(\frac{1}{18.0\text{ cm}}-\frac{1}{-12.0\text{ cm}}\right)$$

$f=16.4$ cm ◊

Reversing the lens does not change what it does to the light.

---

25. The nickel's image in Figure P26.25 has twice the diameter of the nickel and is 2.84 cm from the lens. Determine the focal length of the lens.

Figure P26.25

Solution Looking through the lens, you see the image beyond the lens. Therefore, the image is virtual, with q = –2.84 cm.

Now, $M=\frac{h'}{h}=2=-\frac{q}{p}$,

so $p=-\frac{q}{2}=1.42$ cm

Thus, $$f=\left(\frac{1}{p}+\frac{1}{q}\right)^{-1}=\left(\frac{1}{1.42\text{ cm}}+\frac{1}{-2.84\text{ cm}}\right)^{-1}$$

and $f=2.84$ cm ◊

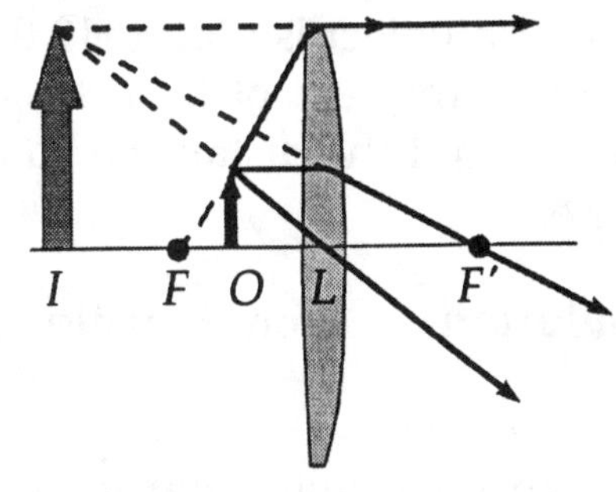

:9. A person looks at a gem with a jeweler's loupe— converging lens that has a focal length of 12.5 cm. The loupe forms a virtual image 30.0 cm from the lens. a) Determine the magnification. Is the image upright or inverted? (b) Construct a ray diagram for this arrangement.

Solution $$\frac{1}{p}+\frac{1}{q}=\frac{1}{f}$$

Thus, $$\frac{1}{p}+\frac{1}{-30.0\text{ cm}}=\frac{1}{12.5\text{ cm}}$$

$$p=8.82\text{ cm}$$

$$M=-\frac{q}{p}=-\frac{(-30.0)}{8.82}=3.40,\quad \text{upright} \quad \lozenge$$

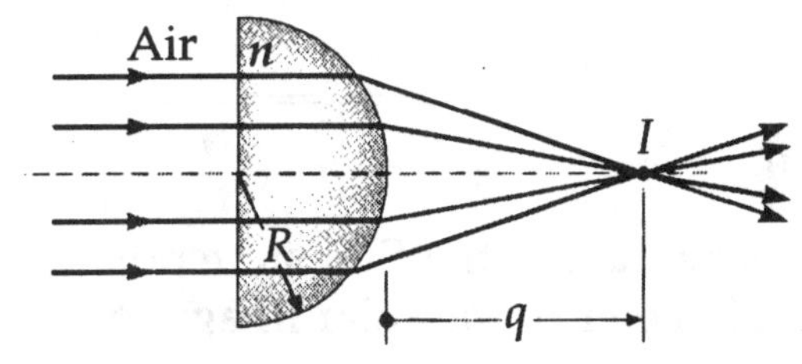

Figure P26.43

43. A parallel beam of light enters a glass hemisphere perpendicular to the flat face, as shown in Figure P26.43. The radius is $R = 6.00$ cm, and the index of refraction is $n = 1.560$. Determine the point at which the beam is focused. (Assume paraxial rays.)

Solution A hemisphere is too thick to be described as a thin lens. The light is undeviated on entry into the flat face. Because of this, we instead consider the light's exit from the second surface, for which $R=-6.00$ cm. The incident rays are parallel, so $p=\infty$.

Then $$\frac{n_1}{p}+\frac{n_2}{q}=\frac{n_2-n_1}{R}$$ becomes $$0+\frac{1}{q}=\frac{(1-1.56)}{-6.00\text{ cm}}$$

and $$q=10.7\text{ cm} \quad \lozenge$$

---

45. An object is placed 12.0 cm to the left of a diverging lens of focal length −6.00 cm. A converging lens of focal length 12.0 cm is placed a distance d to the right of the diverging lens. Find the distance d so that the final image is at infinity. Draw a ray diagram for this case.

Solution From Equation 26.12, $$q_1 = \frac{f_1 p_1}{p_1 - f_1} = \frac{(-6.00 \text{ cm})(12.0 \text{ cm})}{12.0 \text{ cm} - (-6.00 \text{ cm})} = -4.00 \text{ cm}$$

When we require that $q_2 \to \infty$, Equation 26.12 becomes $p_2 = f_2 = 12.0$ cm. Since the object for the converging lens must be 12.0 cm to its left, and since this is the image for the diverging lens which is 4.00 cm to **its** left, the two lenses must be separated by 8.00 cm. Mathematically,

$$p_2 = d - (-4.00 \text{ cm}) \qquad d + 4.00 \text{ cm} = f_2 = 12.0 \text{ cm}, \qquad \text{and} \qquad d = 8.00 \text{ cm} \quad \Diamond$$

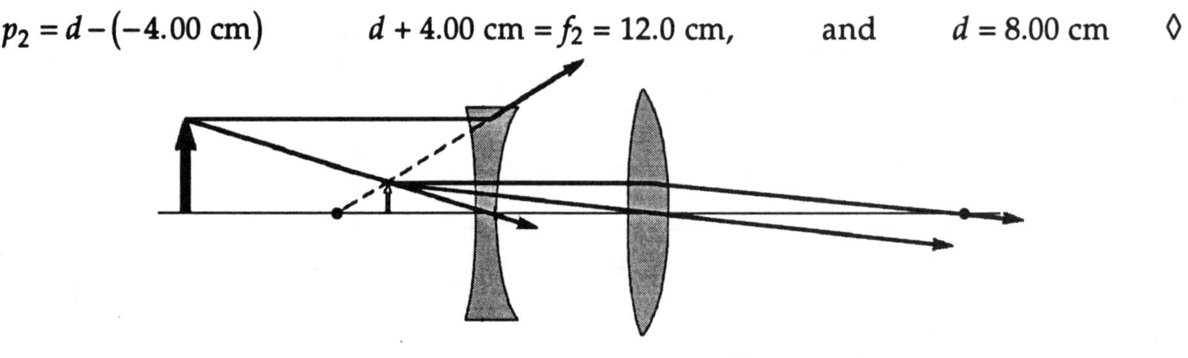

47. The disk of the Sun subtends an angle of 0.533° at the Earth. What are the position and diameter of the solar image formed by a concave spherical mirror of radius 3.00 m?

Solution For the mirror, $f = R/2 = +1.50$ m. In addition, because the distance to the Sun is so much larger than any other figures, we can take $p = \infty$.

The mirror equation, $\frac{1}{p} + \frac{1}{q} = \frac{1}{f}$, then gives a distance from the mirror of $q = f = 1.50$ m. ◊

Now, in $M = -\frac{q}{p} = \frac{h'}{h}$, the magnification is nearly zero, but we can be more precise: the definition of radian measure means that h/p is the angular diameter of the object. Thus the image diameter is

$$h' = -\frac{hq}{p} = (-0.533°)\left(\frac{\pi}{180} \text{ rad/deg}\right)(1.50 \text{ m}) = -0.140 \text{ m} = -1.40 \text{ cm} \quad \Diamond$$

9. In a darkened room, a burning candle is placed .50 m from a white wall. A lens is placed between andle and wall at a location that causes a larger, nverted image to form on the wall. When the lens is noved 90.0 cm toward the wall, another image of the andle is formed. Find (a) the two object distances that roduce the images stated above and (b) the focal length of the lens. (c) Characterize the second image.

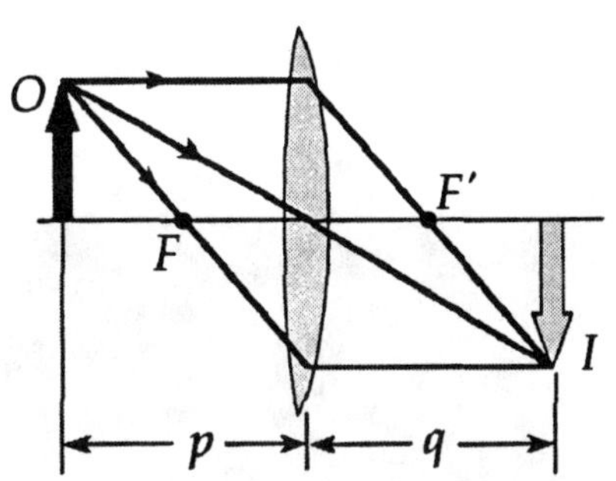

Solution Originally, $p_1 + q_1 = 1.50 \text{ m}.$

n the final situation, $p_2 = p_1 + 0.900 \text{ m},$

and $q_2 = q_1 - 0.900 \text{ m} = (1.50 \text{ m} - p_1) - 0.900 \text{ m} = 0.600 \text{ m} - p_1.$

Our lens equation is $$\frac{1}{p_1} + \frac{1}{q_1} = \frac{1}{f} = \frac{1}{p_2} + \frac{1}{q_2}$$

Substituting, we have $$\frac{1}{p_1} + \frac{1}{1.50 \text{ m} - p_1} = \frac{1}{p_1 + 0.900} + \frac{1}{0.600 - p_1}$$

Adding the fractions, $$\frac{1.50 \text{ m} - p_1 + p_1}{p_1(1.50 \text{ m} - p_1)} = \frac{0.600 - p_1 + p_1 + 0.900}{(p_1 + 0.900)(0.600 - p_1)}$$

Simplified, this becomes $$p_1(1.50 \text{ m} - p_1) = (p_1 + 0.900)(0.600 - p_1)$$

(a) Thus, $$p_1 = \frac{0.540}{1.80} \text{ m} = 0.300 \text{ m} \quad \Diamond$$

$$p_2 = p_1 + 0.900 = 1.20 \text{ m} \quad \Diamond$$

(b) $$\frac{1}{f} = \frac{1}{0.300 \text{ m}} + \frac{1}{1.50 \text{ m} - 0.300 \text{ m}}$$ and $f = 0.240 \text{ m}$ ◊

(c) The second image is real, inverted, and diminished, with $M = \frac{-q_2}{p_2} = -0.250$ ◊

Chapter 27

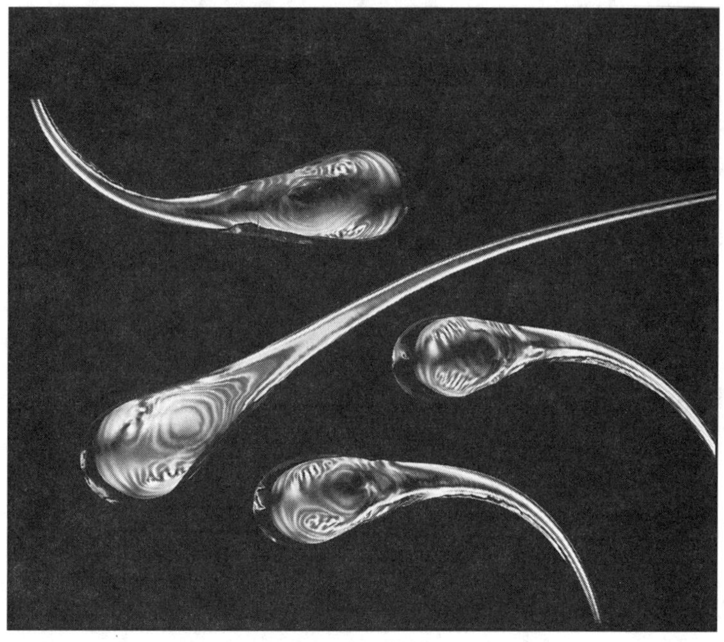

These glass objects, called Prince Rupert drops, are made by dropping molten glass into water. The photograph was made by placing the objects between two crossed polarizers. The patterns observed represent the strain distribution in the glass. (James L. Amos, Peter Arnold, Inc.)

WAVE OPTICS

Chapter 27

WAVE OPTICS

NTRODUCTION

In the previous chapter on geometric optics, we used light rays to examine what ıappens when light passes through a lens or reflects from a mirror. This chapter is :oncerned with wave optics, which deals with the interference and diffraction. These ›henomena cannot be adequately explained with the ray optics of Chapter 26. However, ›y treating light as waves rather than as rays, we can give a satisfactory description of such phenomena.

When light waves pass through a small aperture, a diffraction pattern is observed rather than a sharp spot of light, showing that light spreads beyond the aperture into regions where a shadow would be expected if light traveled in straight lines. Other waves, such as sound waves and water waves, also have this property of being able to bend around corners. This phenomenon, known as diffraction, can be regarded as interference from a great number of coherent wave sources. In other words, diffraction and interference are basically equivalent.

NOTES FROM SELECTED CHAPTER SECTIONS

27.1 Conditions for Interference

In order to observe **sustained** interference in light waves, the following conditions must be met:

- The sources must be coherent; they must maintain a **constant phase** with respect to each other.

- The sources must be **monochromatic**— of a **single wavelength.**

- The **superposition principle** must apply.

27.2 Young's Double-Slit Experiment

A schematic diagram illustrating the geometry used in Young's double-slit experiment is shown in the figure on the following page. The two slits S_1 and S_2 serve as coherent monochromatic sources. The **path difference** $\delta = r_2 - r_1 = d\sin\theta$.

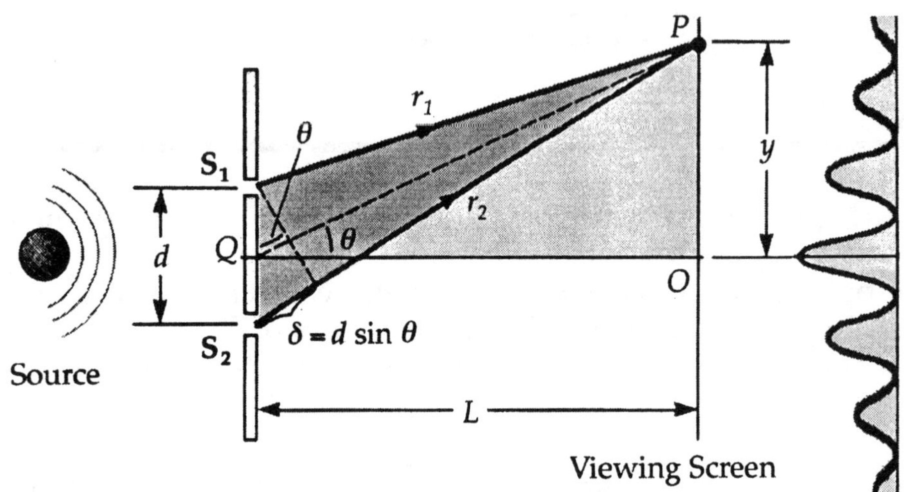

Figure 27.1

27.3 Change of Phase due to Reflection

An electromagnetic wave undergoes a **phase change of 180°** upon reflection from a medium that has **a larger index of refraction** than the one in which it was traveling. There is also a 180° phase change upon reflection from a **conducting surface.**

27.4 Interference in Thin Films

Interference effects in thin films depend on the difference in length of path traveled by the interfering waves as well as any phase changes which may occur due to reflection. It can be shown by the Lloyd's mirror experiment that when light is reflected from a surface of index of refraction greater than the index of the medium in which it is initially traveling, the reflected ray undergoes a 180° (or π radians) phase change. In analyzing interference effects, this can be considered equivalent to the gain or loss of a half wavelength in path difference. Therefore, there are two different cases to consider: (a) a film surrounded by a common medium and (b) a thin film located between two different media. These cases are illustrated on the following page in Figure 27.2.

In case (a), where a phase change occurs at the top surface, the reflected rays will be in phase (constructive interference) if the thickness of the film is an odd number of quarter wavelengths—so that the path lengths will differ by an odd number of half wavelengths.

In case (b), the phase changes due to reflection at **both** the top and bottom surfaces are offsetting and, therefore, constructive interference for the reflected rays will occur when the film thickness is an integer number of half wavelengths—and, therefore, the path difference will be a whole number of wavelengths.

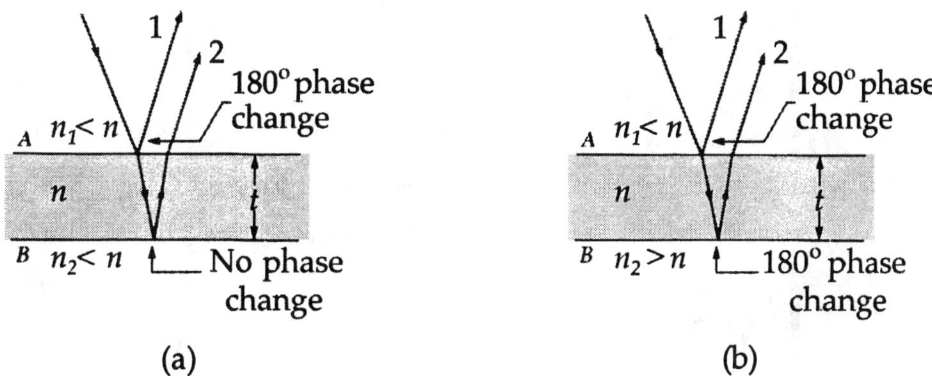

Figure 27.2 (a) Interference of light resulting from reflections at two surfaces of a thin film of thickness t, index of refraction n. (b) A thin film between two different media.

27.5 Diffraction

Fraunhofer diffraction occurs when light rays reaching a point (on a screen) from a diffracting source (edge, slit, etc.) are approximately parallel.

According to **Huygens' principle**, each portion of a slit acts as a source of waves; therefore, light from one portion of a slit can interfere with light from another portion.

In Figure 27.3 (a) on the following page, parallel light rays are shown incident on a slit of width a. In order to analyze the resulting Fraunhofer diffraction pattern, it is convenient to subdivide the width of the slit into many strips of equal width. Each strip can be considered as a source of waves.

The positions of the minima in a single-slit diffraction pattern are determined by the conditions stated in Equation 27.18. The **secondary maxima** (of progressively diminished intensity) **lie approximately halfway between the minima.**

27.6 Resolution of Single-Slit and Circular Apertures

When the central maximum of one image falls on the first minimum of another image, the images are said to be just resolved. This limiting condition of resolution is known as **Rayleigh's criterion.**

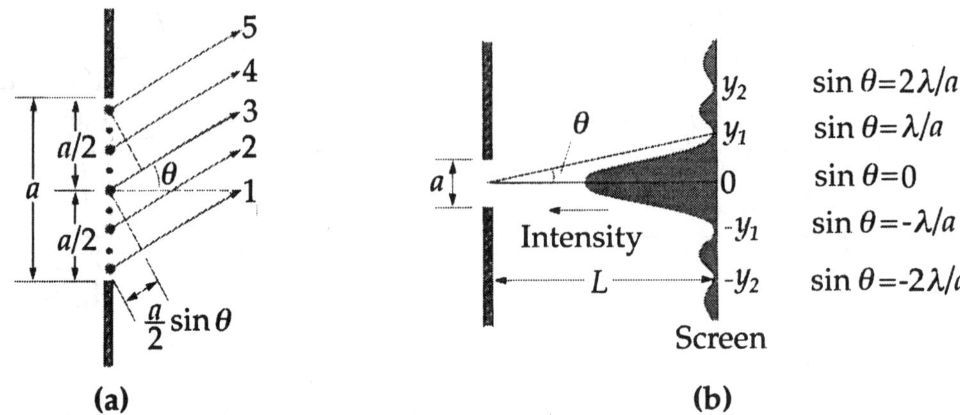

Figure 27.3 (a) Diffraction of light by a narrow slit of width *a*, and (b) the resulting intensity distribution.

EQUATIONS AND CONCEPTS

In the arrangement used for the Young's double-slit experiment (Figure 27.1), two slits separated by a distance, *d*, serve as monochromatic coherent sources. The light intensity at any point on the screen is the resultant of light reaching the screen from both slits.

$$\delta = r_2 - r_1 = d\sin\theta \tag{27.1}$$

$$y = L\tan\theta$$

Also, light from the two slits reaching any point on the screen (except the center) travel unequal path lengths. This difference in length of path is called the path difference.

Bright fringes (constructive interference) will appear at points on the screen for which the path difference is equal to an integral multiple of the wavelength. The positions of bright fringes can also be located by calculating their vertical distance from the center of the screen (y). In each case, the number m is called the order number of the fringe. The central bright fringe ($\theta = 0$, $m = 0$) is called the zeroth-order maximum.

$$\delta = d\sin\theta = m\lambda \qquad (27.2)$$

for $m = 0, \pm 1, \pm 2, \ldots$

$$y_{\text{bright}} \cong \frac{\lambda L}{d} m \quad \text{(for small } \theta\text{)} \qquad (27.5)$$

Dark fringes (destructive interference) will appear at points on the screen which correspond to path differences of an odd multiple of half wavelengths. For these points of destructive interference, waves which leave the two slits in phase arrive at the screen 180° out of phase.

$$\delta = d\sin\theta = \left(m + \tfrac{1}{2}\right)\lambda \qquad (27.3)$$

for $m = 0, \pm 1, \pm 2, \ldots$

$$y_{\text{dark}} \cong \frac{\lambda L}{d}\left(m + \tfrac{1}{2}\right) \quad \text{(for small } \theta\text{)} \qquad (27.6)$$

Two waves which leave the slits initially in phase arrive at the screen **out of phase** by an amount which depends on the path difference.

$$\phi = \frac{2\pi}{\lambda} d\sin\theta \qquad (27.8)$$

$$\phi = \frac{2\pi}{\lambda}\delta$$

Two waves initially in phase and of equal amplitude (E_o) will produce a resultant amplitude at some point on the screen which depends on the phase difference (and therefore on the path difference).

$$E_p = 2E_o \cos\left(\frac{\phi}{2}\right)\sin\left(\omega t + \frac{\phi}{2}\right) \qquad (27.10)$$

The average light intensity (I_{av}) at any point P on the screen is proportional to the square of the amplitude of the resultant wave. **The average intensity can be written:**

- as a function of phase difference ϕ;

$$I_{av} = I_o \cos^2\left(\frac{\phi}{2}\right) \quad (27.11$$

- as a function of the angle (θ) subtended by the screen point at the source midpoint; or

$$I_{av} = I_o \cos^2\left(\frac{\pi d \sin\theta}{\lambda}\right) \quad (27.12$$

- as a function of the vertical distance (y) from the center of the screen.

$$I_{av} \cong I_o \cos^2\left(\frac{\pi d y}{\lambda L}\right) \text{ (for small } \theta\text{)} \quad (27.13)$$

where $I_o \propto 4E_o^2$

The intensity pattern observed on the screen will vary as the number of equally spaced sources is increased; however, the **positions of the principal maxima remain the same.**

In thin-film interference (Figure 27.2), the wavelength of light in the film, λ_n, is not the same as the wavelength in the surrounding medium.

$$\lambda_n = \frac{\lambda}{n} \quad (27.14)$$

The conditions for the interference of light reflected perpendicularly from a thin film can be stated in terms of the thickness (t) and the index of refraction of the film. The conditions expressed by Equations 27.16 and 27.17 are valid when the film is surrounded by a common medium. (See Figure 27.2a)

$$2nt = \left(m + \tfrac{1}{2}\right)\lambda \quad (m = 0, 1, 2, \ldots) \quad (27.16)$$

constructive interference

$$2nt = m\lambda \quad (m = 0, 1, 2, \ldots) \quad (27.17)$$

destructive interference

Figure 27.2b illustrates the case when the index of refraction of the film is greater than that of the medium above the film and less than the index of the medium below the film. In this case a phase change of 180° occurs upon reflection at both the top and bottom surfaces of the film. Under these conditions, constructive interference will be observed when the film thickness is an odd number of half-wavelengths as measured in the film.

The general **condition for destructive interference** in the diffraction pattern of a single slit in terms of θ is a function of the slit width, a, and the wavelength of the incident light.

$$\sin\theta = m\frac{\lambda}{a} \qquad (27.18)$$

$(m = \pm 1,\ \pm 2,\ \pm 3,\ \ldots)$

(Destructive interference)

Rayleigh's criterion states the condition for the resolution of two images due to nearby sources. For a slit, the angular separation between the sources must be greater than the ratio of the wavelength to slit width. In the case of a **circular aperture,** the minimum angular separation depends on D, the diameter of the aperture (or lens).

$$\theta_{\min} = \frac{\lambda}{a} \quad \text{(slit)} \qquad (27.19)$$

$$\theta_{\min} = 1.22\frac{\lambda}{D} \quad \text{(circular aperture)} \qquad (27.20)$$

A grating of equally spaced parallel slits (separated by a distance d) will produce an interference pattern in which there is a series of maxima for each wavelength. Maxima due to wavelengths of different value comprise a spectral order denoted by order number m.

$$d\sin\theta = m\lambda \qquad (27.21)$$

$(m = 0, 1, 2, 3,\ \ldots)$

The **resolving power** of a grating increases as the **number of lines illuminated** is increased and is proportional to the order in which the spectrum is observed.

$$R \equiv \frac{\lambda}{\Delta\lambda} \quad (27.22)$$

$$R = Nm \quad (27.23)$$

When the order number m is substituted from Equation 27.21, $d\sin\theta = m\lambda$, into Equation 27.23, it becomes clear that **the resolution, R, depends on the width of the grating** (Nd). It should also be noted that the angular width $\Delta\theta$ of a spectral line formed by a diffraction grating is inversely proportional to the width of the grating. This statement can be written as $\Delta\theta \propto 1/Nd$.

$$R = \frac{Nd}{\lambda}\sin\theta$$

In the zeroth order (central maximum) all wavelengths are indistinguishable. If in a particular order ($m > 0$) $R = 10\ 000$, the grating will produce a spectrum in which wavelengths differing in value by 1 part in 10 000 can be resolved.

Bragg's law gives the conditions for **constructive interference** of x-rays reflected from the parallel planes of a crystalline solid separated by a distance d. **The angle θ is the angle between the incident beam and the surface.**

$$2d\sin\theta = m\lambda \quad (27.24)$$

$$(m = 1, 2, 3, \ldots)$$

UGGESTIONS, SKILLS, AND STRATEGIES

he following features should be kept in mind while working thin-film interference roblems:

Identify the thin film from which interference effects are being observed.

The type of interference that occurs in a specific problem is determined by the phase relationship between that portion of the wave reflected at the upper surface of the film and that portion reflected at the lower surface of the film.

Phase differences between the two portions of the wave occur because of differences in the distances traveled by the two portions and by phase changes occurring upon reflection.

The wave reflected from the lower surface of the film has to travel a distance equal to twice the thickness of the film before it returns to the upper surface of the film. When **distances alone are considered**, if the extra distance is equal to an integral multiple of λ_n constructive interference will occur; if the extra distance equals $\frac{1}{2}\lambda_n$, $\frac{3}{2}\lambda_n$, and so forth, destructive interference will occur.

Reflections may change such results. When a traveling wave reflects off a surface having a higher index of refraction than the one it is in, a 180° phase shift occurs. This has the same effect as if the wave lost $\frac{1}{2}\lambda_n$ These shifts must be considered in addition to shifts that occur because of the extra distance that one wave travels.

- When distance and phase changes upon reflection are both taken into account, the interference will be constructive if the waves are out of phase by an integral multiple of λ_n. Destructive interference will occur when the total effective path difference is $\frac{1}{2}\lambda_n$, $\frac{3}{2}\lambda_n$, and so forth. For example:

  In a soap bubble (Fig. 27.2a), $n_{\text{air}} < n_{\text{soap}} > n_{\text{air}}$, the condition for constructive interference is $\Delta d = (m + \frac{1}{2})\lambda_n$.

  In an oil slick (Fig. 27.2b), $n_{\text{air}} < n_{\text{oil}} < n_{\text{asphalt}}$, the condition for constructive interference is $\Delta d = m\lambda_n$.

REVIEW CHECKLIST

▷ Describe Young's double-slit experiment to demonstrate the wave nature of light. Account for the phase difference between light waves from the two sources as they arrive at a given point on the screen. State the conditions for constructive and destructive interference in terms of each of the following: path difference, phase difference, distance from the center of the screen, and angle subtended by the observation point at the source mid-point.

▷ Outline the manner in which the superposition principle leads to the correct expression for the intensity distribution on a distant screen due to two coherent sources of equal intensity.

▷ Account for the conditions of constructive and destructive interference in thin films considering both path difference and any expected phase changes due to reflection.

▷ Determine the positions of the minima in a single-slit diffraction pattern. Determine the positions of the principal maxima in the interference pattern of a diffraction grating.

▷ Determine whether or not two sources under a given set of conditions are resolvable as defined by Rayleigh's criterion.

▷ Understand what is meant by the resolving power of a grating, and calculate the resolving power of a grating under specified conditions.

ANSWERS TO SELECTED CONCEPTUAL QUESTIONS

1. What is the necessary condition on the path-length difference between two waves that interfere (a) constructively and (b) destructively?

Answer (a) Two waves interfere constructively if their path difference is either zero or some integral multiple of the wavelength; that is, if the path difference equals $m\lambda$. (b) Two waves interfere destructively if their path difference is an odd multiple of one-half of a wavelength; that is, if the path difference equals $\left(m+\frac{1}{2}\right)\lambda$.

□ □ □ □

3. If Young's double-slit experiment were performed under water, how would the observed interference pattern be affected?

Answer The wavelength of light under water would decrease, since the wavelength of light in a medium is given by $\lambda_n = \lambda_0 / n$ where λ_o is the wavelength in a vacuum and n is the index of refraction of the medium. Since the positions of the bright and dark fringes are proportional to the wavelength, the fringe separation would decrease.

□ □ □ □

4. In Young's double-slit experiment, why do we use monochromatic light? If white light were used, how would the pattern change?

Answer Every color produces its own pattern, with a spacing between the maximums that is characteristic of the wavelength. With several colors, the patterns are superimposed, and it can become difficult to pick out a single maximum. Using monochromatic light can eliminate this problem.

With white light, the central maximum is white. The first side maximum is a full spectrum, with violet on the inside and red on the outside. The second side maximum is a full spectrum also, but red in the second maximum overlaps the violet in the third maximum. At larger angles, the light soon starts mixing to white again, though it is often so faint that you would call it gray.

□ □ □ □

12. Would it be possible to place a nonreflective coating on an airplane to cancel radar waves of wavelength 3 cm?

Answer Yes. In order to do this, first measure the radar-reflectivity of the metal of your airplane. Then, choose a light, durable material that has approximately half the radar-reflectivity of the metal in your plane. Measure its index of refraction, and onto the metal plaster a coating equal in thickness to one-quarter of 3 cm, divided by that index.

This will effectively minimize the radar reflection at 3 cm; however, it will maximize the reflection at 1.5 cm. Thus, while it is possible to do this, it is not effective in a stealth design.

□ □ □ □

13. Why is it so much easier to perform interference experiments with a laser than with an ordinary light source?

Answer If you wish to perform an interference experiment, you need monochromatic light. To obtain it, you must first pass light from an ordinary source through a prism or diffraction grating to disperse different colors into different directions. With a single narrow slit you select a single color and make that light diffract to cover both of the slits for Young's experiment.

Thus you may have trouble lining things up and you will generally have low light power reaching the screen. The laser light is already monochromatic and coherent across the width of the beam.

□ □ □ □

15. Although we can hear around corners, we cannot see around corners. How can you explain this in view of the fact that sound and light are both waves?

Answer Audible sound has wavelengths on the order of meters or centimeters, while visible light has wavelengths on the order of half a micrometer. In this world of breadbox-size objects, λ is comparable to the object size for sound, and sound diffracts around walls and through doorways. But λ / a is much smaller for visible light passing ordinary-size objects or apertures, so light diffracts only through very small angles.

Another way of answering this question would be as follows. We can see by a small angle around a small obstacle or around the edge of a small opening. The side fringes in Figure 27.12 and the Arago spot in the center of Figure 27.13 (in the textbook) show this diffraction. Conversely, we cannot always hear around corners. Out-of-doors, away from reflecting surfaces, have someone a few meters distant face away from you and whisper. The high-frequency, short-wavelength, information-carrying components of the sound do not diffract around his head enough for you to understand his words.

□ □ □ □

20. Describe the change in width of the central maximum of the single-slit diffraction pattern as the width of the slit is made smaller.

Answer Equation 27.18 describes the angles at which you get destructive interference; from it, we can obtain an estimate of the width of the central maximum. For small angles, the equation can be rewritten as $\theta_m = \sin^{-1}(m\lambda / a) \cong m\lambda / a$. Thus, as the width of the slit a decreases, the angle of the first destructive interference θ_1 grows, and the width of the central maximum grows as well.

□ □ □ □

21. Suppose that reflected white light is used to observe a thin, transparent coating on glass as the coating material is gradually deposited by evaporation in a vacuum. Describe possible color changes that might occur during the process of building up the thickness of the coating.

Answer Suppose the coating is intermediate in index of refraction between the vacuum and the glass. When the coating is very thin, light reflected from its top and bottom surfaces will interfere constructively, so the surface will appear bright white. As the thickness reaches one-quarter of the wavelength of violet light in the coating, destructive interference of the violet light will make the surface look red.

Destructive interference will occur successively with blue, green, yellow, orange, and red light, making the surface look red, purple, and then blue. As the coating gets still thicker, we can get constructive interference for violet light, and then for other colors in spectral order. Still thicker coatings will give constructive and destructive interference for several visible wavelengths, so the reflected light will start to look white again.

□ □ □ □

SOLUTIONS TO SELECTED END-OF-CHAPTER PROBLEMS

3. Two radio antennas separated by 300 m, as in Figure P27.3, simultaneously broadcas identical signals at the same wavelength. A radio in a car traveling due north receive the signals. (a) If the car is at the position of the second maximum, what is th wavelength of the signals? (b) How much farther must the car travel to encounter th next minimum in reception? (**Caution:** Do not use the small-angle approximation i this problem.)

Solution Note, with the conditions given, the small angle approximation **does no work well.** That is, sin θ, tan θ, and θ are significantly different. The approach to be usec is outlined below.

(a) At the $m = 2$ maximum,

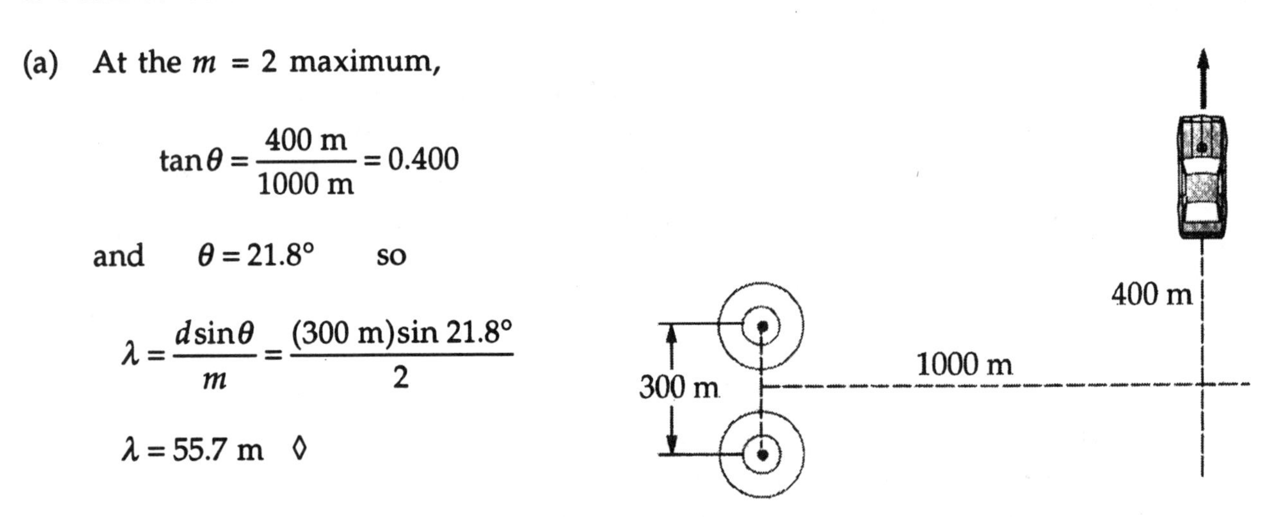

$$\tan\theta = \frac{400\text{ m}}{1000\text{ m}} = 0.400$$

and $\theta = 21.8°$ so

$$\lambda = \frac{d\sin\theta}{m} = \frac{(300\text{ m})\sin 21.8°}{2}$$

$\lambda = 55.7$ m ◊

Figure P27.3

(b) The next minimum encountered is the $m = 2$ minimum, and at that point,

$$d\sin\theta = \left(m+\tfrac{1}{2}\right)\lambda \quad \text{which becomes} \quad d\sin\theta = \tfrac{5}{2}\lambda$$

or $$\sin\theta = \frac{5\lambda}{2d} = \frac{5(55.7\text{ m})}{2(300\text{ m})} = 0.464 \quad \text{and} \quad \theta = 27.7°$$

so $$y = (1000\text{ m})\tan 27.7° = 524\text{ m}$$

Therefore, the car must travel an additional 124 m. ◊

5. Young's double-slit experiment is performed with 589-nm light and a slits-to-screen distance of 2.00 m. The tenth interference minimum is observed 7.26 mm from the central maximum. Determine the spacing of the slits.

Solution In the equation $d\sin\theta = \left(m+\frac{1}{2}\right)\lambda,$

the first minimum is described by $m = 0$
and the tenth is described by $m = 9.$

Thus, for the tenth minimum, $\sin\theta = \frac{\lambda}{d}\left(9+\frac{1}{2}\right)$

For small θ, $\sin\theta \cong \tan\theta,$ and $\tan\theta = \frac{y}{L}$

Solving for d,

$$d = \frac{9.5\lambda}{\sin\theta} = \frac{9.5\lambda L}{y} = \frac{9.5(5890\times10^{-10}\text{ m})(2.00\text{ m})}{7.26\times10^{-3}\text{ m}} = 1.54\times10^{-3}\text{ m}$$

$d = 1.54$ mm ◊

---

7. In Figure 27.1 of this solution manual, let L = 120 cm and d = 0.250 cm. The slits are illuminated with coherent 600-nm light. Calculate the distance y above the central maximum for which the average intensity on the screen is 75.0% of the maximum.

Solution For small θ, $I_{av} = I_o\cos^2\left(\frac{\pi d\sin\theta}{\lambda}\right)$

From the drawing, $\sin\theta \cong \frac{y}{L}$:

$$y = \frac{\lambda L}{\pi d}\cos^{-1}\sqrt{I_{av}/I_o}$$

In addition, since $I_{av} = 0.750 I_o$, we can substitute a value for each variable:

$$y = \frac{\left(6.00\times10^{-7}\text{ m}\right)\left(1.20\text{ m}\right)}{\pi\left(2.50\times10^{-3}\text{ m}\right)}\cos^{-1}\sqrt{0.750} = 0.0480\text{ mm}$$ ◊

---

11. An oil film (n = 1.45) floating on water is illuminated by white light at normal incidence. The film is 280 nm thick. Find (a) the dominant observed color in the reflected light and (b) the dominant color in the transmitted light. Explain your reasoning.

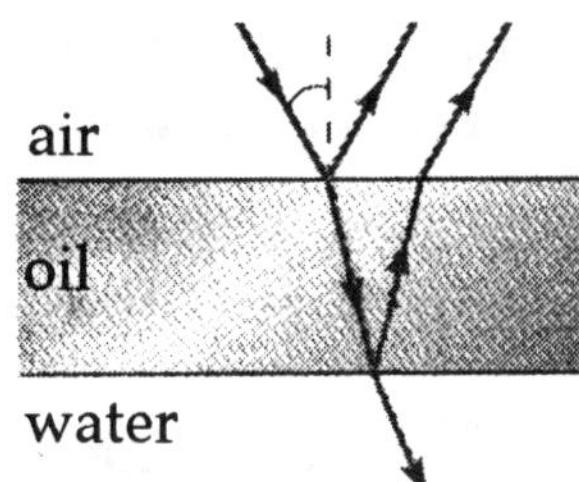

Solution

The light reflected from the top of the oil film undergoes phase reversal. Since 1.45 > 1.33, the light reflected from the bottom undergoes no reversal. For constructive interference of reflected light, we then have

$$2nt = \left(m + \tfrac{1}{2}\right)\lambda \qquad \text{or} \qquad \lambda_m = \frac{2nt}{\left(m + \frac{1}{2}\right)} = \frac{2(1.45)(280 \text{ nm})}{\left(m + \frac{1}{2}\right)}$$

(a) Substituting for m, we have

$m = 0$: λ_0 = 1624 nm (infrared)

$m = 1$: λ_1 = 541 nm (green)

$m = 2$: λ_2 = 325 nm (ultraviolet)

Both infrared and ultraviolet light are invisible to the human eye, so the dominant color is green. ◊

(b) In general, we can say that any light which is not reflected is transmitted. Therefore, subtracting green from white gives red + violet = purple. ◊

We could also note that the reflected light contains little red and violet according to the condition for destructive interference $2nt = m\lambda$.

For m = 1, λ = 812 nm (near infrared)

m = 2, λ = 406 nm (violet)

So these colors make it through to be in the transmitted beam.

13. An air wedge is formed between two glass plates separated at one edge by a very fine wire as in Figure P27.13. When the wedge is illuminated from above by 600-nm light, 30 dark fringes are observed. Calculate the radius of the wire.

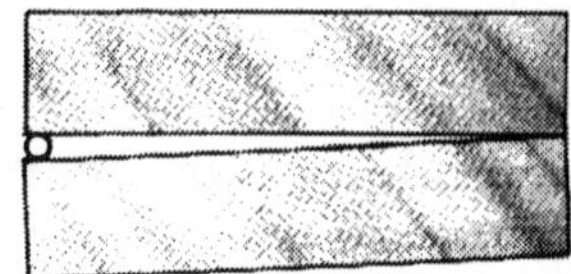

Figure P27.13

Solution Light reflecting from the bottom surface of the top plate undergoes no phase shift, while light reflecting from the top surface of the bottom plate is shifted by π, and must also travel an extra distance $2t$, where t is the thickness of the air wedge.

For destructive interference,

$$2t = m\lambda \quad (\text{where } m = 0, 1, 2, 3, \ldots)$$

The first dark fringe appears where $m = 0$ at the line of contact between the plates. The thirtieth dark fringe gives for the diameter of the wire

$$2t = 29\lambda, \quad \text{so} \quad t = 14.5\lambda.$$

$$r = \tfrac{1}{2}t = \tfrac{1}{2}(14.5)\left(600 \times 10^{-9}\text{ m}\right) = 4.35\ \mu\text{m} \quad \lozenge$$

15. A screen is placed 50.0 cm from a single slit, which is illuminated with 690-nm light. If the distance between the first and third minima in the diffraction pattern is 3.00 mm, what is the width of the slit?

Solution In the equation for single-slit diffraction minima at small angles,

$$\frac{y}{L} \cong \sin\theta = \frac{m\lambda}{a} \qquad \text{(Eq. 27.18)}$$

take differences between the first and third minima, to see that

$$\frac{\Delta y}{L} = \frac{\Delta m\lambda}{a} \quad \text{with } \Delta y = 3.00 \times 10^{-3}\text{ m, and } \Delta m = 3 - 1 = 2$$

The width of the slit is then

$$a = \frac{\lambda L \Delta m}{\Delta y} = \frac{\left(690 \times 10^{-9}\text{ m}\right)(0.500\text{ m})(2)}{3.00 \times 10^{-3}\text{ m}} = 2.30 \times 10^{-4}\text{ m} \quad \lozenge$$

21. A helium-neon laser emits light that has a wavelength of 632.8 nm. The circular aperture through which the beam emerges has a diameter of 0.500 cm. Estimate the diameter of the beam 10.0 km from the laser.

Solution Following Equation 27.20 for diffraction from a circular opening, the beam spreads into a cone of half-angle

$$\theta_{\min} = 1.22\frac{\lambda}{D} = 1.22\frac{\left(632.8\times10^{-9}\text{ m}\right)}{(0.00500\text{ m})} = 1.54\times10^{-4}\text{ rad}$$

The radius of the beam ten kilometers away is, from the definition of radian measure,

$$r_{\text{beam}} = \theta_{\min}\left(1.00\times10^{4}\text{ m}\right) = 1.54\text{ m}$$

and its diameter is $d_{\text{beam}} = 2r_{\text{beam}} = 3.09\text{ m}$ ◊

---

27. The hydrogen spectrum has a red line at 656 nm and a violet line at 434 nm. What is the angular separation between these two spectral lines obtained with a diffraction grating that has 4500 lines/cm?

Solution The grating spacing is $d = (1.00\times10^{-2}\text{ m})/4500 = 2.22\times10^{-6}\text{ m}$

In the first-order spectrum, the angles of diffraction are given by $\sin\theta = \lambda/d$.

$$\sin\theta_1 = \frac{656\times10^{-9}\text{ m}}{2.22\times10^{-6}\text{ m}} = 0.295 \qquad \sin\theta_2 = \frac{434\times10^{-9}\text{ m}}{2.22\times10^{-6}\text{ m}} = 0.195$$

so that $\theta_1 = 17.17°$ and $\theta_2 = 11.26°$

The angular separation is therefore $\Delta\theta = 17.17° - 11.26° = 5.91°$ ◊

In the second-order spectrum, $\Delta\theta = \sin^{-1}(2\lambda_1/d) - \sin^{-1}(2\lambda_2/d) = 13.2°$ ◊

Again, in the third order, $\Delta\theta = \sin^{-1}(3\lambda_1/d) - \sin^{-1}(3\lambda_2/d) = 26.5°$ ◊

Since the red line does not appear in the fourth-order spectrum, the answer is complete.

---

31. A diffraction grating of length 4.00 cm contains 6000 rulings over a width of 2.00 cm. (a) What is the resolving power of this grating in the first three orders? (b) If two monochromatic waves incident on this grating have a mean wavelength of 400 nm, what is their wavelength separation if they are just resolved in the third order?

Solution We assume the two distances are measured in the same direction.

From Eq. 27.23, $R = mN$ where $N = \left(\frac{6000 \text{ lines}}{2.00 \text{ cm}}\right)(4.00 \text{ cm}) = 12000 \text{ lines}$

(a) In the 1st order, $R = (1)(12000 \text{ lines}) = 12000$ ◊

In the 2nd order, $R = (2)(12000 \text{ lines}) = 24000$ ◊

In the 3rd order, $R = (3)(12000 \text{ lines}) = 36000$ ◊

(b) From Equation 27.22, $R = \lambda/\Delta\lambda$

In the 3rd order, $\Delta\lambda = \frac{\lambda}{R} = \frac{4.00 \times 10^{-7} \text{ m}}{3.60 \times 10^{4}} = 1.11 \times 10^{-11} \text{ m} = 0.0111 \text{ nm}$ ◊

---

35. If the interplanar spacing of NaCl is 0.281 nm, what is the predicted angle at which 0.140-nm x-rays are diffracted in a first-order maximum?

Solution The atomic planes in this crystal are shown in Figure 27.25 of the text. The diffraction they produce is described by the Bragg condition, that

$$2d\sin\theta = m\lambda$$

$$\sin\theta = \frac{m\lambda}{2d} = \frac{1(0.140 \times 10^{-9} \text{ m})}{2(0.281 \times 10^{-9} \text{ m})} = 0.249$$

$$\theta = 14.4° \quad ◊$$

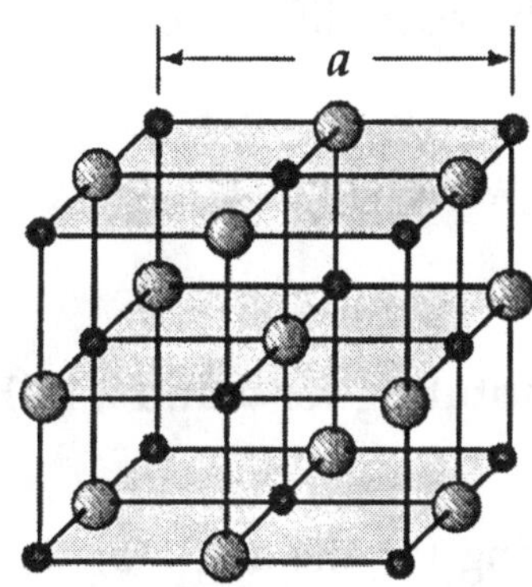

Figure 27.25

---

41. Astronomers observed a 60.0-MHz radio source both directly and by reflection from the sea. If the receiving dish is 20.0 m above sea level, what is the angle of the radio source above the horizon at first maximum?

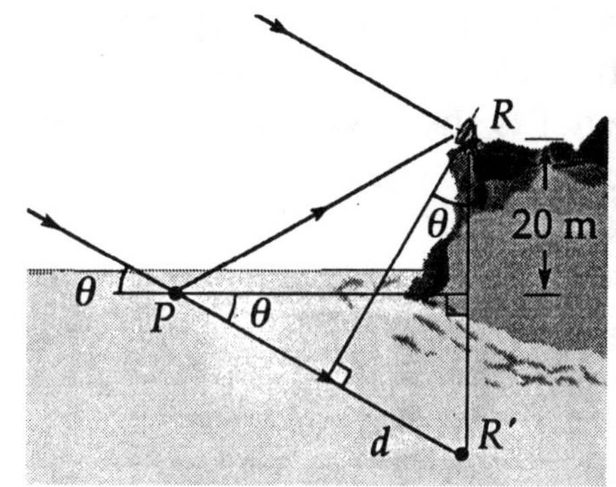

Solution One radio wave reaches the receiver R directly from the distant source at an angle θ above the horizontal. The other wave undergoes phase reversal as it reflects from the water at P.

Constructive interference first occurs for a path difference of

$$d = \frac{\lambda}{2} \qquad (1)$$

The angles θ in the figure are equal because they each form part of a right triangle with a shared angle at R'.

It is equally far from P to R as from P to R', the mirror image of the telescope.

So the path difference is $d = 2(20.0\text{ m})\sin\theta = (40.0\text{ m})\sin\theta$

The wavelength is $\lambda = \frac{c}{f} = \frac{3.00\times10^8\text{ m/s}}{60.0\times10^6\text{ Hz}} = 5.00\text{ m}$

Substituting for d and λ in Equation (1), $(40.0\text{ m})\sin\theta = \frac{5.00\text{ m}}{2}$

Solving for the angle θ,

$$\sin\theta = \frac{5.00\text{ m}}{80.0\text{ m}} \quad \text{and} \quad \theta = 3.58° \quad \Diamond$$

51. Consider the double-slit arrangement shown in Figure P27.50, where the slit separation is d and the slit-to-screen distance is L. A sheet of transparent plastic having an index of refraction n and thickness t is placed over the upper slit. As a result, the central maximum of the interference pattern moves upward a distance y'. Find y'.

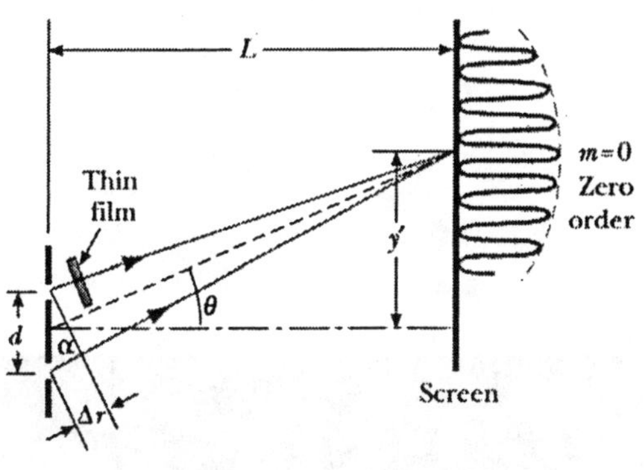

Figure P27.50

Solution The central maximum corresponds to zero phase difference between light taking the two paths marked with arrows in the figure. Thus the added distance Δr traveled by the light from the lower slit must introduce a phase difference equal to that introduced by the plastic film.

Imagine freezing the light at one instant in its propagation. For the lower ray, the number of cycles contained in distance t is t/λ. For the upper ray, the number of cycles completed inside the plastic is

$$\frac{t}{\lambda_n} = \frac{t}{\lambda/n} = \frac{t}{\lambda}$$

The plastic introduces a phase difference

$$\phi = 2\pi\left(\frac{nt}{\lambda} - \frac{t}{\lambda}\right) = \left(\frac{2\pi t}{\lambda}\right)(n-1)$$

Therefore, corresponding difference in optical **path length** Δr can be calculated as

$$\Delta r = \phi\left(\frac{\lambda}{2\pi}\right) = t(n-1)$$

Note that the wavelength of the light does not appear in this equation. If L is large in Figure P27.50, the two rays from the slits are essentially parallel, so the angle θ may be expressed as

$$\tan\theta = \frac{\Delta r}{d} = \frac{y'}{L}$$

Eliminating Δr by substitution,

$$\frac{y'}{L} = \frac{t(n-1)}{d} \quad \text{gives} \quad y' = \frac{t(n-1)L}{d} \quad \Diamond$$

Chapter 28

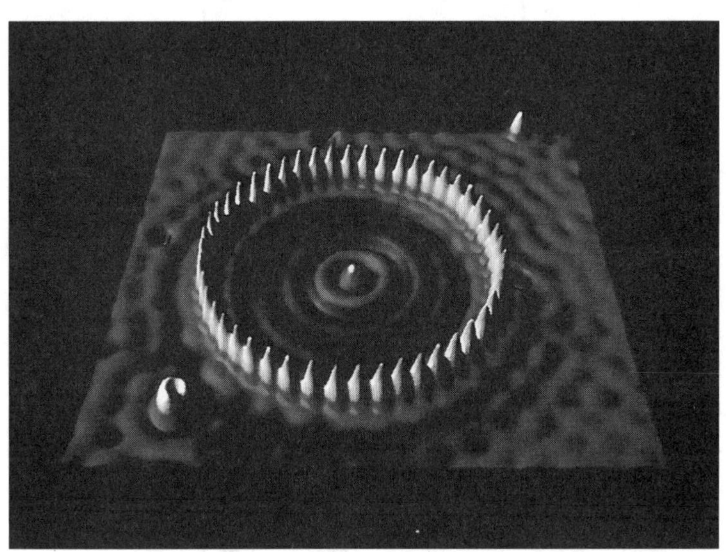

This is a photograph taken from a scanning-tunneling microscope. The image is that of a ring — or corral — of 48 iron atoms on a copper surface. The corral is able to confine surface-electron waves, and may pave the way for future electronic devices. (IBM Corporation Research Division)

QUANTUM PHYSICS

Chapter 28

QUANTUM PHYSICS

NTRODUCTION

In Chapter 9, we noted that Newtonian mechanics must be replaced by Einstein's special theory of relativity when dealing with particles whose speeds are comparable to the speed of light. Although many problems were indeed resolved by the theory of relativity in the early part of the 20th century, many experimental and theoretical problems remained unsolved. Attempts to explain the behavior of matter on the atomic level with the laws of classical physics were totally unsuccessful. Various phenomena, such as blackbody radiation, the photoelectric effect, and the emission of sharp spectra lines by atoms in a gas discharge tube, could not be understood within the framework of classical physics. We shall describe these phenomena because of their importance in subsequent developments.

Another revolution took place in physics between 1900 and 1930. This was the era of a new and more general formulation called **quantum mechanics.** This new approach was highly successful in explaining the behavior of atoms, molecules, and nuclei. As with relativity, the quantum theory requires a modification of our ideas concerning the physical world.

An extensive study of quantum theory is certainly beyond the scope of this book. In this chapter we introduce quantum mechanics as a theory describing the motion of small-mass particles. This scheme, developed from 19253 to 1928 by Schrödinger, Heisenberg, and others, addresses the limitations of the Bohr model and lets us understand a host of phenomena involving atoms, molecules, nuclei, and solids. We begin by describing wave-particle duality, wherein a particle can be viewed as having wave-like properties, and its wavelength can be calculated if its momentum is known. Next, we describe some of the basic features of quantum mechanics and apply these principles to simple, one-dimensional systems. For example, we shall treat the problem of a particle confined to a potential well having infinitely high barriers. We also discuss some simple applications of quantum theory, including the photoelectric effect, the Compton effect, and x-rays.

NOTES FROM SELECTED CHAPTER SECTIONS

28.1 Blackbody Radiation and Planck's Theory

A **black body** is an ideal body that absorbs all radiation incident on it. Any body at some temperature T emits thermal radiation which is characterized by the properties of the body and its temperature. The spectral distribution of blackbody radiation at various temperatures is sketched in the figure. As the temperature increases, the intensity of the radiation (area under the curve) increases, while the peak of the distribution shifts to shorter wavelengths.

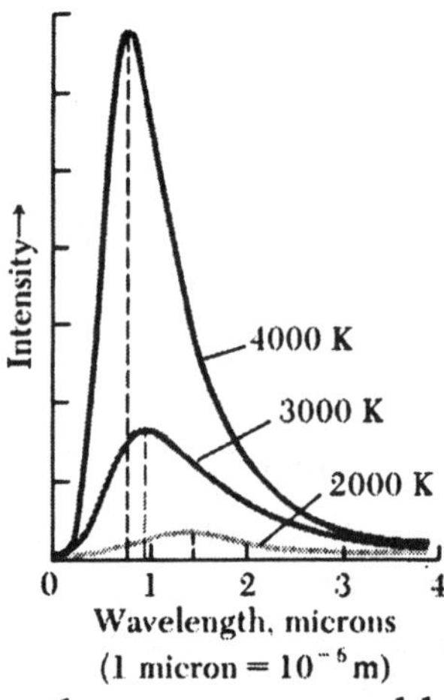

Classical theories failed to explain blackbody radiation. A new theory, proposed by Max Planck, is consistent with this distribution at all wavelengths. Planck made two basic assumptions in the development of this result:

- The oscillators emitting the radiation could only have **discrete energies** given by $E_n = nhf$, where n is a quantum number ($n = 1, 2, 3, \ldots$), f is the oscillator frequency, and h is Planck's constant.

- These oscillators can emit or absorb energy in discrete units called quanta (or photons), where the energy of a light quantum obeys the relation $E = hf$.

Subsequent developments showed that the quantum concept was necessary in order to explain several phenomena at the atomic level, including the photoelectric effect, the Compton effect, and atomic spectra.

28.2 The Photoelectric Effect

When light is incident on certain metallic surfaces, electrons can be emitted from the surfaces. This is called the photoelectric effect, discovered by Hertz. Several features of the photoelectric effect could not be explained with classical physics or with the wave theory of light. In 1905, Einstein provided a successful explanation of the photoelectric effect by extending Planck's quantum concept to include electromagnetic fields.

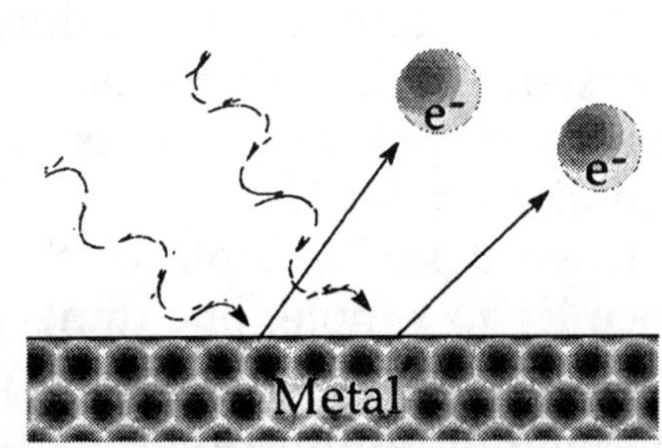

Each of the unusual features of the photoelectric effect can be explained and understood on the basis of the photon theory of light. These observations and their explanations include:

- No electrons are emitted if the incident light frequency falls below some cutoff frequency, f_c, which is characteristic of the material being illuminated. For example, in the case of sodium, $f_c = 5.94 \times 10^{14}$ Hz. This is inconsistent with the wave theory, which predicts that the photoelectric effect should occur at any frequency, provided the light intensity is high enough.

  The photoelectric effect is not observed below a certain cutoff frequency, because the energy of the photon must be greater than or equal to the work function ϕ, the energy required to lift the electron out of the metal. If the energy of the incoming photon is not equal to or greater than ϕ, the electrons will never be ejected from the surface, regardless of the intensity of the light.

- If the light frequency exceeds the cutoff frequency, a photoelectric effect is observed and the number of photoelectrons emitted is proportional to the light intensity. However, the maximum kinetic energy of the photoelectrons is independent of light intensity, a fact that cannot be explained by the concepts of classical physics.

  That K_{max} is independent of the light intensity can be understood with the following argument: If the light intensity is doubled, the number of photons is doubled, which doubles the number of photoelectrons emitted. However, their kinetic energy, which equals $hf - \phi$, depends only on the light frequency and the work function, not on the light intensity.

- The maximum kinetic energy of the photoelectrons increases with increasing light frequency. That K_{max} increases with increasing frequency is easily understood with Equation 28.5.

- Finally, electrons are emitted from the surface almost instantaneously (less than 10^{-9} s after the surface is illuminated), even at low light intensities. Classically, one would expect that the electrons would require some time to absorb the incident radiation before they acquire enough kinetic energy to escape from the metal.

  The almost instantaneous electron emission is consistent with the particle theory of light, in which the incident energy appears in small packets and there is a one-to-one interaction between photons and electrons. This is in contrast to having the energy of the light distributed uniformly over a large area.

28.3 The Compton Effect

The Compton effect involves the scattering of an x-ray by an electron. The scattered x-ray undergoes a **change** in wavelength $\Delta\lambda$, called the Compton shift which cannot be explained using classical concepts. By treating the x-ray as a photon (the quantum concept), the scattering process between the photon and electron predicts a shift in photon (x-ray) wavelength given by Equation 28.7, where θ is the angle between the incident and scattered x-ray and m_e is the mass of the electron. The formula is in excellent agreement with experimental results.

The figure to the right represents the data for Compton scattering of x-rays from graphite at $\theta = 90.0°$. In this case, the Compton shift is $\Delta\lambda = 0.0024$ nm and λ_o is the wavelength of the incident x-ray beam.

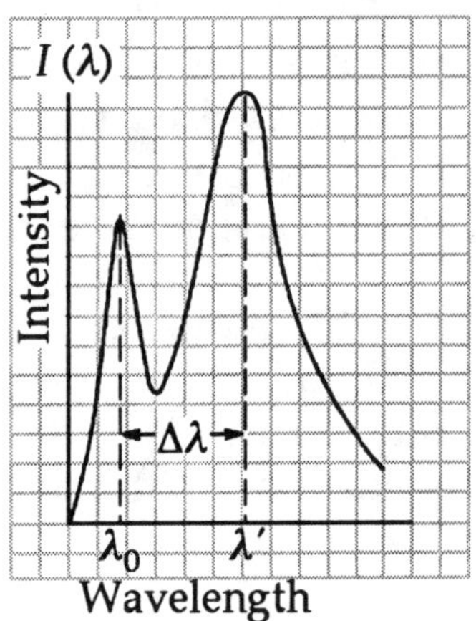

28.4 Photons and Electromagnetic Waves

The results of some experiments are better described on the basis of the photon model of light; other experimental outcomes are better described in terms of the wave model. The photon theory and the wave theory complement each other—light exhibits both wave and photon characteristics.

28.5 The Wave Properties of Particles

De Broglie postulated that a particle in motion has wave properties and a corresponding wavelength inversely proportional to the particle's momentum.

28.7 The Uncertainty Principle

Quantum theory predicts that it is fundamentally impossible to make simultaneous measurements of a particle's position and momentum with infinite accuracy.

28.8 An Interpretation of Quantum Mechanics

The wave function is a complex valued quantity, the absolute square of which gives the probability per volume of finding a particle at a given point at some instant; the wave function contains all the information that can be known about the particle.

The **Schrödinger equation** describes the manner in which matter waves change in time and space.

The average experimental value of a quantity such as position or energy is called the **expectation value** of the quantity.

28.9 A Particle in a Box

A particle confined to a line segment and represented by a well-defined de Broglie wave function is represented by a sinusoidal wave. The allowed states of the system are called **stationary states** since they represent **standing waves.**

The minimum energy which the particle can have is called the **zero-point energy.**

28.10 The Schrödinger Equation

The basic problem in wave mechanics is to determine a solution to the Schrödinger equation. The solution will provide the allowed wave functions and energy levels of the system.

28.11 Tunneling Through a Barrier

When a particle is incident onto a barrier, the height of which is greater than the energy of the particle, there is a finite probability that the particle will penetrate the barrier. In this process, called **tunneling,** part of the incident wave is transmitted and part is reflected.

EQUATIONS AND CONCEPTS

The Wien displacement law properly describes the peak in the energy spectrum emitted by a blackbody radiator.

$$\lambda_{max}T = 0.2898 \times 10^{-2}\ \text{m}\cdot\text{K} \quad (28.1$$

A **black body** is an ideal body that absorbs all radiation incident on it. Any body at some temperature T emits thermal radiation which is characterized by the properties of the body and its temperature. As the temperature of a blackbody increases, the intensity of the radiation increases, while the peak of the distribution shifts to shorter wavelengths.

Vibrating molecules are characterized by discrete energy levels called quantum states. The positive integer n is called a quantum number.

$$E_n = nhf \quad (28.2)$$

$$h = 6.626 \times 10^{-34}\ \text{J}\cdot\text{s}$$

Molecules emit or absorb energy in discrete units of light energy called quanta. The energy of a quantum or photon corresponds to the energy difference between adjacent quantum states.

$$E = hf \quad (28.3)$$

When light is incident on certain metallic surfaces, electrons can be emitted from the surfaces. This is the photoelectric effect, discovered by Hertz. One cannot explain many of the features of the photoelectric effect using classical concepts. However, in 1905, Einstein provided a successful explanation of the effect by extending Planck's quantum concept to include electromagnetic fields. In his model, Einstein assumed that light consists of a stream of particles called **photons** whose energy is given by $E = hf$, where h is Planck's constant and f is their frequency.

'he maximum kinetic energy of the jected photoelectron also depends on the vork function of the metal, ϕ, which is ypically a few eV. This model is in excellent agreement with experimental esults, including the prediction of a cutoff or threshold) wavelength above which no photoelectric effect is observed.

$$K_{max} = hf - \phi \quad (28.5)$$

$$\lambda_c = \frac{hc}{\phi} \quad (28.6)$$

The Compton effect involves the scattering of an x-ray by an electron. The scattered x-ray undergoes a change in wavelength called the Compton shift, which cannot be explained using classical concepts.

$$\lambda' - \lambda_o = \frac{h}{m_e c}(1 - \cos\theta) \quad (28.7)$$

By treating the x-ray as a photon (the quantum concept), the scattering process between the photon and electron predicts a shift in photon (x-ray) wavelength, where θ is the angle between the incident and scattered x-ray and m_e is the mass of the electron. The formula is in excellent agreement with experimental results.

The quantity $h/(m_e c)$ = 0.00243 nm is called the Compton wavelength.

According to the de Broglie hypothesis, a material particle should have an associated wavelength λ which depends on its momentum.

$$\lambda = \frac{h}{mv} \quad (28.9)$$

Matter waves can be represented by a wave function, ψ, which, in general, depends on position and time. In the expression for the part of the wave function which depends only on position, and which describes a freely moving particle, k is the wave number.

$$\psi(x) = A\sin\left(\frac{2\pi x}{\lambda}\right) \quad (28.13)$$

The normalization condition is a statement of the requirement that the particle exists at some point (along the x axis in the one-dimensional case) at all times.

$$\int_{-\infty}^{\infty} |\psi|^2\, dx = 1 \quad (28.14)$$

Although it is not possible to specify the position of a particle exactly, it is possible to calculate the probability P_{ab} of finding the particle within an interval $a \le x \le b$.

$$P_{ab} = \int_a^b |\psi|^2\, dx \quad (28.15)$$

The average of many measured values for the position of a particle is called the expectation value of the coordinate x.

$$\langle x \rangle \equiv \int_{-\infty}^{\infty} x|\psi|^2\, dx \quad (28.16)$$

The uncertainty principle states that if a measurement of position is made with a precision Δx and a **simultaneous** measurement of momentum is made with a precision Δp, then the product of the two uncertainties can never be smaller than $\hbar/2$.

$$\Delta x \Delta p_x \ge \frac{\hbar}{2} = \frac{h}{4\pi} \quad (28.11)$$

he allowed wave functions for a particle a rigid box of width L are sinusoidal.

$$\psi(x) = A\sin\left(\frac{n\pi x}{L}\right) \quad (28.17)$$

$n = 1, 2, 3, \ldots$

'he energy of a particle in a box is uantized and the least energy which the article can have is called the zero-point nergy.

$$E_n = \left(\frac{h^2}{8mL^2}\right)n^2 \quad (28.18)$$

$n = 1, 2, 3, \ldots$

'he time independent Schrödinger quation for a particle confined to moving long the x-axis (total energy E is constant) llows in principle the determination of he wave functions and energies of the llowed states if the potential energy 'unction is known.

$$\frac{d^2\psi}{dx^2} = -\frac{2m}{\hbar^2}(E - U)\psi \quad (28.19)$$

REVIEW CHECKLIST

▷ Describe the account of blackbody radiation proposed by Planck.

▷ Describe the Einstein model for the photoelectric effect, and the predictions of the fundamental photoelectric effect equation for the maximum kinetic energy of photoelectrons. Recognize that Einstein's model of the photoelectric effect involves the photon concept ($E = hf$), and that the basic features of the photoelectric effect are consistent with this model.

▷ Describe the Compton effect (the scattering of x-rays by electrons) and be able to use the formula for the Compton shift. Recognize that the Compton effect can only be explained using the photon concept.

▷ Discuss the wave properties of particles, the de Broglie wavelength concept, and the dual nature of both matter and light.

▷ Describe the concept of wave function for the representation of matter waves and state in equation form the normalization condition and expectation value of the coordinate.

▷ Discuss the manner in which the uncertainty principle makes possible a better understanding of the dual wave-particle nature of light and matter.

ANSWERS TO SELECTED CONCEPTUAL QUESTIONS

6. Why does the existence of a cutoff frequency in the photoelectric effect favor particle theory of light rather than a wave theory?

Answer Wave theory predicts that the photoelectric effect should occur at an frequency, provided that the light intensity is high enough. As is implied by th question, this is in contradiction to experimental results.

□ □ □ □

10. The brightest star in the constellation Lyra is the bluish star Vega, whereas th brightest star in Bootes is the reddish star Arcturus. How do you account for th difference in color of the two stars?

Answer In general, the stars having higher surface temperature produce photon with higher average energy. Thus, blue stars have a higher surface temperature than rec stars.

□ □ □ □

11. An x-ray photon is scattered by an electron. What happens to the frequency of the scattered photon relative to that of the incident photon?

Answer The x-ray photon transfers some of its energy to the electron. Thus, its energy, and therefore its frequency, must be decreased.

□ □ □ □

16. If the photoelectric effect is observed for one metal, can you conclude that the effect will also be observed for another metal under the same conditions? Explain.

Answer No. Suppose that the incident light frequency at which you first observed the photoelectric effect is above the cutoff frequency of the first metal, but less than the cutoff frequency of the second metal. In that case, the photoelectric effect would not be observed at all in the second metal.

□ □ □ □

OLUTIONS TO SELECTED END-OF-CHAPTER PROBLEMS

. The human eye is most sensitive to 560-nm light. What temperature black body vould radiate most intensely at this wavelength?

olution We use Wien's law: $\lambda_{max}T = 0.290\times10^{-2}\ \text{m}\cdot\text{K}$

hus, $$T = \frac{2.90\ \text{mm}\cdot\text{K}}{560\times10^{-6}\ \text{mm}} = 5180\ \text{K} \quad \Diamond$$

elated Commentary: This is close to the temperature of the surface of the Sun (which is, in turn, a pretty good black body). Living things on Earth evolved having sensitivity to electromagnetic waves around this wavelength because there is such a lot of it bouncing around, carrying information.)

---

3. Calculate the energy, in electron volts, of a photon the frequency of which is (a) 6.20 × 10^{14} Hz, (b) 3.10 GHz, (c) 46.0 MHz. (d) Determine the corresponding wavelengths for these photons.

Solution $E = hf$

(a) $E = (6.63\times10^{-34}\ \text{J}\cdot\text{s})(6.20\times10^{14}\ \text{Hz}) = 4.11\times10^{-19}\ \text{J} = 2.57\ \text{eV} \quad \Diamond$

(b) $E = (6.63\times10^{-34}\ \text{J}\cdot\text{s})(3.10\times10^{9}\ \text{Hz}) = 2.06\times10^{-24}\ \text{J} = 12.8\ \mu\text{eV} \quad \Diamond$

(c) $E = (6.63\times10^{-34}\ \text{J}\cdot\text{s})(46.0\times10^{6}\ \text{Hz}) = 3.05\times10^{-26}\ \text{J} = 1.90\ \times10^{-7}\text{eV} \quad \Diamond$

(d) $\lambda_a = c/f = (3.00\times10^{8}\ \text{m/s})/(6.20\times10^{14}\ \text{s}^{-1}) = 4.84\times10^{-7}\ \text{m} = 484\ \text{nm}$ ◊ (Blue light)

$\lambda_b = (3.00\times10^{8}\ \text{m/s})/(3.10\times10^{9}\ \text{s}^{-1}) = 0.0968\ \text{m} = 9.68\ \text{cm}$ ◊ (Microwave)

$\lambda_c = (3.00\times10^{8}\ \text{m/s})/(46.0\times10^{6}\ \text{s}^{-1}) = 6.52\ \text{m}$ ◊ (Radio, in the public LO band)

---

7. Molybdenum has a work function of 4.20 eV. (a) Find the cutoff wavelength an[d] threshold frequency for the photoelectric effect. (b) Calculate the stopping potential if th[e] incident light has a wavelength of 180 nm.

Solution We write Einstein's equation as $eV_s = hf - \phi$. The cutoff wavelength an[d] threshold frequency describe light barely able to produce photoelectrons, with zero kineti[c] energy.

(a) $0 = hf_c - \phi = \dfrac{hc}{\lambda_c} - \phi$: $\lambda_c = \dfrac{hc}{\phi} = \dfrac{(6.63 \times 10^{-34}\ \text{J}\cdot\text{s})(3.00 \times 10^{8}\ \text{m/s})}{(4.20\ \text{eV})(1.602 \times 10^{-19}\ \text{J/eV})} = 296\ \text{nm}$ ◊

$$f = \frac{c}{\lambda} = \frac{3.00 \times 10^{8}\ \text{m/s}}{296 \times 10^{-9}\ \text{m}} = 1.01 \times 10^{15}\ \text{Hz} \quad ◊$$

(b) $\dfrac{hc}{\lambda} = \phi + eV_s$

$$\frac{(6.63 \times 10^{-34}\ \text{J}\cdot\text{s})(3.00 \times 10^{8}\ \text{m/s})}{180 \times 10^{-9}\ \text{m}} = (4.20\ \text{eV})(1.60 \times 10^{-19}\ \text{J/eV}) + (1.60 \times 10^{-19}\ \text{C})V_s$$

Thus, $V_s = 2.71\ \text{V}$ ◊

13. A 0.00160-nm photon scatters from a free electron. For what (photon) scattering angle will the kinetic energy of the recoiling electron and the energy of the scattered photon be the same?

Solution The energy of the incoming photon is

$$E_o = \frac{hc}{\lambda} = \frac{(6.63 \times 10^{-34}\ \text{J}\cdot\text{s})(3.00 \times 10^{8}\ \text{m/s})}{0.00160 \times 10^{-9}\ \text{m}} = 1.24 \times 10^{-13}\ \text{J}$$

Since the outgoing photon and the electron each have half of this energy in kinetic form,

$$E' = 6.22 \times 10^{-14}\ \text{J} \quad \text{and} \quad \lambda' = \frac{hc}{E'} = 3.20 \times 10^{-12}\ \text{m}$$

The shift in wavelength is $\Delta\lambda = \lambda' - \lambda = 1.60 \times 10^{-12}\ \text{m}$

But by Equation 28.7, $\Delta\lambda = \lambda_c(1 - \cos\theta)$

so $\cos\theta = 1 - \dfrac{\Delta\lambda}{\lambda_c} = 1 - \dfrac{1.60 \times 10^{-12}\ \text{m}}{0.00243 \times 10^{-9}\ \text{m}} = 0.342$ and $\theta = 70.0°$ ◊

7. Calculate the de Broglie wavelength for an electron that has kinetic energy (a) 50.0 eV nd (b) 50.0 keV.

olution

a) We can find the electron speed first, and then the wavelength.

$$K = \tfrac{1}{2}mv^2 = (50.0\ \text{eV})(1.60\times10^{-19}\ \text{C}/\text{e}^-)(1\ \text{J}/\text{C}\cdot\text{V})$$

$$v = \sqrt{2\frac{(80.0\times10^{-19}\ \text{J})}{(9.11\times10^{-31}\ \text{kg})}} = 4.19\times10^{6}\ \text{m}/\text{s}$$

$$\lambda = \frac{h}{p} = \frac{h}{mv} = \frac{6.63\times10^{-34}\ \text{J}\cdot\text{s}}{(9.11\times10^{-31}\ \text{kg})(4.19\times10^{6}\ \text{m}/\text{s})} = 0.174\ \text{nm} \quad \Diamond$$

(b) Or we can skip computing the speed:

$$K = \left(50.0\times10^{3}\ \text{eV}\right)\left(1.60\times10^{-19}\ \text{J}/\text{eV}\right)$$

Since $K = \dfrac{mv^2}{2} = \dfrac{p^2}{2m}$, $\quad p = \sqrt{2m_eK} = \sqrt{2\left(9.11\times10^{-31}\right)8.00\times10^{-15}}$

$$p = 1.21\times10^{-22}\ \text{kg}\cdot\text{m}/\text{s}$$

$$\lambda = \frac{h}{p} = 5.49\times10^{-12}\ \text{m} \quad \Diamond$$

Slightly more accurate are the relativistic answers:

From $\quad \left(mc^2 + K\right)^2 = p^2c^2 + m^2c^4,$

We find $\quad p = 1.23\times10^{-22}\ \text{kg}\cdot\text{m}/\text{s}$

and $\quad \lambda = h/p = 5.37\times10^{-12}\ \text{m} \quad \Diamond$

21. Robert Hofstadter won the 1961 Nobel Prize in physics for his pioneering work i scattering 20.0-GeV electrons from nuclei. (a) What is the γ-factor for a 20.0-GeV electron where $\gamma = \left(1 - v^2/c^2\right)^{-1/2}$? What is the momentum of the electron in $\text{kg}\cdot\text{m/s}$ (b) What is the wavelength of a 20.0-GeV electron and how does it compare with the siz of a nucleus?

Solution (a) From $E = \gamma m_e c^2$,

$$\gamma = \frac{20.0\times10^3\ \text{MeV}}{0.511\ \text{MeV}} = 39100 \quad \Diamond$$

For these extreme-relativistic electrons, with $m_e c^2 << pc$,

$$E^2 = p^2c^2 + m^2c^4 \cong p^2c^2.$$

$$p = \frac{E}{c} = \frac{(2.00\times10^4\ \text{MeV})(1.60\times10^{-13}\ \text{J/MeV})}{3.00\times10^8\ \text{m/s}} = 1.07\times10^{-17}\ \text{kg}\cdot\text{m/s} \quad \Diamond$$

(b) $$\lambda = \frac{h}{p} = \frac{6.63\times10^{-34}\ \text{J}\cdot\text{s}}{1.07\times10^{-17}\ \text{kg}\cdot\text{m/s}} = 6.22\times10^{-17}\ \text{m} \quad \Diamond$$

Since the size of a nucleus is on the order of 10^{-14} m, the 20-GeV electrons would be small enough to go through the nucleus.

---

23. The resolving power of a microscope depends on the wavelength used. If one wished to use a microscope to "see" an atom, a resolution of approximately 1.00×10^{-11} m would have to be obtained. (a) If electrons are used (in an electron microscope), what minimum kinetic energy is required for the electrons? (b) If photons are used, what minimum photon energy is needed to obtain the required resolution?

Solution (a) Since the de Broglie wavelength is $\lambda = \dfrac{h}{p}$,

$$p_e = \frac{h}{\lambda} = \frac{6.63\times10^{-34}\ \text{J}\cdot\text{s}}{1.00\times10^{-11}\ \text{m}} = 6.63\times10^{-23}\ \text{kg}\cdot\text{m/s}$$

$$K_e = \frac{p_e^{\,2}}{2m} = \frac{(6.63\times10^{-23}\ \text{kg}\cdot\text{m/s})^2}{2(9.11\times10^{-31}\ \text{kg})} = 2.41\times10^{-15}\ \text{J} = 15.1\ \text{keV} \quad \Diamond$$

For better accuracy, you can use the relativistic equation

$\left(mc^2 + K\right)^2 = p^2c^2 + m^2c^4$ to find that $K = 14.9\ \text{keV}$. ◊

(b) For photons:

$$E = hf = \frac{hc}{\lambda} = \frac{(6.63\times10^{-34}\ \text{J}\cdot\text{s})(3.00\times10^{8}\ \text{m/s})}{1.00\times10^{-11}\ \text{m}} = 1.99\times10^{-14}\ \text{J} = 124\ \text{keV} \quad \Diamond$$

For the photon, this wavelength λ = 0.100 Å is in the x-ray range of the electromagnetic spectrum.

---

25. Neutrons traveling at 0.400 m/s are directed through a double slit having a 1.00-mm separation. An array of detectors is placed 10.0 m from the slit. (a) What is the de Broglie wavelength of the neutrons? (b) How far off axis is the first zero-intensity point on the detector array? (c) When a neutron reaches a detector, can we say which slit the neutron passed through? Explain.

Solution

(a) $\lambda = \frac{h}{mv} = \frac{6.63\times10^{-34}\ \text{J}\cdot\text{s}}{(1.67\times10^{-27}\ \text{kg})(0.400\ \text{m/s})} = 9.93\times10^{-7}\ \text{m} \quad \Diamond$

(b) The condition for destructive interference in a multiple-slit experiment is $d\sin\theta = \left(m+\frac{1}{2}\right)\lambda$ with $m = 0$ for the first minimum. Then,

$$\theta = \sin^{-1}\left(\frac{\lambda}{2d}\right) = 0.0284° \qquad \frac{y}{L} = \tan\theta$$

$$y = L\tan\theta = (10.0\ \text{m})(\tan 0.0284°) = 4.96\ \text{mm} \quad \Diamond$$

(c) We cannot say the neutron passed through one slit. We can only say it passed through the pair of slits, as a water wave does to produce an interference pattern.

---

27. An electron (m_e = 9.11×10^{-31} kg) and a bullet (m = 0.0200 kg) each have a speed of 500 m/s, accurate to within 0.0100%. Within what limits could we determine the position of the objects?

Solution

For the electron, the uncertainty in momentum is

$$\Delta p = m_e \Delta v = (9.11 \times 10^{-31}\ \text{kg})(500\ \text{m/s})(1.00 \times 10^{-4})$$

$$\Delta p = 4.56 \times 10^{-32}\ \text{kg} \cdot \text{m/s}$$

The minimum uncertainty in position is then

$$\Delta x = \frac{h}{4\pi \Delta p} = \frac{6.63 \times 10^{-34}\ \text{J} \cdot \text{s}}{4\pi(4.56 \times 10^{-32}\ \text{kg} \cdot \text{m/s})} = 1.16\ \text{mm} \quad \Diamond$$

For the bullet,

$$\Delta p = m\Delta v = (0.0200\ \text{kg})(500\ \text{m/s})(1.00 \times 10^{-4}) = 1.00 \times 10^{-3}\ \text{kg} \cdot \text{m/s}$$

$$\Delta x = \frac{h}{4\pi \Delta p} = 5.28 \times 10^{-32}\ \text{m} \quad \Diamond$$

Quantum mechanics describes all objects, but the quantum fuzziness in position is unobservably small for the bullet and large for the small-mass electron.

31. A free electron has a wave function $\psi(x) = A\sin\left(5.00 \times 10^{10}\, x\right)$ where x is in meters. Find (a) the de Broglie wavelength, (b) the momentum, and (c) the energy in electron volts.

Solution

(a) The wave function $\psi(x) = A\ \sin(5.00 \times 10^{10}\, x)$ will go through one full cycle between $x_1 = 0$ and $(5.00 \times 10^{10})x_2 = 2\pi$. The wavelength is then

$$\lambda = x_2 - x_1 = \frac{2\pi}{5.00 \times 10^{10}\ \text{m}^{-1}} = 1.26 \times 10^{-10}\ \text{m} \quad \Diamond$$

To say the same thing, we can inspect $A\ \sin((5.00 \times 10^{10}\ \text{m}^{-1})x)$ to see that the wave number is $k = 5.00 \times 10^{10}\ \text{m}^{-1}$.

(b) Since $\lambda = \frac{h}{p}$, the momentum is

$$p = \frac{h}{\lambda} = \frac{6.63 \times 10^{-34}\ \text{J} \cdot \text{s}}{1.26 \times 10^{-10}\ \text{m}} = 5.28 \times 10^{-24}\ \text{kg} \cdot \text{m/s} \quad \Diamond$$

(c) The electron's kinetic energy is

$$K = \tfrac{1}{2}mv^2 = \frac{p^2}{2m}$$

Solving,

$$K = \frac{(5.28 \times 10^{-24}\ \text{kg} \cdot \text{m/s})^2}{2(9.11 \times 10^{-31}\ \text{kg})}\left(\frac{1\ \text{eV}}{1.60 \times 10^{-19}\ \text{J}}\right) = 95.5\ \text{eV} \quad \Diamond$$

Its relativistic total energy is 511 keV + 95.5 eV.

35. An electron is contained in a one-dimensional box of width 0.100 nm. (a) Draw an energy-level diagram for the electron for levels up to $n = 4$. (b) Find the wavelengths of all photons that can be emitted by the electron in making transitions that will eventually get it from the $n = 4$ state to the $n = 1$ state.

Solution (a) We can draw a diagram that parallels our treatment of standing mechanical waves. In each state, we measure the distance d from one node to another (N to N), and base our solution upon that:

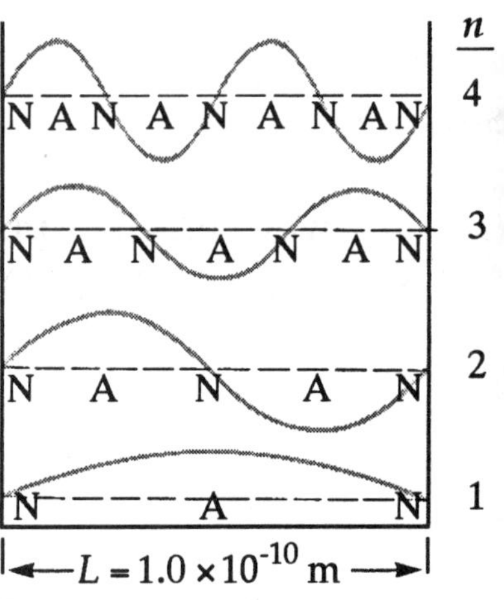

Since $d_{N\text{ to }N} = \dfrac{\lambda}{2}$ and $\lambda = \dfrac{h}{p}$, $p = \dfrac{h}{\lambda} = \dfrac{h}{2d}$

Next, $$K = \frac{p^2}{2m} = \frac{h^2}{8md^2} = \frac{1}{d^2}\left(\frac{\left(6.63\times10^{-34}\text{ J}\cdot\text{s}\right)^2}{8(9.11\times10^{-31}\text{ kg})}\right)$$

Evaluating, $$K = \frac{6.03\times10^{-38}\text{ J}\cdot\text{m}^2}{d^2} = \frac{3.77\times10^{-19}\text{ eV}\cdot\text{m}^2}{d^2}$$

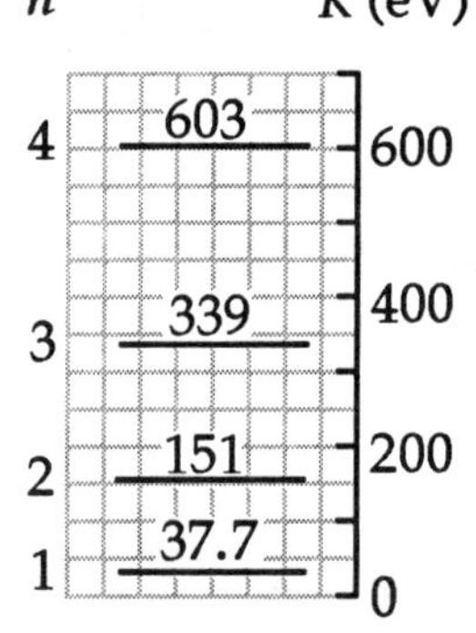

In state 1, $d = 1.00\times10^{-10}$ m $K_1 = 37.7$ eV

In state 2, $d = 5.00\times10^{-11}$ m $K_2 = 151$ eV

In state 3, $d = 3.33\times10^{-11}$ m $K_3 = 339$ eV

In state 4, $d = 2.50\times10^{-11}$ m $K_4 = 603$ eV

These energy levels are shown in the diagram to the right. ◊

(b) When the charged, massive electron inside the box makes a downward transition from one energy level to another, a chargeless, massless photon comes out of the box, carrying the difference in energy, ΔE. Its wavelength is

$$\lambda = \frac{c}{f} = \frac{hc}{\Delta E} = \frac{\left(6.63\times10^{-34}\text{ J}\cdot\text{s}\right)\left(3.00\times10^{8}\text{ m/s}\right)}{\Delta E\left(1.602\times10^{-19}\text{ J/eV}\right)} = \frac{1.24\times10^{-6}\text{ eV}\cdot\text{m}}{\Delta E}$$

Transition	$4\to3$	$4\to2$	$4\to1$	$3\to2$	$3\to1$	$2\to1$
ΔE (eV)	264	452	565	188	302	113
Wavelength (nm)	4.71	2.75	2.20	6.60	4.12	11.0

The wavelengths of light released for each transition are given in the table above. ◊

1. Show that the wave function $\psi = Ae^{i(kx-\omega t)}$ is a solution to the Schrödinger equation (Eq. 28.19), where $k = 2\pi/\lambda$ and $U = 0$.

Solution

From $\psi = A\,e^{i(kx-\omega t)}$, (1)

we evaluate $\dfrac{d\psi}{dx} = ikAe^{i(kx-\omega t)}$

and $\dfrac{d^2\psi}{dx^2} = -k^2Ae^{i(kx-\omega t)}$ (2)

We substitute Equations (1) and (2) into the Schrödinger equation, so that

$$\frac{d^2\psi}{dx^2} = -\frac{2m}{\hbar^2}(E-U)\psi \quad \text{(Eq. 28.19)}$$

becomes $-k^2Ae^{i(kx-\omega t)} = -\dfrac{2m}{\hbar^2}(K)Ae^{i(kx-\omega t)}$ (3)

The wave function $\psi = Ae^{i(kx-\omega t)}$ is a solution to the Schrödinger equation if Equation (3) is true.

Both sides depend on A, x, and t in the same way, so we can cancel several terms, and determine that we have a solution if:

$$k^2 = \frac{2m}{\hbar^2}K$$

But this is true for a nonrelativistic particle with mass, since

$$\frac{2m}{\hbar^2}K = \frac{2m}{(h/2\pi)^2}\left(\tfrac{1}{2}mv^2\right) = \frac{4\pi^2m^2v^2}{h^2} = \left(\frac{2\pi p}{h}\right)^2 = \left(\frac{2\pi}{\lambda}\right)^2 = k^2$$

Therefore, the given wave function does satisfy Equation 28.19. ◊

43. An electron with kinetic energy $E = 5.00$ eV is incident on a barrier with thickness $L = 0.200$ nm and height $U = 10.0$ eV (Fig. P28.43). What is the approximate probability that the electron (a) will tunnel through the barrier and (b) will be reflected?

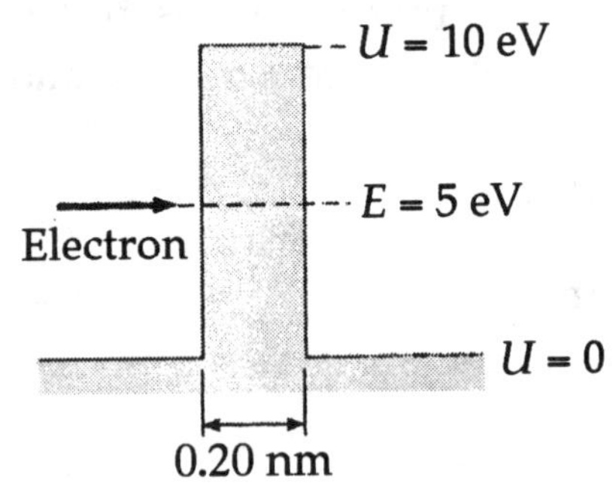

Figure P28.43

Solution

We apply Equation 28.24, $T \cong e^{-2KL}$

with transmission coefficient: $K = \dfrac{\sqrt{2m(U-E)}}{\hbar}$

Substituting, $K = \dfrac{\sqrt{2(9.11\times10^{-31}\text{ kg})(10.0\text{ eV} - 5.00\text{ eV})(1.60\times10^{-19}\text{ J/eV})}}{6.63\times10^{-34}\text{ J}\cdot\text{s}/2\pi}$

so that $K = 1.14\times10^{10}\text{ m}^{-1}$

(a) For transmission, $T \cong e^{-2(1.14\times10^{10}\text{ m}^{-1})(2.00\times10^{-10}\text{ m})} = e^{-4.58} = 0.0103$ ◊

(b) If the electron does not tunnel, it is reflected, with probability $1 - 0.010 = 0.990$ ◊

Related Commentary: A typical scanning - tunneling electron microscope (STM) can be built for less than \$5000, and can be tied directly to a home computer.

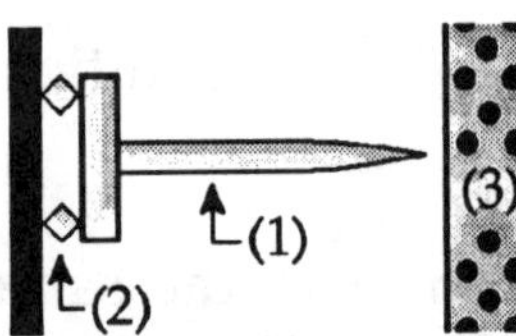

The STM essentially consists of a needle (1) that is mounted on a few piezo-electric crystals (2). When a voltage is induced across the piezo-electric crystals, the crystals change their shape, and the needle moves.

The STM charges the needle, and moves the tip of the needle to within a few angstroms of the test sample (3); the remaining gap provides the energy barrier that is required for tunneling. When the electron cloud of one of the needle's atoms is close to the electron cloud of one of the sample's atoms, a relatively large number of electrons will tunnel from one to the other, and generate a current. The current is then measured by a computer. As the needle moves across the surface of the sample, an image of the atoms is generated.

51. The table below shows data obtained in a photoelectric experiment. (a) Using these data, make a graph similar to Figure 28.8 that plots as a straight line. From the graph, determine (b) an experimental value for Planck's constant (in joule-seconds) and (c) the work function (in electron volts) for the surface. (Two significant figures for each answer are sufficient.)

Wavelength (nm)	Maximum Kinetic Energy of Photoelectrons (eV)
588	0.67
505	0.98
445	1.35
399	1.63

Solution Convert each wavelength to a frequency using the relation $\lambda f = c$, where c is the speed of light:

$$\lambda_1 = 588 \times 10^{-9}\text{ m} \qquad f_1 = 5.10 \times 10^{14}\text{ Hz}$$

$$\lambda_2 = 505 \times 10^{-9}\text{ m} \qquad f_2 = 5.94 \times 10^{14}\text{ Hz}$$

$$\lambda_3 = 445 \times 10^{-9}\text{ m} \qquad f_3 = 6.74 \times 10^{14}\text{ Hz}$$

$$\lambda_4 = 399 \times 10^{-9}\text{ m} \qquad f_4 = 7.52 \times 10^{14}\text{ Hz}$$

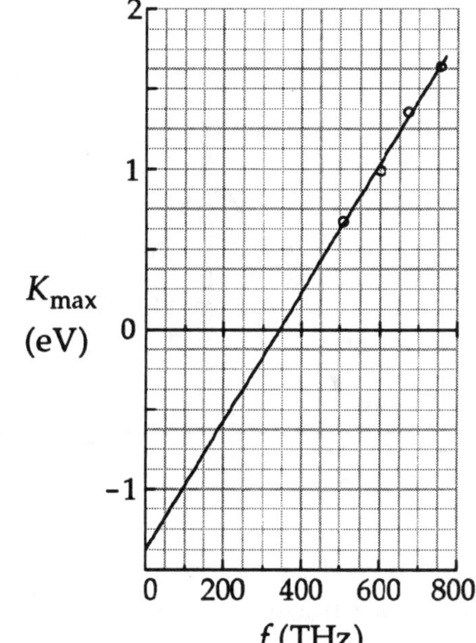

(a) Plot each point on a energy vs. frequency graph, as shown at the right. Extend a line through the set of 4 points, as far as the -y intercept.

(b) Our basic equation is $K_{max} = hf - \phi$. Therefore, Planck's constant should be equal to the slope of the K-f graph, which can be found from a least-squares fit or from reading the graph as:

$$h_{exp} = \frac{\text{Rise}}{\text{Run}} = \frac{1.25\text{ eV} - 0.25\text{ eV}}{6.5 \times 10^{14}\text{ Hz} - 4.0 \times 10^{14}\text{ Hz}} = 4.0 \times 10^{-15}\text{ eV} \cdot \text{s} = 6.4 \times 10^{-34}\text{ J} \cdot \text{s} \quad \Diamond$$

From the scatter on the graph, we estimate the uncertainty as 3%.

(c) From the same equation, the work function for the surface is the negative of the y-intercept of the graph. You can extend the graph until the line intercepts the y axis, and then check your accuracy with a calculation similar to the one below:

$$\phi = -\left(K_{max} - h_{exp} f\right) = -\left(0.67\text{ eV} - \left(4.0 \times 10^{-15}\text{ eV} \cdot \text{s}\right)\left(5.10 \times 10^{14}\text{ s}^{-1}\right)\right) = 1.4\text{ eV} \quad \Diamond$$

55. An electron is represented by the time-independent wave function

$$\psi(x) = \begin{cases} Ae^{-\alpha x} & \text{for } x > 0 \\ Ae^{+\alpha x} & \text{for } x < 0 \end{cases}$$

(a) Sketch the wave function as a function of x. (b) Sketch the probability that the electron is found between x and $x + dx$. (c) Why could this be a physically reasonable wave function? (d) Normalize the wave function. (e) Determine the probability of finding the electron somewhere in the range

$$x_1 = -\frac{1}{2\alpha} \quad \text{to} \quad x_2 = \frac{1}{2\alpha}$$

Solution

(a)

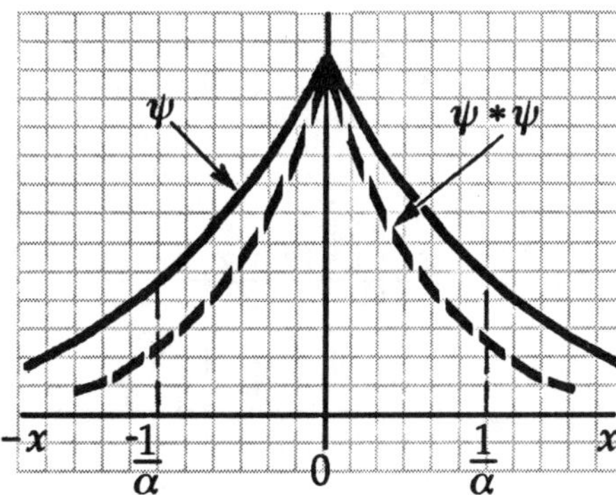

(b) $P(x \to x + dx) = |\psi|^2\, dx = A^2 e^{-2\alpha x}\, dx$,

for $x > 0$

$P(x \to x + dx) = |\psi|^2\, dx = A^2 e^{+2\alpha x}\, dx$,

for $x < 0$ ◊

(c) ψ is continuous; $\psi \to 0$ as $x \to \infty$; it mimics an electron bound at $x = 0$. ◊

(d) Since ψ is symmetric, $\int_{-\infty}^{\infty} \psi^* \psi\, dx = 2\int_0^{\infty} \psi^* \psi\, dx = 1$

or $2A^2 \int_0^{\infty} e^{-2\alpha x}\, dx = 1$ or $\left(\frac{2A^2}{-2\alpha}\right)\left(e^{-\infty} - e^0\right) = 1$

This gives $A = \sqrt{\alpha}$ ◊

(e) $P_{(-1/2\alpha)\to(1/2\alpha)} = 2\int_{x=0}^{1/2\alpha} (\sqrt{\alpha})^2 e^{-2\alpha x}\, dx = \left(\frac{2\alpha}{-2\alpha}\right)\left(e^{-2\alpha/2\alpha} - 1\right) = \left(1 - e^{-1}\right) = 0.632$ ◊

Chapter 29

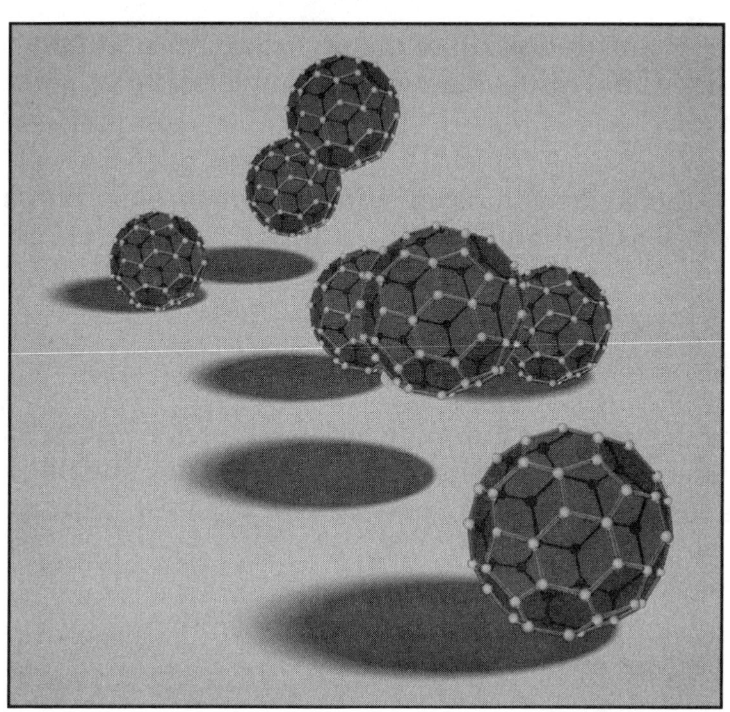

This is an artist's rendition of "bucky-balls," named for R. Buckminster Fuller, inventor of the geodesic dome. This new form of carbon, C_{60}, has many intriguing and useful properties: mechanical — as super-strong microscopic balls, electrical — due to its superconductivity, and chemical — as an inert "prison" for other atoms and ions.

ATOMIC PHYSICS

Chapter 29

ATOMIC PHYSICS

INTRODUCTION

A large portion of this chapter is an application of quantum mechanics to the study of the hydrogen atom. Understanding the hydrogen atom, the simplest atomic system, is especially important for several reasons:

- Much of what is learned about the hydrogen atom with its single electron can be extended to such single-electron ions as He^+ and Li^{2+}.
- The hydrogen atom is an ideal system for performing precise tests of theory against experiment and for improving our overall understanding of atomic structure.
- The quantum numbers used to characterize the allowed states of hydrogen can be used to describe the allowed states of more complex atoms. This enables us to understand the periodic table of the elements, one of the greatest triumphs of quantum mechanics.
- The basic ideas about electrons in atoms provide background for studying nuclei.

NOTES FROM SELECTED CHAPTER SECTIONS

29.1 Early Models of the Atom

A need for modification of the Bohr theory became apparent when improved spectroscopic techniques were used to examine the spectral lines of hydrogen. It was found that many of the lines in the Balmer and other series were not single lines at all. Instead, each was a group of lines spaced very close together. An additional difficulty arose when it was observed that, in some situations, certain single spectral lines were split into three closely spaced lines when the atoms were placed in a strong magnetic field.

29.2 The Hydrogen Atom Revisited

In the three-dimensional problem of the hydrogen atom, three quantum numbers are required for each stationary state, corresponding to the three independent degrees of freedom for the electron.

The three quantum numbers which emerge from the theory are represented by the symbols n, ℓ, and m_ℓ. The quantum number n is called the **principal quantum number**; ℓ is called the **orbital quantum number**; and m_ℓ is called the **orbital magnetic quantum number.**

There are certain important relationships between these quantum numbers, as well as certain restrictions on their values. These restrictions are:

The three quantum numbers are integers.
The values of n can range from 1 to ∞.
The values of ℓ can range from 0 to $n - 1$.
The values of m_ℓ can range from $-\ell$ to ℓ.

For historical reasons, **all states with the same principal quantum number are said to form a shell.** These shells are identified by the letters K, L, M, . . . , which designate the states for which $n = 1, 2, 3, \ldots$. Likewise, **the states having the same values of n and ℓ are said to form a subshell.** The letters $s, p, d, f, g, h, \ldots$ are used to designate the states for which $\ell = 0, 1, 2, 3, \ldots$.

29.3 The Spin Magnetic Quantum Number

The **spin magnetic quantum number** m_s accounts for the two closely spaced energy states corresponding to the two possible orientations of the electron spin.

29.4 The Wave Functions for Hydrogen

Since the wave function of the electron in the hydrogen atom depends only on the radial distance r, the wave functions that describe the s states are **spherically symmetric.**

29.5 The "Other" Quantum Numbers

The allowed energy levels for the electron in a hydrogen atom are determined by the value of the **principal quantum number,** n. There are three "other" quantum numbers which also serve to characterize the wave function of the hydrogen electron. The **orbital quantum number,** ℓ, restricts the orbital angular momentum to discrete values. The **magnetic orbital quantum number,** m_ℓ, specifies the allowed directions of the angular momentum with respect to an external magnetic field. The **spin magnetic quantum number,** m_s, restricts the spin angular momentum of the hydrogen electron to have only two possible orientations in space. The spin contribution to the magnetic moment is twice the contribution of the orbital motion.

29.6 The Exclusion Principle and the Periodic Table

The **Pauli exclusion principle** states that no two electrons can exist in identical quantum states. This means that no two electrons in a given atom can be characterized by the same set of quantum numbers at the same time.

Hund's rule states that when an atom has orbitals of equal energy, the order in which they are filled by electrons is such that a maximum number of electrons will have unpaired spins.

29.7 Atomic Spectra: Visible and X-Ray

An atom will emit electromagnetic radiation if an electron in an excited state makes a transition to a lower energy state. The set of wavelengths observed for a species by such processes is called an **emission spectrum**. Likewise, atoms with electrons in the ground-state configuration can also absorb electromagnetic radiation at specific wavelengths, giving rise to an **absorption spectrum.** Such spectra can be used to identify the elements in gases.

Since the orbital angular momentum of an atom changes when a photon is emitted or absorbed (that is, as a result of a transition) and since angular momentum must be conserved, we conclude that **the photon involved in the process must carry angular momentum.**

X-rays are emitted by atoms when an electron undergoes a transition from an outer shell into an electron vacancy in one of the inner shells. Transitions into a vacant state in the K shell give rise to the K series of spectral lines; in a similar way, transitions into a vacant state in the L shell create the L series of lines, and so on. The x-ray spectrum of a metal target consists of a set of sharp characteristic lines superimposed on a broad, continuous spectrum.

29.8 Atomic Transitions

Stimulated absorption occurs when light with photons of an energy which matches the energy separation between two atomic energy levels is absorbed by an atom.

An atom in an excited state has a certain probability of returning to its original energy state. This process is called **spontaneous emission**.

When a photon with an energy equal to the excitation energy of an excited atom is incident on the atom, it can increase the probability of de-excitation. This is called **stimulated emission**, and results in a second photon of energy equal to that of the incident photon.

29.9 Lasers and Holography

The following three conditions must be satisfied in order to achieve laser action:

- The system must be in a state of population inversion (that is, more atoms in an excited state than in a ground state).
- The excited state of the system must be a **metastable state**, which means its lifetime must be long compared with the usually short lifetimes of excited states. When such is the case, stimulated emission will likely occur before spontaneous emission.
- The emitted photons must be confined in the system long enough to allow them to stimulate further emission from other excited atoms. This is achieved by the use of reflecting mirrors at the ends of the system. One end is made totally reflecting, and the other is slightly transparent to allow the laser beam to escape.

EQUATIONS AND CONCEPTS

The **potential energy** of the hydrogen atom depends on the distance of the electron from the nucleus.

$$U(r) = -\frac{k_e e^2}{r} \quad (29.1)$$

When numerical values of the constants are used, the energy level values for hydrogen can be expressed in units of electron volts (eV).

$$E_n = -\frac{13.6}{n^2}\ \text{eV} \quad (29.2)$$

The lowest energy state or ground state corresponds to the principal quantum number $n = 1$. The energy level approaches $E = 0$ as n approaches infinity. This is the **ionization energy** for the atom.

The simplest wave function for hydrogen is the one which describes the 1*s* state and depends only on the radial distance *r*. The parameter a_0 is the Bohr radius.

$$\psi_{1s}(r) = \frac{1}{\sqrt{\pi a_o^3}} e^{-r/a_o} \quad (29.3)$$

$$a_o = \frac{\hbar^2}{m_e k_e e^2} = 0.0529\ \text{nm} \quad (11.21)$$

The radial probability density for the 1*s* state of hydrogen is defined as the probability of finding the electron in a spherical shell of radius *r* and thickness *dr*.

$$P_{1s}(r) = \left(\frac{4r^2}{a_o^3}\right) e^{-2r/a_o} \quad (29.7)$$

The value of the quantum number ℓ determines the magnitude of the electron's **orbital angular momentum,** L.

$$L = \sqrt{\ell(\ell+1)}\hbar \quad (29.10)$$

$$(\ell = 0, 1, 2, 3, \ldots, n-1)$$

When an atom is placed in an external magnetic field, the projection (or component) of the orbital angular momentum L_z along the direction of the magnetic field is quantized.

$$L_z = m_\ell \hbar \quad (29.11)$$

$$(m_\ell = -\ell, -\ell+1, -\ell+2, \ldots, \ell)$$

If an orbiting electron is placed in a weak magnetic field directed along the z axis, the projection of the angular momentum vector of the electron (**L**) along the z axis can have only discrete values.

$$\cos\theta = \frac{m_\ell}{\sqrt{\ell(\ell+1)}} \quad (29.12)$$

In addition to orbital angular momentum, the electron has an intrinsic angular momentum or **spin angular momentum** S, as if due to spinning on its axis. This spin angular momentum is described by a single quantum number s whose value can only be $\frac{1}{2}$.

$$S = \sqrt{s(s+1)}\hbar \quad (29.13)$$

$$S = \frac{\sqrt{3}}{2}\hbar$$

The spin angular momentum is space quantized with respect to the direction of an external magnetic field (along the z direction). The spin magnetic quantum number m_s can have values of $\pm\frac{1}{2}$.

$$S_z = m_s\hbar = \pm\frac{1}{2}\hbar \quad (29.14)$$

The **spin magnetic moment** μ_s is related to the spin angular momentum.

$$\mu_s = \left(-\frac{e}{m_e}\right)\mathbf{S} \quad (29.15)$$

The shielding effect of the nuclear charge by inner-core electrons must be taken into account when calculating the allowed energy levels of multiple-electron atoms. The atomic number is replaced by an effective atomic number, Z_{eff}, which depends on the values of n and ℓ.

$$E_n = -\frac{13.6 {Z_{eff}}^2}{n^2} \text{eV} \qquad (29.19)$$

Z_{eff} is the atomic number, modified by the number of shielding electrons: for K-shell electrons, $Z_{eff} = Z - 1$; For an M shell electron, $Z_{eff} = Z - 9$.

$$Z_{eff} = Z - n_s$$

where

n_s = the number of shielding electrons

SUGGESTIONS, SKILLS, AND STRATEGIES

Before reading this chapter, it may be helpful to review Section 5 of Chapter 11. This section deals with the Bohr model of the atom, and makes it easier to understand the modern concept of the atomic structure.

After reading the chapter in your text, review the significance of each of the quantum numbers that is used to describe the various electronic states of electrons in an atom. Next, review the set of allowed values for each of the quantum numbers.

In addition to the principal quantum number n (which can range from 1 to ∞), other quantum numbers are necessary to specify completely the possible energy levels in the hydrogen atom and also in more complex atoms.

All energy states with the same principal quantum number, n, form a shell. These shells are identified by the spectroscopic notation K, L, M, . . . corresponding to $n = 1, 2, 3, \ldots$.

The orbital quantum number ℓ [which can range from 0 to $(n - 1)$], determines the allowed value of orbital angular momentum. All energy states having the same values of n and ℓ form a subshell. The letter designations $s, p, d, f, \ldots$ correspond to values of $\ell = 0, 1, 2, 3, \ldots$.

The magnetic orbital quantum number m_ℓ (which can range from $-\ell$ to ℓ) determines the possible orientations of the electron's orbital angular momentum vector in the presence of an external magnetic field.

The spin magnetic quantum number m_s, can have only two values, $m_s = -\frac{1}{2}$ and $m_s = \frac{1}{2}$, which in turn correspond to the two possible directions of the electron's intrinsic spin.

Chapter 29

REVIEW CHECKLIST

▷ Understand the significance of the wave function and the associated radial probability density for the ground state of hydrogen.

▷ For each of the quantum numbers, n (the principal quantum number), ℓ (the orbital quantum number), m_ℓ (the orbital magnetic quantum number), and m_s (the spin magnetic quantum number):

- qualitatively describe what each implies concerning atomic structure;
- state the allowed values which may be assigned to each, and the number of allowed states which may exist in a particular atom corresponding to each quantum number.

▷ Associate the customary shell and subshell spectroscopic notations with allowed combinations of quantum numbers n and ℓ. Calculate the possible values of the orbital angular momentum, L, corresponding to a given value of the principal quantum number.

▷ Describe how allowed values of the magnetic orbital quantum number, m_ℓ, may lead to a restriction on the orientation of the orbital angular momentum vector in an external magnetic field. Find the allowed values for L_z (the component of the angular momentum along the direction of an external magnetic field) for a given value of L.

▷ State the Pauli exclusion principle and describe its relevance to the periodic table of the elements. Show how the exclusion principle leads to the known electronic ground state configuration of the light elements.

ANSWERS TO SELECTED CONCEPTUAL QUESTIONS

2. Must an atom first be ionized before it can emit light?

Answer When an electron in an atom shifts to an lower state, it emits light. When the electron absorbs light, it shifts to a higher state. The electron must first absorb energy and shift to an excited state ($n > 1$), before it can emit light by shifting to a lower state, but the atom need not be ionized ($n = \infty$).

□ □ □ □

5. When a hologram is produced, the system (including light source, object, beam splitter, and so on) must be held motionless within a quarter of a wavelength. Why?

Answer The hologram is an interference pattern between light scattered from the object and the reference beam. If anything moves by a distance comparable to the wavelength of the light (or more), the pattern will wash out. The effect is just like making the slits vibrate in Young's experiment.

□ □ □ □

6. Why is an inhomogeneous magnetic field used in the Stern-Gerlach experiment?

Answer The gradient of a magnetic field is equal to the rate of change in the magnetic field with respect to the distance along its direction of most rapid increase. Since the deflecting force on a neutral atom with a magnetic moment is proportional to the gradient of the magnetic field, atoms having oppositely directed magnetic dipole moments in a **nonuniform** field therefore experience net forces in opposite directions. This causes them to be displaced in opposite directions as the atoms pass through the magnetic field. In a **uniform** field, the gradient of the magnetic field is zero, so the deflecting force is zero.

□ □ □ □

. Could the Stern-Gerlach experiment be performed with ions rather than neutral atoms? Explain.

Answer Practically speaking, the answer would be no. Since ions have a net charge, the magnetic force $q\mathbf{v} \times \mathbf{B}$ would deflect the beam, making it very difficult to separate the ions with different orientations of their magnetic moments.

□ □ □ □

8. Describe some experiments that support the conclusion that the spin quantum number for electrons can only have the values $\pm 1/2$.

Answer The Stern-Gerlach experiment using hydrogen atoms (one unpaired electron) verifies that the electron spin equals 1/2. Electron spin resonance experiments on atoms or ions with one unpaired electron also support this conclusion.

□ □ □ □

9. Discuss some of the consequences of the exclusion principle.

Answer If the Pauli exclusion principle were not valid, the elements and their chemical behavior would be grossly different because every electron would end up in the lowest energy level of the atom. All matter would therefore be nearly alike in its chemistry and composition, since the shell structures of each element would be identical. Most materials would have a much higher density, and the spectra of atoms and molecules would be very simple, resulting in the existence of less color in the world.

□ □ □ □

10. Why do lithium, potassium, and sodium exhibit similar chemical properties?

Answer The three elements have similar electronic configurations, with filled inner shells, plus a single electron in an *s* orbital. Since atoms typically interact through their unfilled outer shells, and the outer shell of each of these atoms is similar, the chemical interactions of the three atoms is also similar.

□ □ □ □

11. From Table 29.4, we find that the ionization energies for Li, Na, K, Rb, and Cs ar 5.390, 5.138, 4.339, 4.176, and 3.893 eV, respectively. Explain why these values are to b expected in terms of the atomic structures.

Answer These elements all have their valence electron in an *s* state. The outermos electron is relatively loosely bound, so the ionization energies of these metals are low compared to other atoms. Comparing these elements with one another, we may say tha a higher atomic number implies a slightly larger atom; a larger average distance betwee the outermost electron and the center of the atom implies weaker attraction and a lowe ionization energy.

□ □ □ □

14. It is easy to understand how two electrons (one spin up, one spin down) can fill the 1*s* shell for a helium atom. How is it possible that eight more electrons can fit into the $2s$, $2p$ level to complete the $1s^2 2s^2 2p^6$ shell for a neon atom?

Answer Each of the eight electrons must have just one quantum number different from each of the others. They can differ (in m_s) by being spin-up or spin-down. They can differ (in ℓ) in angular momentum and in the general shape of the wave function (look at the $2s$ and $2p$ graphs in Figure 29.8). Those electrons with $\ell = 1$ can differ (in m_ℓ) in orientation of angular momentum, as shown by Figure 29.9.

□ □ □ □

17. The efficiencies of most solid-state lasers are on the order of 1% to 2%. Although the laser output is monochromatic and highly directional, can you use Figures 29.22 and 29.23 to determine why the energy input must exceed laser energy output by a factor of 50 to 100? Explain your answer.

Answer Only energy in a narrow spectral band from the flash lamp will cause transitions from E_1 to E_3. Additionally, not all the light emitted by the flash lamp is absorbed in the ruby rod.

□ □ □ □

SOLUTIONS TO SELECTED END-OF-CHAPTER PROBLEMS

. A general expression for the energy levels of one-electron atoms and ions is

$$E_n = -\left(\frac{\mu k_e^2 q_1^2 q_2^2}{2\hbar^2}\right)\frac{1}{n^2}$$

where k_e is the Coulomb constant, q_1 and q_2 are the charges of the two particles, and μ is the reduced mass, given by $\mu = m_1 m_2 / (m_1 + m_2)$. In Problem 2, we found that the wavelength for the $n = 3$ to $n = 2$ transition of the hydrogen atom is 656.3 nm (visible red light). What are the wavelengths for this same transition in (a) positronium, which consists of an electron and a positron, and (b) singly ionized helium? (**Note**: A positron is a positively charged electron.)

Solution For hydrogen, $\mu = \dfrac{m_{\text{proton}} m_e}{m_{\text{proton}} + m_e} \cong m_e$

The photon energy is $E_3 - E_2$

Its wavelength is experimentally found to be $\lambda = \dfrac{c}{f} = \dfrac{hc}{E_3 - E_2} = 656.3 \text{ nm},$

(a) For positronium, $\mu = \dfrac{m_e m_e}{m_e + m_e} = \dfrac{m_e}{2},$

so the energy of each level is one half as large as in hydrogen, which we could call "protonium." The photon energy is inversely proportional to its wavelength, so

$$\lambda_{32} = 2(656 \text{ nm}) = 1312 \text{ nm} \quad \text{(in the infrared region)} \quad \Diamond$$

(b) For He^+, $\mu \approx m_e$, $q_1 = e$, and $q_2 = 2e$, so each energy is $2^2 = 4$ times larger than hydrogen. Then,

$$\lambda_{32} = \left(\frac{656}{4}\right) \text{nm} = 164 \text{ nm} \quad \text{(in the ultraviolet region)} \quad \Diamond$$

7. The wave function for an electron in the 2p state of hydrogen is

$$\psi_{2p} = \frac{1}{\sqrt{3}(2a_o)^{3/2}} \frac{r}{a_o} e^{-r/2a_o}$$

What is the most likely distance from the nucleus to find an electron in the 2p state? (See Fig. 29.8.)

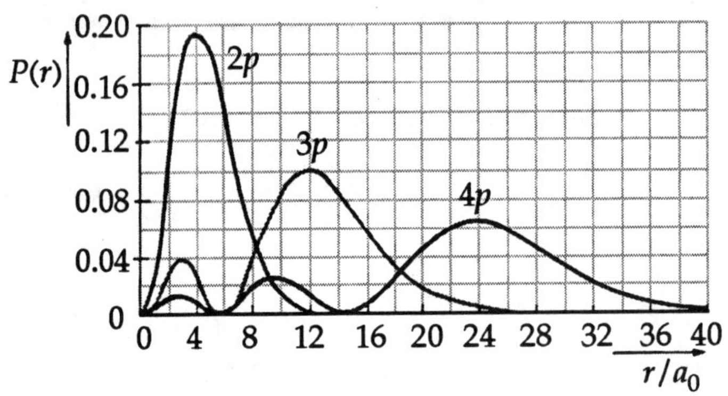

Figure 29.8

Solution

The radial probability density function is given by

$$P = 4\pi r^2 |\psi|^2 = 4\pi r^2 \left(\frac{r^2}{24a_o^5}\right) e^{-r/a_o}$$

The most likely position for the electron corresponds to the point at which $\frac{dP}{dr} = 0$ (where P reaches its **maximum** value).

$$\frac{dP}{dr} = \frac{4\pi}{24a_o^5}\left(4r^3 e^{-r/a_o} + r^4\left(-\frac{1}{a_o}\right)e^{-r/a_o}\right) = 0$$

Eliminating the constants, $$r^3\left(4 + r\left(-\frac{1}{a_o}\right)\right)e^{-r/a_o} = 0$$

If $r = 0$, the left-hand term goes to zero. If $r = \infty$, the exponential term goes to zero. However, in both these cases, the figure shows that P is a **minimum**. The maximum value occurs when the middle term is zero:

$$4 + r\left(-\frac{1}{a_o}\right) = 0 \quad \text{so that} \quad r = 4a_o \quad \Diamond$$

Show that the 1*s* wave function for an electron in hydrogen,

$$\psi(r) = \frac{1}{\sqrt{\pi a_o^3}} e^{-r/a_o}$$

atisfies the radially symmetric Schrödinger equation, $-\frac{\hbar^2}{2m}\left(\frac{d^2\psi}{dr^2} + \frac{2}{r}\frac{d\psi}{dr}\right) - \frac{k_e e^2}{r}\psi = E\psi.$

olution

'o solve this problem, we substitute the wave function and its derivatives into the chrödinger equation, and then start simplifying the resulting equation. If we find that he resulting equation is true, then we know that the Schrödinger equation is satisfied.

$$\frac{d\psi}{dr} = \frac{1}{\sqrt{\pi a_o^3}} \frac{d}{dr}\left(e^{-r/a_o}\right) = -\frac{1}{\sqrt{\pi a_o^3}}\left(\frac{1}{a_o}\right)e^{-r/a_o} = -\frac{\psi}{a_o} \qquad (1)$$

ikewise,

$$\frac{d^2\psi}{dr^2} = -\frac{1}{\sqrt{\pi a_o^5}} \frac{d}{dr} e^{-r/a_o} = \frac{1}{\sqrt{\pi a_o^7}} e^{-r/a_o} = \frac{1}{a_o^2}\psi \qquad (2)$$

Substituting (1) and (2) into the Schrödinger equation, and noting that *m* is the electron's mass,

$$-\frac{\hbar^2}{2m_e}\left(\frac{1}{a_o^2} - \frac{2}{a_o r}\right)\psi - \frac{k_e e^2}{r}\psi = E\psi \qquad (3)$$

Substituting $\hbar^2 = m_e k_e e^2 a_o$ (Eq. 11.21), and canceling m_e and ψ,

$$-\frac{k_e e^2 a_o}{2}\left(\frac{1}{a_o^2} - \frac{2}{a_o r}\right) - \frac{k_e e^2}{r} = E$$

We find that the second and third terms add to zero, leaving the equation that was given for the ground state energy of hydrogen:

$$E = -\frac{k_e e^2}{2a_o}$$

This is true, so the Schrödinger equation is satisfied. ◊

---

15. How many sets of quantum numbers are possible for an electron for which (a) n = 1, (b) $n = 2$, (c) $n = 3$, (d) $n = 4$, and (e) $n = 5$? Check your results to show that they agree with the general rule that the number of sets of quantum numbers is equal to $2n^2$.

Solution (a) For $n = 1$, $\ell = 0$, $m_\ell = 0$, $m_s = \pm\frac{1}{2}$:

n	ℓ	m_ℓ	m_s
1	0	0	$-\frac{1}{2}$
1	0	0	$+\frac{1}{2}$

This yields $2n^2 = 1(1)^2 = 2$ sets ◊

(b) For $n = 2$, we have

n	ℓ	m_ℓ	m_s
2	0	0	$\pm\frac{1}{2}$
2	1	−1	$\pm\frac{1}{2}$
2	1	0	$\pm\frac{1}{2}$
2	1	1	$\pm\frac{1}{2}$

This yields $2n^2 = 2(2)^2 = 8$ sets ◊

Note that the number is twice the number of m_ℓ values. Also, for each ℓ there are $(2\ell + 1)$ different m_ℓ values. Finally, ℓ can take on values ranging from 0 to $n - 1$. So the general expression is

$$s = \sum_{0}^{n-1} 2(2\ell+1)$$

The series is an arithmetic progression 2 + 6 + 10 + 14, the sum of which is

$$s = \frac{n}{2}[2a + (n-1)d] \quad \text{where } a = 2,\ d = 4$$

$$s = \frac{n}{2}[4 + (n-1)4] = 2n^2$$

(c) $n = 3$: $2(1) + 2(3) + 2(5) = 2 + 6 + 10 = 18$ $\qquad 2n^2 = 2(3)^2 = 18$ ◊

(d) $n = 4$: $2(1) + 2(3) + 2(5) + 2(7) = 32$ $\qquad 2n^2 = 2(4)^2 = 32$ ◊

(e) $n = 5$: $32 + 2(9) = 32 + 18 = 50$ $\qquad 2n^2 = 2(5)^2 = 50$ ◊

3. (a) Scanning through Table 29.4 in order of increasing atomic number, note that the lectrons fill the subshells in such a way that those subshells with the lowest values of $n + \ell$ are filled first. If two subshells have the same value of $n + \ell$, the one with the ower value of n is filled first. Using these two rules, write the order in which the ubshells are filled through $n + \ell = 7$. (b) Predict the chemical valence for the elements hat have atomic numbers 15, 47, and 86, and compare your predictions with the actual alences.

Solution

(a)

$n + \ell$	1	2	3	4	5	6	7
subshell	1*s*	2*s*	2*p*, 3*s*	3*p*, 4*s*	3*d*, 4*p*, 5*s*	4*d*, 5*p*, 6*s*	4*f*, 5*d*, 6*p*, 7*s*

(b) Z = 15: Filled subshells: 1*s*, 2*s*, 2*p*, 3*s* (12 electrons)

Valence subshell: 3 electrons in 3*p* subshell

Prediction: Valence +3 or –5

Element is phosphorus: Valence +3 or –5

Z = 47: Filled subshells: 1*s*, 2*s*, 2*p*, 3*s*, 3*p*, 4*s*, 3*d*, 4*p*, 5*s* (38 electrons)

Outer subshell: 9 electrons in 4*d* subshell

Prediction: Valence –1

Element is silver, Valence +1 (Prediction fails)

Z = 86: Filled shells: 1*s*, 2*s*, 2*p*, 3*s*, 3*p*, 4*s*, 3*d*, 4*p*, 5*s*, 4*d*, 5*p*, 6*s*, 4*f*, 5*d*, 6*p*

Outer subshell: Full

Prediction: Inert gas

Element is Radon, Inert gas (Prediction works)

27. Use the method illustrated in Example 29.7 to calculate the wavelength of the x-ray emitted from a molybdenum target ($Z = 42$) when an electron moves from the L shell ($n = 2$) to the K shell ($n = 1$).

Solution Following Example 29.7, suppose the electron is originally in the L shell with just one other electron in the K shell between it and the nucleus, so it moves in a field of effective charge $(42 - 1)e$. Its energy is then $E_L = -(42 - 1)^2\, 13.6 \text{ eV}/4$. In its final state we estimate the screened charge holding it in orbit as again $(42 - 1)e$, so its energy is $E_K = -(42 - 1)^2\, 13.6$ eV. The photon energy emitted is the difference.

$$E_\gamma = \tfrac{3}{4}(42 - 1)^2(13.6 \text{ eV}) = 1.71 \times 10^4 \text{ eV} = 2.74 \times 10^{-15} \text{ J}$$

Then $f = E/h = 4.14 \times 10^{18}$ Hz and $\lambda = c/f = 0.725$ Å ◊

31. A ruby laser delivers a 10.0-ns pulse of 1.00 MW average power. If the photons have a wavelength of 694.3 nm, how many are contained in the pulse?

Solution The pulse's energy is $E = (1.00 \times 10^6 \text{ W})(1.00 \times 10^{-8} \text{ s}) = 1.00 \times 10^{-2} \text{ J}$

The energy of each photon is $E_\gamma = hf = \dfrac{hc}{\lambda} = \dfrac{(6.626 \times 10^{-34})(3.00 \times 10^8)}{694.3 \times 10^{-9}} \text{ J} = 2.86 \times 10^{-19} \text{ J}$

So $N = \dfrac{E}{E_\gamma} = \dfrac{1.00 \times 10^{-2}}{2.86 \times 10^{-19}} = 3.49 \times 10^{16}$ photons ◊

33. Show that the average value of r for the 1s state of hydrogen has the value $3a_0/2$. (**Hint:** Use Eq. 29.7.)

Solution The average value (expectation value) of r is $\langle r \rangle = \int_0^\infty r P_{1s}(r)\,dr$

where $P_{1s}(r) = \left(4r^2/a_0^3\right)e^{-2r/a_0}$: $\langle r \rangle = \dfrac{4}{a_0^3}\int_0^\infty r^3 e^{-2r/a_0}\,dr$

Letting $x = 2r/a_0$, we find: $\langle r \rangle = \tfrac{1}{4}a_0 \int_0^\infty x^3 e^{-x}\,dx$

Integrating by parts gives $\langle r \rangle = \tfrac{3}{2}a_0$ ◊

5. If a muon (a negatively charged particle having a mass 206 times the electron's mass) is captured by a lead nucleus, $Z = 82$, the resulting system behaves like a one-electron atom. (a) What is the "Bohr radius" for a muon captured by a lead nucleus? (**Hint:** Follow the treatment in Section 11.5) (b) Using Equation 29.2 with e replaced by Ze, calculate the ground state energy of a muon captured by a lead nucleus. (c) What is the transition energy for a muon descending from the $n = 2$ to the $n = 1$ level in a muonic lead atom?

Solution (a) We expand the treatment of the Bohr atom in Section 11.5 to a particle of charge $-e$ and mass m orbiting a much more massive nucleus of charge Ze. Angular momentum is quantized according to $mvr = n\hbar$.

$n = 1, 2, 3, \ldots$ and $\Sigma F = ma$ means $$\frac{k_e Ze^2}{r^2} = \frac{mv^2}{r}$$

We eliminate v with $v = \dfrac{n\hbar}{mr}$: $$\frac{k_e Ze^2}{r^2} = \frac{mn^2\hbar^2}{m^2 r^3}$$

Thus, $r = \dfrac{n^2\hbar^2}{mk_e Ze^2}$, and the generalized Bohr radius is $$\frac{\hbar^2}{mk_e Ze^2}$$

With $m = 206\ m_e$ and $Z = 82$, the radius for muonic lead is $a_\mu = \dfrac{a_0}{206 \times 82} = 3.13\ \text{fm}$ ◊

(b) The energy is

$$\tfrac{1}{2}mv^2 - \frac{k_e Ze^2}{r} = \frac{\frac{1}{2}k_e Ze^2}{r} - \frac{k_e Ze^2}{r} = -\frac{k_e Ze^2}{2r} = -\frac{k_e Ze^2}{2\left(\dfrac{n^2\hbar^2}{mk_e Ze^2}\right)} = -\frac{mk_e^2 Z^2 e^4}{2\hbar^2 n^2} = \frac{k_e Ze^2}{2a_\mu n^2}$$

For the ground-state energy, we take $n = 1$:

$$E_{\mu 1} = -\frac{mk_e^2 Z^2 e^4}{2\hbar^2} = -206 \times 82^2 \times 13.6\ \text{eV} = -18.8\ \text{MeV} \quad ◊$$

(c) $E_{\mu 2} = \dfrac{-18.8}{4} = -4.7\ \text{MeV}$

$\Delta E_{\mu 2 \to \mu 1} = 14.1\ \text{MeV}$ ◊

---

41. (a) Find the most probable position for an electron in the 2*s* state of hydrogen. (Hin Let $x = r/a_0$, find an equation for x, and show that $x = 5.236$ is a solution to this equation. (b) Show that the wave function given by Equation 29.8 is normalized.

Solution We use Equation 29.8: $\psi_{2s}(r) = \frac{1}{4}\left(2\pi a_0^{\ 3}\right)^{-1/2}\left(2-\frac{r}{a_0}\right)e^{-r/2a_0}$

By Equation 29.6, the radial probability distribution function is

$$P(r) = 4\pi r^2\psi^2 = \frac{1}{8}\left(\frac{r^2}{a_0^{\ 3}}\right)\left(2-\frac{r}{a_0}\right)^2 e^{-r/a_0}$$

Its extremes are given by

(a) $$\frac{dP(r)}{dr} = \frac{1}{8}\left[\frac{2r}{a_0^{\ 3}}\left(2-\frac{r}{a_0}\right)^2 - \frac{2r^2}{a_0^{\ 3}}\left(\frac{1}{a_0}\right)\left(2-\frac{r}{a_0}\right) - \frac{r^2}{a_0^{\ 3}}\left(2-\frac{r}{a_0}\right)^2\left(\frac{1}{a_0}\right)\right]e^{-r/a_0} = 0$$

We do not need the hint if we factor in the following manner:

$$\frac{1}{8}\left(\frac{r}{a_0^{\ 3}}\right)\left(2-\frac{r}{a_0}\right)\left[2\left(2-\frac{r}{a_0}\right) - \frac{2r}{a_0} - \frac{r}{a_0}\left(2-\frac{r}{a_0}\right)\right]e^{-r/a_0} = 0$$

Therefore, $[\ldots\ldots] = 4 - \frac{6r}{a_0} + \left(\frac{r}{a_0}\right)^2 = 0$ which has solutions $r = \left(3 \pm \sqrt{5}\right)a_0$

$\left[\text{The roots of } \frac{dP}{dr} = 0 \text{ at } r = 0,\ r = 2a_0, \text{ and } r = \infty \text{ are minima } (P(r) = 0).\right]$

We substitute the two roots into $P(r)$:

When $r = \left(3-\sqrt{5}\right)a_0 = 0.764a_0$, then $P(r) = \frac{0.0519}{a_0}$

When $r = \left(3+\sqrt{5}\right)a_0 = 5.24a_0$, then $P(r) = \frac{0.191}{a_0}$

Therefore, the most probable value of r is $\left(3+\sqrt{5}\right)a_0 = 5.24a_0$ ◊

(b) $\int_0^\infty P(r)dr = \int_0^\infty \frac{1}{8}\left(\frac{r^2}{a_o^3}\right)\left(2-\frac{r}{a_o}\right)^2 e^{-r/a_o}\,dr$ (Eq. 29.8)

Let $u = \frac{r}{a_o}$, and $dr = a_o\,du$, so that $\int_0^\infty P(r)dr = \int_0^\infty \frac{1}{8}u^2(4-4u+u^2)e^{-u}\,du$

$$\int_0^\infty P(r)dr = \int_0^\infty \tfrac{1}{8}(u^4-4u^3+4u^2)e^{-u}\,du$$

Using a table of integrals, or integrating by parts repeatedly,

$$\int_0^\infty P(r)dr = -\tfrac{1}{8}(u^4+4u^2+8u+8)e^{-u}\Big]_0^\infty = 1 \text{ as desired } \Diamond$$

---

45. For hydrogen in the 1s state, what is the probability of finding the electron farther than $2.50\,a_o$ from the nucleus?

Solution The radial probability distribution function is $P(r) = 4\pi r^2|\psi|^2$

With $\psi_{1s} = \left(\pi a_o^3\right)^{-1/2} e^{-r/a_o}$, it is $P(r) = 4r^2 a_o^{-3} e^{-2r/a_o}$

The required probability is then $P = \int_{2.5a_o}^\infty P(r)dr = \int_{2.5a_o}^\infty \frac{4r^2}{a_o^3} e^{-2r/a_o}\,dr$

Let $z = \frac{2r}{a_o}$, $dz = \frac{2dr}{a_o}$: $P = \frac{1}{2}\int_5^\infty z^2 e^{-z}\,dz$

$$P = -\tfrac{1}{2}(z^2+2z+2)e^{-z}\Big|_5^\infty$$

$$P = -\tfrac{1}{2}[0] + \tfrac{1}{2}(25+10+2)e^{-5} = \left(\frac{37}{2}\right)(0.00674)$$

$$P = 0.125 \quad \Diamond$$

---

47. According to classical physics, an accelerated charge e radiates at a rate

$$\frac{dE}{dt} = -\frac{1}{6\pi \epsilon_0} \frac{e^2 a^2}{c^3}$$

(a) Show that an electron in a classical hydrogen atom (see Fig. 29.3) spirals into the nucleus at a rate

$$\frac{dr}{dt} = -\frac{e^4}{12\pi^2 \epsilon_0^2 r^2 m_e^2 c^3}$$

(b) Find the time it takes the electron to reach $r = 0$, starting from $r_0 = 2.00 \times 10^{-10}$ m.

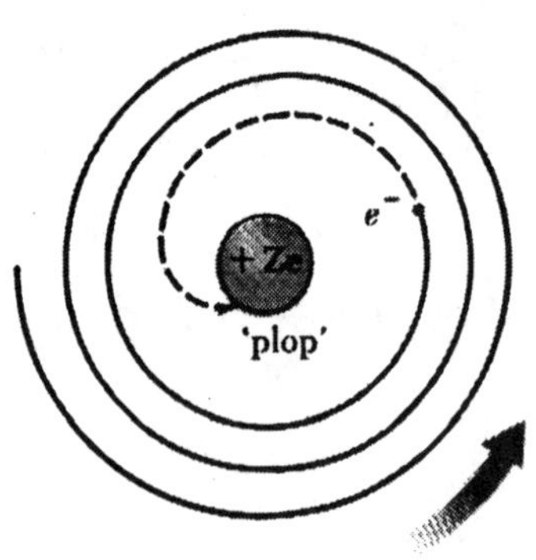

Figure 29.3

Solution According to a classical model, the electron moving in a circular orbit about the proton in the hydrogen atom experiences a force $k_e e^2/r^2$; and from Newton's second law, $F = ma$, its acceleration is $k_e e^2/m_e r^2$.

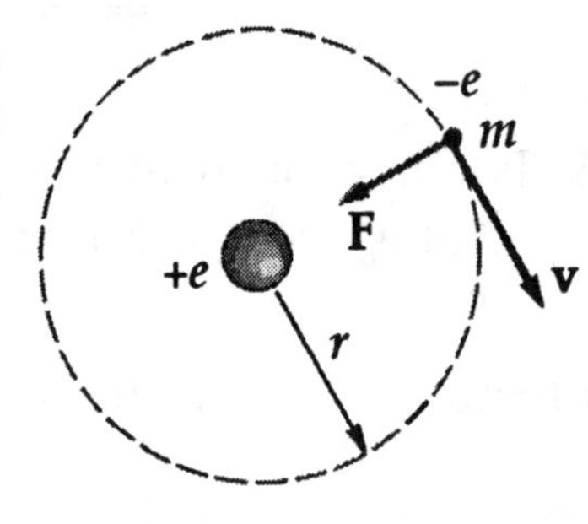

(a) Using the fact that the Coulomb constant $k_e = \dfrac{1}{4\pi \epsilon_0}$,

$$a = \frac{v^2}{r} = \frac{k_e e^2}{m_e r^2} = \frac{e^2}{4\pi \epsilon_0 m_e r^2} \quad (1)$$

From the Bohr model of the atom (Chapter 11), we can write the total energy of the atom as

$$E = -\frac{k_e e^2}{2r} = -\frac{e^2}{8\pi \epsilon_0 r} \quad \text{so that} \quad \frac{dE}{dt} = \frac{e^2}{8\pi \epsilon_0 r^2} \frac{dr}{dt} = -\frac{1}{6\pi \epsilon_0} \frac{e^2 a^2}{c^3} \quad (2)$$

Substituting (1) into (2) for a, solving for $\dfrac{dr}{dt}$, and simplifying gives

$$\frac{dr}{dt} = -\frac{4r^2}{3c^3}\left(\frac{e^2}{4\pi \epsilon_0 m_e r^2}\right)^2 = -\frac{e^4}{12\pi^2 \epsilon_0^2 r^2 m_e^2 c^3} \quad \Diamond$$

b) We can express $\frac{dr}{dt}$ in the simpler form: $$\frac{dr}{dt} = -\frac{A}{r^2} = -\frac{3.15 \times 10^{-21}}{r^2}$$

Thus, $$-\int_{2\times10^{-10}\text{ m}}^{0} r^2 dr = 3.15 \times 10^{-21} \int_0^{T} dt$$

and $$T = \left(3.17 \times 10^{20}\right) \frac{r^3}{3}\Bigg]_0^{2\times10^{-10}\text{ m}} = 8.46 \times 10^{-10}\text{ s} = 0.846\text{ ns} \quad \Diamond$$

We know that atoms 'last' much longer than 0.8 ns; thus, classical physics does not hold (fortunately) for atomic systems.

---

49. Light from a certain He-Ne laser has a power output of 1.00 mW and a cross-sectional area of 10.0 mm^2. The entire beam is incident on a metal target that requires 1.50 eV to remove an electron from its surface. (a) Perform a classical calculation to determine how long it takes one atom in the metal to absorb 1.50 eV from the incident beam. (**Hint:** Assume the surface area of an atom is 1.00 × 10^{-20} m^2, and first calculate the energy incident on each atom per second.) (b) Compare the (wrong) answer obtained in part (a) to the actual response time for photoelectric emission $\left(\sim 10^{-9}\text{ s}\right)$, and discuss the reasons for the large discrepancy.

Solution (a) The intensity of the beam is $$I = \frac{P}{A} = \frac{1.00 \times 10^{-3}\text{ W}}{10.0 \times 10^{-6}\text{ m}^2} = 100\text{ W/m}^2$$

The power incident on our one atom is

$$P_a = IA = \left(100\text{ W/m}^2\right)\left(1.00 \times 10^{-20}\text{ m}^2\right) = 1.00 \times 10^{-18}\text{ W}$$

To absorb 1.50 eV = 2.40 × 10^{-19} J, the atom must be in the beam for time

$$t = \frac{E}{P} = \frac{2.40 \times 10^{-19}\text{ J}}{1.00 \times 10^{-18}\text{ J/s}} = 0.240\text{ s} \quad \Diamond$$

(b) The classical answer is too large by a factor of about a billion because the beam does not carry smeared-out energy but energy in photon lumps. The atoms do not all have to wait together for energy to accumulate, like school teachers for retirement. In the first nanosecond of absorption, some atom will get hit with a photon and release an electron. The photocurrent begins right away.

Chapter 30

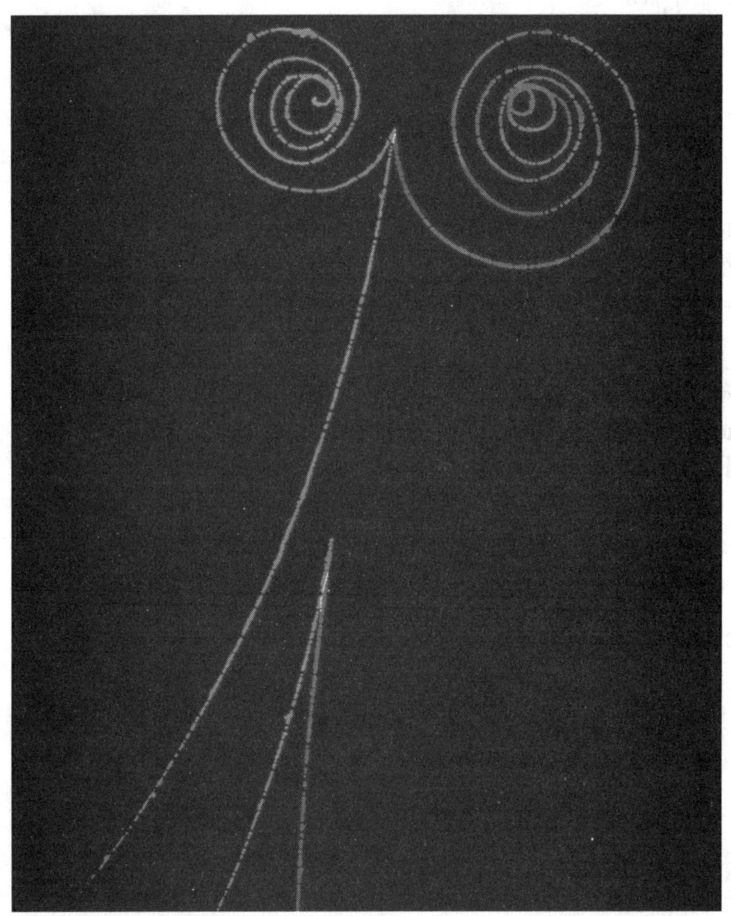

This is a bubble chamber photograph of electron (green) and positron (red) tracks produced by energetic gamma rays. The highly curved tracks at the top are due to an electron-positron pair that bend in opposite directions in the magnetic field. *(Lawrence Berkeley Laboratory / Science Photo Library, Photo Researchers, Inc.)*

NUCLEAR PHYSICS

Chapter 30

NUCLEAR PHYSICS

INTRODUCTION

In this chapter we discuss the properties and structure of the atomic nucleus. We start by describing the basic properties of nuclei, and then discuss nuclear forces and binding energy, nuclear models, and the phenomenon of radioactivity. We also discuss nuclear reactions and the various processes by which nuclei decay.

NOTES FROM SELECTED CHAPTER SECTIONS

30.1 Some Properties of Nuclei

Important quantities in the description of nuclear properties are:

- The **atomic number**, Z, which equals the number of protons in the nucleus.
- The **neutron number**, N, which equals the number of neutrons in the nucleus.
- The **mass number**, A, which equals the number of nucleons (neutrons plus protons) in the nucleus.

The nuclei of all atoms of a particular element contain the same number of protons but often contain different numbers of neutrons. Nuclei that are related in this way are called isotopes. The isotopes of an element have the same Z value but different N and A values.

The **atomic mass unit,** u, is defined such that the mass of one atom of the isotope ^{12}C is exactly 12 u.

Experiments have shown that most nuclei are approximately spherical and all **have nearly the same density**. The stability of nuclei is due to the **nuclear force.** This is a **short range, attractive** force which acts between all nuclear particles. Nuclei have **intrinsic angular momentum** which is quantized by the **nuclear spin quantum number** which may be integer or half integer.

The **magnetic moment of the nucleus** is measured in terms of the **nuclea magneton.** When placed in an external magnetic field, nuclear magneti moments precess with a frequency called the **Larmor precessional frequency.**

30.2 Binding Energy

The total mass of a nucleus is always less than the sum of the masses of it individual nucleons. The **binding energy** of the nucleus is its mass difference multiplied by c^2.

30.3 Radioactivity

There are three processes by which a radioactive substance can undergo decay:

- alpha (α) decay, where the emitted particles are ^{4}He nuclei,
- beta (β) decay, in which the emitted particles are either electrons or positrons,
- gamma (γ) decay, in which the emitted rays are high-energy photons.

A positron is a particle similar to the electron in all respects except that it has a charge of $+e$, and is the antimatter twin of the electron. In beta decay, the electron is denoted by a e^-, while the positron is designated with a e^+.

The three types of radiation have quite different penetrating powers. Alpha particles barely penetrate a sheet of paper, beta particles can penetrate a few millimeters of aluminum, and gamma rays can penetrate several centimeters of lead.

30.4 The Decay Processes

Alpha decay can occur because, according to quantum mechanics, some nuclei have energy barriers that can be penetrated by the alpha particles (the tunneling process). Beta decay is energetically more favorable for those nuclei having a large excess of neutrons. A nucleus can undergo beta decay in two ways. It can emit either an electron (e^-) and an antineutrino ($\bar{\nu}$), or a positron (e^+) and a neutrino (ν). In the electron-capture process, the nucleus of an atom absorbs one of its own electrons (usually from the K shell) and emits a neutrino.

The neutrino has the following properties:

- It has zero electric charge.
- It has a mass smaller than that of the electron, and in fact its mass may be zero (although recent experiments suggest that this may not be true).
- It has a spin of 1/2, which satisfies the law of conservation of angular momentum.
- It interacts very weakly with matter and is therefore very difficult to detect.

In gamma decay, a nucleus in an excited state decays to its ground state and emits a gamma ray.

The disintegration energy is the energy released as a result of the decay process.

30.5 Natural Radioactivity

Radioactive nuclei are generally classified into two groups: (1) unstable nuclei found in nature, which give rise to what is called **natural radioactivity**, and (2) nuclei produced in the laboratory through nuclear reactions, which exhibit **artificial radioactivity.**

30.6 Nuclear Reactions

Nuclear reactions are events in which collisions change the identity or properties of nuclei. The total energy released as a result of a nuclear reaction is called the **reaction energy,** Q_r.

An **endothermic reaction** is one in which Q_r is negative, and the minimum energy for which the reaction will occur is called the **threshold energy.**

EQUATIONS AND CONCEPTS

Most nuclei are approximately spherical in shape and have an average radius which is proportional to the cube root of the mass number, or total number of nucleons. This means that the volume is proportional to A and that all nuclei have nearly the same density.

$$r = r_0 A^{1/3} \tag{30.1}$$

$$r_0 = 1.2 \text{ fm}$$

The nucleus has an angular momentum and a corresponding nuclear magnetic moment associated with it. The nuclear magnetic moment is measured in terms of a unit of moment called the nuclear magneton μ_n.

$$\mu_n \equiv \frac{e\hbar}{2m_p} = 5.05 \times 10^{-27} \text{ J/T} \tag{30.3}$$

The binding energy of any nucleus can be calculated in terms of the mass of a neutral hydrogen atom, the mass of a neutron, and the atomic mass of the associated compound nucleus.

$$E_b = \left(ZM(H) + Nm_n - M\left({}^{A}_{Z}\text{X}\right)\right) \times 931.5 \frac{\text{MeV}}{\text{u}} \tag{30.4}$$

$$(E_b \text{ in MeV})$$

The number of radioactive nuclei in a given sample which undergoes decay during a time interval Δt depends on the number of nuclei present. The number of decays depends also on the decay constant λ which is characteristic of a particular isotope.

$$\frac{dN}{dt} = -\lambda N \tag{30.5}$$

The number of nuclei in a radioactive sample decreases exponentially with time. The plot of number of nuclei N versus elapsed time t is called a decay curve.

$$N = N_o e^{-\lambda t} \tag{30.6}$$

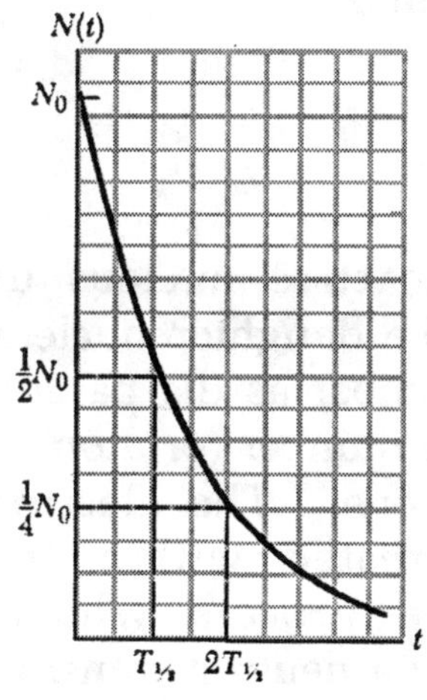

The decay rate R or activity of a sample of radioactive nuclei is defined as the number of decays per second.

$$R = \left|\frac{dN}{dt}\right| = R_o e^{-\lambda t} \tag{30.7}$$

The half-life $T_{1/2}$ is the time required for half of a given number of radioactive nuclei to decay.

$$T_{1/2} = \frac{0.693}{\lambda} \tag{30.8}$$

When a nucleus decays by **alpha emission**, the parent nucleus loses two neutrons and two protons. In order for alpha emission to occur, the mass of the parent nucleus must be greater than the combined mass of the daughter nucleus and the emitted alpha particle. The mass difference is converted into energy and appears as kinetic energy shared (unequally) by the alpha particle and the daughter nucleus.

$${}^{A}_{Z}\text{X} \rightarrow {}^{A-4}_{Z-2}\text{Y} + {}^{4}_{2}\text{He} \tag{30.9}$$

$${}^{238}_{92}\text{U} \rightarrow {}^{234}_{90}\text{Th} + {}^{4}_{2}\text{He} \tag{30.10}$$

The disintegration energy Q can be calculated in MeV when the masses are expressed in u.

$$Q = \left(M_X - M_Y - M_{^4\text{He}}\right)(931.5\text{ MeV/u}) \quad (30.1\text{:}$$

When a radioactive nucleus undergoes **beta decay**, the daughter nucleus has the same mass number as the parent nucleus, but the charge number (or atomic number) increases by one. The electron that is emitted is created within the parent nucleus by a process which can be represented by a neutron transformed into a proton and an electron. The total energy released in beta decay is greater than the combined kinetic energies of the electron and the daughter nucleus. This difference in energy is associated with a third particle called a neutrino.

$$^A_Z\text{X} \rightarrow {}^{A}_{Z+1}\text{Y} + \text{e}^- + \overline{\nu} \quad (30.14)$$

$$\text{n} \rightarrow \text{p} + \text{e}^- + \overline{\nu} \quad (30.18)$$

Nuclei which undergo alpha or beta decay are often left in an excited energy state. The nucleus returns to the ground state by emission of one or more photons. Gamma-ray emission results in no change in mass number or atomic number.

$$^A_Z\text{X}^* \rightarrow {}^A_Z\text{X} + \gamma \quad (30.20)$$

Nuclear reactions can occur when target nuclei are bombarded with energetic particles. In these reactions the structure, identity, or properties of the target nuclei are changed.

$$\text{a} + \text{X} \rightarrow \text{Y} + \text{b} \quad (30.23)$$

The quantity of energy required to balance the equation representing a nuclear reaction (e.g. Eq. 30.23) is called the Q_r value of the reaction. The Q_r value can be calculated in terms of the total mass of the reactants minus the total mass of the products. Q_r is positive in the case of exothermic reactions and negative for endothermic reactions.

$$Q_r = (M_a + M_X - M_Y - M_b)c^2 \qquad (30.24)$$

TABLE 30.2 Various Decay Pathways

Alpha Decay	${}^{A}_{Z}X \rightarrow {}^{A-4}_{Z-2}X + {}^{4}_{2}He$
Beta Decay (e^-)	${}^{A}_{Z}X \rightarrow {}_{Z+1}^{A}X + e^- + \bar{\nu}$
Beta Decay (e^+)	${}^{A}_{Z}X \rightarrow {}_{Z-1}^{A}X + e^+ + \nu$
Electron Capture	${}^{A}_{Z}X + {}^{0}_{-1}e \rightarrow {}_{Z-1}^{A}X + \nu$
Gamma Decay	${}^{A}_{Z}X^* \rightarrow {}^{A}_{Z}X + \gamma$

SUGGESTIONS, SKILLS, AND STRATEGIES

The rest energy of a particle is given by $E = mc^2$. It is therefore often convenient to express the unified mass unit in terms of its equivalent energy, 1 u = 1.660559×10^{-27} kg or 1 u = 931.5 MeV/c^2. When masses are expressed in units of u, energy values are then $E = m(931.5 \text{ MeV/u})$.

Equation 30.6 can be solved for the particular time t after which the number of remaining nuclei will be some specified fraction of the original number N_0. This can be done by taking the natural log of each side of Equation 30.6 to find

$$t = \frac{1}{\lambda}\ln\left(\frac{N_0}{N}\right)$$

REVIEW CHECKLIST

▷ Use the appropriate nomenclature in describing the static properties of nuclei.

▷ Discuss nuclear stability in terms of the strong nuclear force and a plot of N vs. Z.

▷ Account for nuclear binding energy in terms of the Einstein mass-energy relationship. Describe the basis for energy released by fission and fusion in terms of the shape of the curve of binding energy per nucleon vs. mass number.

▷ Identify each of the components of radiation that are emitted by the nucleus through natural radioactive decay and describe the basic properties of each. Write out typical equations to illustrate the processes of transmutation by alpha and beta decay and explain why the neutrino must be considered in the analysis of beta decay.

▷ State and apply to the solution of related problems, the formula which expresses decay rate as a function of the decay constant and the number of radioactive nuclei. Describe the process of carbon dating as a means of determining the age of ancient objects.

▷ Calculate the Q_r value of given nuclear reactions and determine the threshold energy of endothermic reactions.

ANSWERS TO SELECTED CONCEPTUAL QUESTIONS

3. Why do heavier elements require more neutrons in order to maintain stability?

Answer The protons, although held together by the nuclear force, are repulsed by the electrostatic force. However, if enough protons were placed together in a nucleus, the electrostatic force would overcome the nuclear force, which is based on the number of particles, and cause the nucleus to fission.

The addition of neutrons prevents such fission. The neutron does not increase the electrostatic force, being electrically neutral, but does contribute to the nuclear force.

□ □ □ □

5. Why do nearly all the naturally occurring isotopes lie above the $N = Z$ line in Figure 30.3?

Answer Extra neutrons are required to overcome the increasing electrostatic repulsion of the protons.

□ □ □ □

7. What fraction of a radioactive sample has decayed after two half-lives have elapsed?

Answer After the first half-life, half the original sample remains. After the second half-life $\left(\frac{1}{2}\right)\left(\frac{1}{2}\right) = \frac{1}{4}$ of the original sample remains, and three quarters of a radioactive sample has decayed after two half-lives.

Alternatively, we could calculate the remaining quantity using Equation 30.6:

$$N = N_o e^{-\lambda t} = N_o e^{-\lambda(2T_{1/2})} = N_o e^{-\lambda(2)(0.693/\lambda)}$$

$$N = N_o e^{-1.39} = N_o(0.25)$$

□ □ □ □

8. Two samples of the same radioactive nuclide are prepared. Sample A has twice the initial activity of sample B. How does the half-life of A compare with the half-life of B? After each has passed through five half-lives, what is the ratio of their activities?

Answer Since the two samples are of the same radioactive nuclide, they have the same half-life; the 2:1 difference in activity is due to a 2:1 difference in the mass of each sample. After 5 half lives, each will have decreased in mass by a power of $2^5 = 32$. However, since this simply means that the mass of each is 32 times smaller, the ratio of the masses will still be (2/32):(1/32), or 2:1. Therefore, the ratio of their activities will **always** be 2:1.

□ □ □ □

12. If a nucleus such as ^{226}Ra initially at rest undergoes alpha decay, which has more kinetic energy after the decay, the alpha particle or the daughter nucleus?

Answer The alpha particle and the daughter nucleus carry equal and opposite momentum. Since kinetic energy can be written as $p^2/2m$, the less massive alpha particle has much more of the decay energy than the recoiling nucleus.

□ □ □ □

16. Suppose it could be shown that the cosmic ray intensity at the Earth's surface was much greater 10 000 years ago. How would this difference affect what we accept as valid carbon-dated values of the age of ancient samples of once-living matter?

Answer If the cosmic ray intensity at the Earth's surface was much greater 10 000 years ago, a greater fraction of the Earth's carbon dioxide would contain the heavy nuclide ^{14}C at that time, than now. Thus, there would initially be a greater fraction of ^{14}C in the organic artifacts, and we would believe that the artifact was more recent than it actually was.

For example, suppose that the actual ratio of atmospheric ^{14}C to ^{12}C, two half-lives (11460 years) ago was 2.6×10^{-12}. The current ratio of isotopes would be

$$\left(2.6\times10^{-12}\right)\left(\tfrac{1}{2}\right)\left(\tfrac{1}{2}\right)=0.65\times10^{-12}$$

We, believing the initial ratio to be 1.3×10^{-12}, would see the same current ratio, but would think that the artifact had died only one half-life (5730 years) ago.

$$\left(1.3\times10^{-12}\right)\left(\tfrac{1}{2}\right)=0.65\times10^{-12}$$

□ □ □ □

SOLUTIONS TO SELECTED END-OF-CHAPTER PROBLEMS

7. Certain stars at the end of their lives are thought to collapse, combining their protons and electrons to form a neutron star. Such a star could be thought of as a gigantic atomic nucleus. If a star of mass equal to that of the Sun ($M = 1.99 \times 10^{30}$ kg) collapsed into neutrons ($m_n = 1.67 \times 10^{-27}$ kg), what would the radius be? (**Hint:** $r = r_0 A^{1/3}$.)

Solution The number of nucleons in a star of one solar mass is

$$A = \frac{1.99 \times 10^{30} \text{ kg}}{1.67 \times 10^{-27} \text{ kg}} = 1.20 \times 10^{57}$$

Therefore, $r = r_o A^{1/3} = (1.2 \times 10^{-15} \text{ m})\sqrt[3]{1.20 \times 10^{57}} = 12.7 \text{ km}$ ◊

15. The isotope $^{139}_{57}$La is stable. A radioactive isobar (see Problem 14) of this lanthanum isotope, $^{139}_{59}$Pr, is located below the line of stable nuclei in Figure 30.3 and decays by e^+ emission. Another radioactive isobar of ^{139}La, $^{139}_{55}$Cs decays by e^- emission and is located above the line of stable nuclei in Figure 30.3. (a) Which of these three isobars has the highest neutron-to-proton ratio? (b) Which has the greatest binding energy per nucleon? (c) Which do you expect to be heavier, ^{139}Pr or ^{139}Cs?

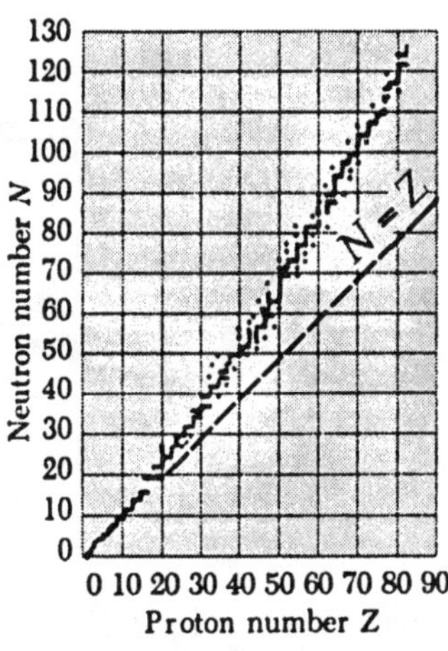

Figure 30.3

Solution

(a) For $^{139}_{59}$Pr the neutron number is 139 – 59 = 80. For $^{139}_{55}$Cs the neutron number is 84, so this cesium isotope has the greatest neutron-to-proton ratio. ◊

(b) Binding energy per nucleon measures stability so it is greatest for the stable nucleus, the lanthanum isotope. Note also that it has a magic number of neutrons, 82. ◊

(c) Cs–139 has 55 protons and 84 neutrons. Pr–139 has 59 protons and 80 neutrons. When we plot both of them onto Figure 30.3, we see that Cesium is a little further away from the center of the zone of stable nuclei. Being less stable goes with being able to lose more energy in decay, and with having more mass. We therefore expect Cesium to be heavier. ◊

17. A sample of radioactive material contains 1.00×10^{15} atoms and has an activity o. 6.00×10^{11} Bq. What is its half-life?

Solution $$\frac{dN}{dt} = -\lambda N$$

$$\lambda = \frac{1}{N}\left(-\frac{dN}{dt}\right) = \left(1.00 \times 10^{15} \text{ atoms}\right)^{-1}\left(6.00 \times 10^{11} \text{ s}^{-1}\right) = 6.00 \times 10^{-4} \text{ s}^{-1}$$

$$T_{1/2} = \frac{(\ln 2)}{\lambda} = 1160 \text{ s} \quad \Diamond$$

(This is also 19.3 minutes)

---

21. A freshly prepared pure sample of a certain radioactive isotope has an activity of 10.0 mCi. After 4.00 h, its activity is 8.00 mCi. (a) Find the decay constant and half-life. (b) How many atoms of the isotope were contained in the freshly prepared sample? (c) What is the sample's activity 30.0 h after it is prepared?

Solution

From the rate equation, $R = R_0 e^{-\lambda t}$, we can solve for λ to get

(a) $$\lambda = \frac{1}{t}\ln\left(\frac{R_0}{R}\right) = \left(\frac{1}{4.00 \text{ h}}\right)\ln\left(\frac{10.0 \text{ mCi}}{8.00 \text{ mCi}}\right) = 5.58 \times 10^{-2} \text{ h}^{-1} = 1.55 \times 10^{-5} \text{ s}^{-1} \quad \Diamond$$

$$T_{1/2} = \frac{\ln 2}{\lambda} = 12.4 \text{ h} \quad \Diamond$$

(b) $R_0 = 10.0 \text{ mCi} = (10.0 \times 10^{-3})(3.70 \times 10^{10} \text{ decays/s}) = 3.70 \times 10^{8} \text{ decays/s}$

Since $R_0 = \lambda N_0$, we find $$N_0 = \frac{R_0}{\lambda} = \frac{3.70 \times 10^{8} \text{ decays/s}}{1.55 \times 10^{-5} \text{ s}^{-1}} = 2.39 \times 10^{13} \text{ atoms} \quad \Diamond$$

(c) $$R = R_0 e^{-\lambda t} = (10.0 \text{ mCi})e^{-(5.58\times10^{-2} \text{ h}^{-1})(30.0 \text{ h})} = 1.88 \text{ mCi} \quad \Diamond$$

---

27. Enter the correct isotope symbol in each open square in Figure P30.27, which shows the sequences of decays starting with uranium-235 and ending with the stable isotope lead-207.

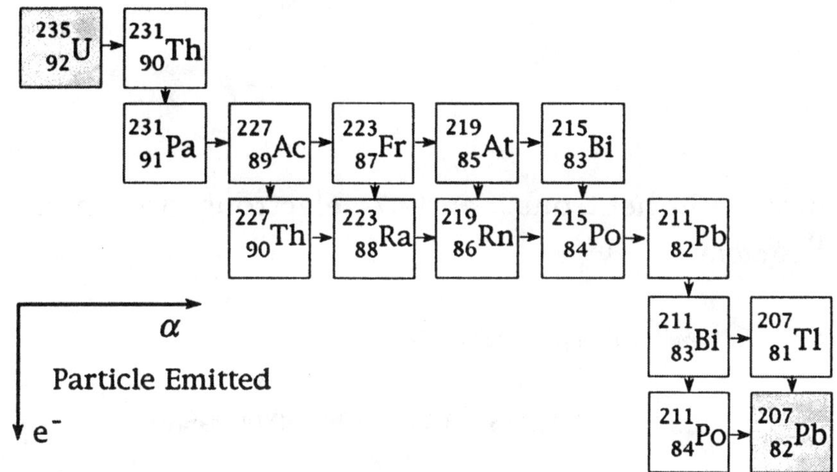

Figure P30.27 (modified)

Solution Whenever an $\alpha = {}^4_2He$ is emitted, Z drops by 2 and A by 4. Whenever a $e^- = {}^{\ 0}_{-1}e^-$ is emitted, Z increases by 1 and A is unchanged. We find the chemical name by looking up Z in a periodic table. The values in the shaded boxes (^{235}U and ^{207}Pb) were given; all others have been filled in as part of the solution.

---

29. Find the energy released in the alpha decay $\quad {}^{238}_{92}U \to {}^{234}_{90}Th + {}^4_2He$

You will find the following mass values useful:

$$M\left({}^{238}_{92}U\right) = 238.050\ 786\ \text{u}$$

$$M\left({}^{234}_{90}Th\right) = 234.043\ 583\ \text{u}$$

$$M\left({}^4_2He\right) = 4.002\ 603\ \text{u}$$

Solution

$$Q = (M_U - M_{Th} - M_{He})(931.5\ \text{MeV/u})$$

$$Q = (238.050786 - 234.043583 - 4.002603)(931.5) = 4.28\ \text{MeV} \quad \lozenge$$

---

31. The nucleus ${}^{15}_{8}\text{O}$ decays by electron capture. Write (a) the basic nuclear process and (b) the decay process referring to neutral atoms. (c) Determine the energy of the neutrino. Disregard the daughter's recoil.

Solution

(a) $e^- + \text{p} \rightarrow \text{n} + \nu$ ◊

(b) Add 7 protons, 7 neutrons, and 7 electrons to each side to give
$^{15}\text{O atom} \rightarrow {}^{15}\text{N atom} + \nu$ ◊

(c) From Table A.3, $m(^{15}\text{O}) = m(^{15}\text{N}) + Q/c^2$

$$\Delta m = 15.003065 - 15.000109 = 0.002956 \text{ u}$$

$$Q = (931.5 \text{ MeV/u})(0.002956 \text{ u}) = 2.75 \text{ MeV} \quad ◊$$

---

35. Natural gold has only one isotope, ${}^{197}_{79}\text{Au}$. If natural gold is irradiated by a flux of slow neutrons, electrons are emitted. (a) Write the appropriate reaction equation. (b) Calculate the maximum energy of the emitted beta particles. The mass of ${}^{198}_{80}\text{Hg}$ is 197.96675 u.

Solution The ${}^{197}_{79}\text{Au}$ will absorb a neutron to become ${}^{198}_{79}\text{Au}$, which emits an e^- to become ${}^{198}_{80}\text{Hg}$.

(a) For nuclei, the reaction is: ${}^{197}_{79}\text{Au nucleus} + {}^{1}_{0}\text{n} \rightarrow {}^{198}_{80}\text{Hg nucleus} + {}^{0}_{-1}\text{e}^- + \bar{\nu}$ ◊

Adding 79 electrons to both sides: ${}^{197}_{79}\text{Au atom} + {}^{1}_{0}\text{n} \rightarrow {}^{198}_{80}\text{Hg atom} + \bar{\nu}$

(b) From Table A.3, $196.96656 + 1.008665 = 197.96675 + 0 + Q/c^2$

$$Q = \Delta mc^2 = (0.008475 \text{ u})(931.5 \text{ MeV/u})$$

$$Q = 7.89 \text{ MeV} \quad ◊$$

---

9. A by-product of some fission reactors is the isotope $^{239}_{94}\text{Pu}$, an alpha emitter having a half-life of 24 000 years:

$$^{239}_{94}\text{Pu} \rightarrow {}^{235}_{92}\text{U} + \alpha$$

Consider a sample of 1.00 kg of pure $^{239}_{94}\text{Pu}$ at t = 0. Calculate (a) the number of $^{239}_{94}\text{Pu}$ nuclei present at t = 0 and (b) the initial activity in the sample. (c) How long does the sample have to be stored if a "safe" activity level is 0.100 Bq?

Solution

The decay constant λ appears in every equation about radioactivity, so it is convenient to compute it first:

$$\lambda = \frac{0.693}{T_{1/2}} = \frac{0.693}{(2.40\times10^{4}\text{ y})(3.156\times10^{7}\text{ s/y})} = 9.15\times10^{-13}\text{ s}^{-1}$$

(a) Number of nuclei = $\dfrac{\text{mass parent}}{\text{mass of 1 nucleus}} = \dfrac{1.00\text{ kg}}{(239\text{ u})(1.66\times10^{-27}\text{ kg/u})} = 2.52\times10^{24}$ ◊

(b) $R_o = \lambda N_o = (9.15\times10^{-13}\text{ s}^{-1})(2.52\times10^{24}) = 2.31\times10^{12}\text{ decays/s} = 2.31\times10^{12}\text{ Bq}$ ◊

(c) $R = R_o e^{-\lambda t}$ or $e^{-\lambda t} = \dfrac{R}{R_0} = \dfrac{0.100\text{ Bq}}{2.31\times10^{12}\text{ Bq}} = 4.34\times10^{-14}$

$\lambda t = -\ln(4.34\times10^{-14}) = 30.8$ and

$$t = \frac{30.8}{\lambda} = \frac{30.8}{9.15\times10^{-13}\text{ s}^{-1}} = 3.36\times10^{13}\text{ s} = 1.07\times10^{6}\text{ y}$$

t = 1.07 million years ◊

45. (a) Find the radius of the $^{12}_{6}C$ nucleus. (b) Find the force of repulsion between a proton at the surface of a $^{12}_{6}C$ nucleus and the remaining five protons. (c) How much work (in MeV) has to be done to overcome this electrostatic repulsion in order to put the last proton into the nucleus? (d) Repeat parts (a), (b), and (c) for $^{238}_{92}U$.

Solution

(a) $$r = r_o A^{1/3} = 1.2A^{1/3}\ \text{fm}$$

When $A = 12$, $$r = 1.2(12)^{1/3}\ \text{fm} = 2.75\ \text{fm} \quad \lozenge$$

(b) Since the proton interacts with Z − 1 other protons, which we assume to be distributed uniformly throughout the nucleus, we find (where $Z = 6$ and $r = 2.75$ fm):

$$F = k_e \frac{(Z-1)e^2}{r^2} = \left(8.99 \times 10^9\ \frac{\text{N}\cdot\text{m}^2}{\text{C}^2}\right)\frac{(5)(1.60 \times 10^{-19}\ \text{C})^2}{(2.75 \times 10^{-15}\ \text{m})^2} = 152\ \text{N} \quad \lozenge$$

(c) $$U = k_e \frac{q_1 q_2}{r} = k_e \frac{(Z-1)e^2}{r}$$

Solving, $$U = \left(8.99 \times 10^9\ \frac{\text{N}\cdot\text{m}^2}{\text{C}^2}\right)\frac{(5)(1.60 \times 10^{-19}\ \text{C})^2}{2.75 \times 10^{-15}\ \text{m}}$$

thus $$U = 4.19 \times 10^{-13}\ \text{J} = 2.62\ \text{MeV} \quad \lozenge$$

(d) For $^{238}_{92}U$, we take $A = 238$ and $Z = 92$ to find

$$r = 7.44\ \text{fm}, \quad \lozenge$$

$$F = 379\ \text{N}, \quad \lozenge$$

and $$U = 17.6\ \text{MeV} \quad \lozenge$$

7. Carbon detonations are powerful nuclear reactions that temporarily tear apart the cores of massive stars late in their lives. These blasts are produced by carbon fusion, which requires a temperature of about 6×10^8 K to overcome the strong Coulomb repulsion between carbon nuclei. (a) Estimate the repulsive energy barrier to fusion, using the required ignition temperature for carbon fusion. (In other words, what is the average kinetic energy of a carbon nucleus at 6×10^8 K?) (b) Calculate the energy (in MeV) released in each of these "carbon-burning" reactions:

$$^{12}\text{C} + {}^{12}\text{C} \rightarrow {}^{20}\text{Ne} + {}^{4}\text{He}$$

$$^{12}\text{C} + {}^{12}\text{C} \rightarrow {}^{24}\text{Mg} + \gamma$$

(c) Calculate the energy (in kWh) given off when 2.00 kg of carbon completely fuses according to the first reaction.

Solution

(a) At 6×10^8 K, each carbon nucleus has thermal energy of

$$\tfrac{3}{2}k_B T = (1.5)(8.62 \times 10^{-5}\text{ eV/K})(6 \times 10^8\text{ K}) = 8 \times 10^4\text{ eV} \quad \Diamond$$

(b) Energy released $= \left[2m(\text{C}^{12}) - m(\text{Ne}) - m(\text{He}^4)\right]c^2$

$$= (24.000000 - 19.992439 - 4.002603)(931.5)\text{ MeV} = 4.62\text{ MeV} \quad \Diamond$$

Energy released $= \left[2m(\text{C}^{12}) - m(\text{Mg}^{24})\right](931.5)\text{ MeV/u}$

$$= (24.000000 - 23.985042)(931.5)\text{ MeV} = 13.9\text{ MeV} \quad \Diamond$$

(c) Energy released = the energy of reaction of the number of carbon nuclei in a 2.00-kg sample, which corresponds to

$$\Delta E = \left(\frac{2000\text{ g}}{12\text{ g/mol Carbon}}\right)\left(\frac{6.02 \times 10^{23}\text{ atoms}}{\text{mol Carbon}}\right)\left(\frac{1\text{ fusion}}{2\text{ atoms}}\right)\left(\frac{4.62\text{ MeV}}{\text{fusion}}\right)\left(\frac{1\text{ kWh}}{2.25 \times 10^{19}\text{ MeV}}\right)$$

$$\Delta E = 10.3 \times 10^6\text{ kWh} \quad \Diamond$$

49. The decay of an unstable nucleus by alpha emission is represented by Equation 30.9. The disintegration energy Q given by Equation 30.12 must be shared by the alpha particle and the daughter nucleus in order to conserve both energy and momentum in the decay process. (a) Show that Q and K_α, the kinetic energy of the alpha particle, are related by the expression

$$Q = K_\alpha\left(1+\frac{M_\alpha}{M}\right)$$

where M is the mass of the daughter nucleus. (b) Use the result of part (a) to find the energy of the alpha particle emitted in the decay of ^{226}Ra. (See Example 30.5 for the calculation of Q.)

Solution

(a) Let us assume that the parent nucleus (mass M_p) is initially at rest, and let us denote the masses of the daughter nucleus and alpha particle by M_d and M_α, respectively. Applying the equations of conservation of momentum and energy for the alpha decay process gives

$$M_d v_d = M_\alpha v_\alpha \qquad (1)$$

$$M_p c^2 = M_d c^2 + M_\alpha c^2 + \tfrac{1}{2}M_\alpha {v_\alpha}^2 + \tfrac{1}{2}M_d {v_d}^2 \qquad (2)$$

The disintegration energy Q is given by

$$Q = (M_p - M_d - M_\alpha)c^2 = \tfrac{1}{2}M_\alpha {v_\alpha}^2 + \tfrac{1}{2}M_d {v_d}^2 \qquad (3)$$

Eliminating v_d from Equations (1) and (3) gives

$$Q = \tfrac{1}{2}M_\alpha {v_\alpha}^2 + \tfrac{1}{2}M_d\left(\frac{M_\alpha}{M_d}v_\alpha\right)^2 = \tfrac{1}{2}M_\alpha {v_\alpha}^2 + \tfrac{1}{2}\frac{{M_\alpha}^2}{M_d}{v_\alpha}^2$$

$$= \tfrac{1}{2}M_\alpha {v_\alpha}^2\left(1+\frac{M_\alpha}{M_d}\right) = K_\alpha\left(1+\frac{M_\alpha}{M_d}\right)$$

(b) $K_\alpha = \dfrac{Q}{1+\dfrac{M_\alpha}{M_d}} = \dfrac{4.87 \text{ MeV}}{1+\dfrac{4}{222}} = 4.78 \text{ MeV}$ ◊

Chapter 31

A technician works on one of the particle detectors at CERN, the European center for particle physics near Geneva, Switzerland. Electrons and positrons accelerated to an energy of 50 GeV collide in a circular tunnel 2 km in circumference, located 100 m underground. (David Parker / Science Photo Library, Photo Researchers, Inc.)

PARTICLE PHYSICS AND COSMOLOGY

Chapter 31

PARTICLE PHYSICS AND COSMOLOGY

INTRODUCTION

In this chapter, we examine the properties and classifications of the various known subatomic particles and the fundamental interactions that govern their behavior. We also discuss the current theory of elementary particles, in which all matter is believed to be constructed from only two families of particles, quarks and leptons. Finally, we discuss how clarifications of such models might help scientists understand cosmology, which deals with the evolution of the Universe.

NOTES FROM SELECTED CHAPTER SECTIONS

31.1 The Fundamental Forces in Nature

There are four fundamental forces in nature: **strong** (hadronic), **electromagnetic**, **weak**, and **gravitational**. The strong force is the force between nucleons that keeps the nucleus together. The weak force is responsible for beta decay. The electromagnetic and weak forces are now considered to be manifestations of a single force called the **electroweak** force.

The fundamental forces are described in terms of particle or quanta exchanges which **mediate** the forces. The electromagnetic force is mediated by photons, which are the quanta of the electromagnetic field. Likewise, the strong force is fundamentally mediated by field particles called **gluons**, the weak force is mediated by particles called the W and Z **bosons**, and the gravitational force is mediated by quanta of the gravitational field called **gravitons**.

31.2 Positrons and Other Antiparticles

An antiparticle and a particle have the same mass, but opposite charge. Furthermore, other properties may have opposite values such as lepton number and baryon number. It is possible to produce particle-antiparticle pairs in nuclear reactions if the available energy is greater than $2mc^2$, where m is the mass of the particle.

Pair production is a process in which a gamma ray with an energy of at least 1.02 MeV interacts with a nucleus, and an electron-positron pair is created.

Pair annihilation is an event in which an electron and a positron can annihilate to produce two gamma rays, each with an energy of at least 0.511 MeV.

Chapter 31

31.3 Mesons and the Beginning of Particle Physics

The interaction between two particles can be represented in a diagram called a **Feynman diagram.**

31.4 Classification of Particles

All particles (other than photons and gravitons) can be classified into two categories: **hadrons** and **leptons.**

There are two classes of hadrons: **mesons** and **baryons** which are grouped according to their masses and spins. It is believed that hadrons are composed of units called **quarks** which are fundamental in nature.

Leptons have no structure or size and are therefore considered to be truly elementary particles.

31.5 Conservation Laws

In all reactions and decays, quantities such as energy, linear momentum, angular momentum, electric charge, baryon number, and lepton number are strictly conserved. Certain particles have properties called **strangeness** and **charm**. These unusual properties are conserved only in those reactions and decays that occur via the strong and electromagnetic forces.

Whenever a nuclear reaction or decay occurs, the sum of the baryon numbers before the process must equal the sum of the baryon numbers after the process.

The sum of the electron-lepton numbers before a reaction or decay must equal the sum of the electron-lepton numbers after the reaction or decay.

31.6 Strange Particles and Strangeness

Whenever the strong force or the electromagnetic force causes a reaction or decay to occur, the sum of the strangeness numbers before the process must equal the sum of the strangeness numbers after the process.

31.9 Quarks

Recent theories in elementary particle physics have postulated that all hadrons are composed of smaller units known as **quarks**. Quarks have a fractional electric charge and a baryon number of 1/3. There are six flavors of quarks, up (u), down (d), strange (s), charmed (c), top (t), and bottom (b). All baryons contain three quarks, while all mesons contain one quark and one antiquark.

According to the theory of **quantum chromodynamics**, quarks have a property called **color**, and the strong force between quarks is referred to as the **color force**.

EQUATIONS AND CONCEPTS

Pions and muons are very unstable particles. A decay sequence is shown in Equation 31.1.

$$\pi^- \to \mu^- + \bar{\nu}_\mu \qquad (31.1)$$

$$\mu^- \to e^- + \nu_\mu + \bar{\nu}_e$$

Hubble's law states a linear relationship between the velocity of a galaxy and its distance R from the Earth. The constant H is called the **Hubble parameter.**

$$v = HR \qquad (31.7)$$

$$H = 17 \times 10^{-3} \text{ m/(s} \cdot \text{light-year)}$$

Each baryon and meson is composed of quarks and antiquarks; the identity of each particle is based upon the combination of quarks involved. When a particle decays, a quark can annihilate with its corresponding antiquark; alternatively, a quark-antiquark pair can be spontaneously formed. However, in any interaction, the baryon number is always conserved.

Quark, Antiquark	Charge	Baryon Number
$u, \bar{u}$	$+\frac{2}{3}e, -\frac{2}{3}e$	$\frac{1}{3}, -\frac{1}{3}$
$d, \bar{d}$	$-\frac{1}{3}e, +\frac{1}{3}e$	$\frac{1}{3}, -\frac{1}{3}$
$s, \bar{s}$	$-\frac{1}{3}e, +\frac{1}{3}e$	$\frac{1}{3}, -\frac{1}{3}$
$c, \bar{c}$	$+\frac{2}{3}e, -\frac{2}{3}e$	$\frac{1}{3}, -\frac{1}{3}$
$b, \bar{b}$	$-\frac{1}{3}e, +\frac{1}{3}e$	$\frac{1}{3}, -\frac{1}{3}$
$t, \bar{t}$	$+\frac{2}{3}e, -\frac{2}{3}e$	$\frac{1}{3}, -\frac{1}{3}$

REVIEW CHECKLIST

▷ Be aware of the four fundamental forces in nature and the corresponding field particles or quanta via which these forces are mediated.

▷ Understand the concepts of the antiparticle, pair production, and pair annihilation.

▷ Know the broad classification of particles and the characteristic properties of the several classes (relative mass value, spin, decay mode).

NSWERS TO SELECTED CONCEPTUAL QUESTIONS

Describe the quark model of hadrons, including the properties of quarks.

nswer In the quark model, all hadrons are composed of smaller units called uarks. Quarks have a fractional electric charge and a baryon number of 1/3. There are x flavors of quarks; up (u) down (d), strange (s), charmed (c), top (t), and bottom (b). All aryons contain three quarks, and all mesons contain one quark and one anti-quark. ection 31.9 has a more detailed discussion of the quark model.

□ □ □ □

. Describe the properties of baryons and mesons and the important differences etween them.

Answer There are two types of hadrons called baryons and mesons. Hadrons nteract primarily through the strong force and are not elementary particles, being composed of either three quarks (baryons), or a quark and an antiquark (mesons). Baryons have a nonzero baryon number with a spin of either 1/2 or 3/2. Mesons have a baryon number of zero, and a spin of either 0 or 1.

□ □ □ □

5. Particles known as resonances have very short lifetimes, of the order of 10^{-23} s. From this information, would you guess that they are hadrons or leptons? Explain.

Answer Resonances are hadrons. Such particles decay into other strongly interacting particles such as protons, neutrons, and pions with very short lifetimes. In fact, they decay so quickly that they cannot be detected directly. Decays which occur via the weak force have lifetimes of 10^{-13} s or longer; particles that decay via the strong force have lifetimes in the range of 10^{-16} s to 10^{-19} s.

□ □ □ □

6. Kaons all decay into final states that contain no protons or neutrons. What is th baryon number of kaons?

Answer All stable particles other than protons and neutrons have baryon numbe zero. Since the baryon number must be conserved, and the final states of the kaon deca contain no protons or neutrons, the baryon number of all kaons must be **zero**.

□ □ □ □

7. The Ξ^o particle decays by the weak interaction according to the decay mod $\Xi^o \rightarrow \Lambda^o + \pi^o$. Would you expect this decay to be fast or slow? Explain.

Answer This decay should be slow, since decays which occur via the weak interaction typically take 10^{-10} s or longer to occur.

□ □ □ □

9. Identify the particle decays listed in Table 31.2 that occur by the electromagnetic interaction. Justify your answer.

Answer The decays of the neutral pion, eta, and neutral sigma occur by the electromagnetic interaction. These are the three shortest lifetimes in Table 31.2. All produce photons, which are the quanta of the electromagnetic force. All conserve strangeness.

□ □ □ □

10. Two protons in a nucleus interact via the strong interaction. Are they also subject to the weak interaction?

Answer Yes, but the strong interaction predominates.

□ □ □ □

2. An antibaryon interacts with a meson. Can a baryon be produced in such an nteraction? Explain.

nswer Unless the particles have enough kinetic energy to produce a aryon-antibaryon pair, the answer is **No**. Antibaryons have a baryon number of -1; aryons have a baryon number of +1; mesons have a baryon number of 0. If such an nteraction were to occur, and produce a baryon, the baryon number would not be onserved.

□ □ □ □

4. How many quarks are there in (a) a baryon, (b) an antibaryon, (c) a meson, (d) an antimeson? How do you account for the fact that baryons have half-integral spins and mesons have spins of 0 or 1? (**Hint:** Quarks have spin 1/2.)

Answer All baryons and antibaryons consist of three quarks. All mesons and antimesons consist of two quarks. Since quarks have spins of 1/2, it follows that all baryons (which consist of three quarks) must have half-integral spins, and all mesons (which consist of two quarks) must have spins of 0 or 1.

□ □ □ □

15. In the theory of quantum chromodynamics, quarks come in three colors. How would you justify the statement that "all baryons and mesons are colorless"?

Answer Each flavor of quark can have three colors, designated as red, green, and blue. Antiquarks are colored antired, antigreen, and antiblue. Baryons consist of three quarks, each having a different color. Mesons consist of a quark of one color and an antiquark with a corresponding anticolor. Thus, baryons and mesons are "white", or colorless.

□ □ □ □

SOLUTIONS TO SELECTED END-OF-CHAPTER PROBLEMS

5. One of the mediators of the weak interaction is the Z^o boson, with a mass of 96 GeV/c^2. Use this information to find an approximate value for the range of the weak interaction.

Solution The rest energy of the Z^o boson is E_o = 96 GeV. The maximum time a virtual Z^0 boson can exist is found from $\Delta E \Delta t \geq \hbar / 2$, or

$$\Delta t \approx \frac{\hbar}{2\Delta E} = \frac{1.055 \times 10^{-34} \text{ J} \cdot \text{s}}{2(96 \text{ GeV})(1.60 \times 10^{-10} \text{ J/GeV})} = 3.43 \times 10^{-27} \text{ s}$$

The maximum distance it can travel in this time is

$$d = c(\Delta t) = (3.00 \times 10^8 \text{ m/s})(3.43 \times 10^{-27} \text{ s}) \sim 10^{-18} \text{ m} \quad \Diamond$$

The distance d is an approximate value for the range of the weak interaction.

---

9. A neutral pion at rest decays into two photons according to

$$\pi^o \rightarrow \gamma + \gamma$$

Find the energy, momentum, and frequency of each photon.

Solution Since the pion is at rest and momentum is conserved, the two gamma-rays must have equal momenta in opposite directions. So, they must share equally in the energy of the pion.

$$m_{\pi^0} = 135.0 \text{ MeV}/c^2 \qquad \text{(Table 31.2)}$$

Therefore,

$$E_\gamma = 67.5 \text{ MeV} = 1.08 \times 10^{-11} \text{ J} \quad \Diamond$$

$$p = \frac{E}{c} = \frac{67.5 \text{ MeV}}{3.00 \times 10^8 \text{ m/s}} = 3.61 \times 10^{-20} \text{ kg} \cdot \text{m/s} \quad \Diamond$$

$$f = \frac{E}{h} = 1.63 \times 10^{22} \text{ Hz} \quad \Diamond$$

---

1. Name one possible decay mode (see Table 31.2) for Ω^+, $\overline{K}s^o$, $\overline{\Lambda}^o$, and $\overline{n}$.

Solution The particles in this problem are the antiparticles of those listed in Table 31.2. Therefore, the decay modes include the antiparticles of those shown in the decay modes in Table 31.2:

$$\Omega^+ \rightarrow \overline{\Lambda}^o + K^+$$

$$\overline{K}^o \rightarrow \pi^+ + \pi^- \quad \text{or} \quad \pi^o + \pi^o$$

$$\overline{\Lambda}^o \rightarrow \overline{p} + \pi^+$$

$$\overline{n} \rightarrow \overline{p} + e^+ + \nu_e \quad \Diamond$$

---

15. The following reactions or decays involve one or more neutrinos. Supply the missing neutrinos (ν_e, ν_μ, or ν_τ).

(a) $\pi^- \rightarrow \mu^- + ?$

(b) $K^+ \rightarrow \mu^+ + ?$

(c) $? + p^+ \rightarrow n + e^+$

(d) $? + n \rightarrow p^+ + e^-$

(e) $? + n \rightarrow p^+ + \mu^-$

(f) $\mu^- \rightarrow e^- + ? + ?$

Solution

(a) $\pi^- \rightarrow \mu^- + \overline{\nu}_\mu$ $\qquad L_\mu:\ 0 \rightarrow 1 - 1$

(b) $K^+ \rightarrow \mu^+ + \nu_\mu$ $\qquad L_\mu:\ 0 \rightarrow -1 + 1$

(c) $\overline{\nu}_e + p^+ \rightarrow n + e^+$ $\qquad L_e:\ -1 + 0 \rightarrow 0 - 1$

(d) $\nu_e + n \rightarrow p^+ + e^-$ $\qquad L_e:\ 1 + 0 \rightarrow 0 + 1$

(e) $\nu_\mu + n \rightarrow p^+ + \mu^-$ $\qquad L_\mu:\ 1 + 0 \rightarrow 0 + 1$

(f) $\mu^- \rightarrow e^- + \overline{\nu}_e + \nu_\mu$ $\qquad L_\mu:\ 1 \rightarrow 0 + 0 + 1$ and $L_e:\ 0 \rightarrow 1 - 1 + 0$ $\Diamond$

---

17. Determine which of the reactions below can occur. For those that cannot occur determine the conservation law (or laws) violated.

(a) $p \to \pi^+ + \pi^0$

(b) $p + p \to p + p + \pi^0$

(c) $p + p \to p + \pi^+$

(d) $\pi^+ \to \mu^+ + \nu_\mu$

(e) $n \to p + e^- + \overline{\nu}_e$

(f) $\pi^+ \to \mu^+ + n$

Solution

(a) $p \to \pi^+ + \pi^0$ — Baryon number is violated: $1 \to 0 + 0$

(b) $p + p \to p + p + \pi^0$ — This reaction can occur.

(c) $p + p \to p + \pi^+$ — Baryon number is violated: $1 + 1 \to 1 + 0$

(d) $\pi^+ \to \mu^+ + \nu_\mu$ — This reaction can occur.

(e) $n \to p + e^- + \overline{\nu}_e$ — This reaction can occur.

(f) $\pi^+ \to \mu^+ + n$ — Violates baryon number: $0 \to 0 + 1$ and violates muon-lepton number: $0 \to -1 + 0$ ◊

19. Determine whether or not strangeness is conserved in the following decays and reactions:

(a) $\Lambda^o \rightarrow p + \pi^-$

(b) $\pi^- + p \rightarrow \Lambda^o + K^o$

(c) $\overline{p} + p \rightarrow \overline{\Lambda}^o + \Lambda^o$

(d) $\pi^- + p \rightarrow \pi^- + \Sigma^+$

(e) $\Xi^- \rightarrow \Lambda^o + \pi^-$

(f) $\Xi^0 \rightarrow p + \pi^-$

Solution

We look up the strangeness quantum numbers in Table 31.2.

(a) $\Lambda^o \rightarrow p + \pi^-$
Strangeness: $-1 \rightarrow 0 + 0$
(-1 does not equal 0, so strangeness is not conserved)

(b) $\pi^- + p \rightarrow \Lambda^o + K^o$
Strangeness: $0 + 0 \rightarrow -1 + 1$
(0 = 0 and strangeness is conserved)

(c) $\overline{p} + p \rightarrow \overline{\Lambda}^o + \Lambda^o$
Strangeness: $0 + 0 \rightarrow +1 - 1$
(0 = 0 and strangeness is conserved)

(d) $\pi^- + p \rightarrow \pi^- + \Sigma^+$
Strangeness: $0 + 0 \rightarrow 0 - 1$
(0 does not equal –1 so strangeness is not conserved)

(e) $\Xi^- \rightarrow \Lambda^o + \pi^-$
Strangeness: $-2 \rightarrow -1 + 0$
(–2 does not equal –1 so strangeness is not conserved)

(f) $\Xi^0 \rightarrow p + \pi^-$
Strangeness: $-2 \rightarrow 0 + 0$
(–2 does not equal 0 so strangeness is not conserved) ◊

25. Analyze each reaction in terms of constituent quarks:

(a) $\pi^- + p \rightarrow K^o + \Lambda^o$

(b) $\pi^+ + p \rightarrow K^+ + \Sigma^+$

(c) $K^- + p \rightarrow K^+ + K^o + \Omega^-$

(d) $p + p \rightarrow K^o + p + \pi^+ + ?$

In the last reaction, identify the mystery particle.

Solution We look up the quark constituents of the particles in Table 31.4.

(a) $d\bar{u}$ + uud → $d\bar{s}$ + uds

(b) $\bar{d}u$ + uud → $u\bar{s}$ + uus

(c) $\bar{u}s$ + uud → $u\bar{s}$ + $d\bar{s}$ + sss

(d) uud + uud → $d\bar{s}$ + uud + $u\bar{d}$ + uds A uds is either a Λ^o or a Λ^o ◊

---

28. A distant quasar is moving away from Earth at such high speed that the violet 434-nm hydrogen line is observed at 650 nm, in the red portion of the spectrum. (a) How fast is the quasar receding? (See the hint in Problem 27.) (b) Using Hubble's law, determine the distance from Earth to this quasar.

Solution

(a) $$\frac{\lambda'}{\lambda} = \frac{650 \text{ nm}}{434 \text{ nm}} = 1.498 = \sqrt{\frac{1+\frac{v}{c}}{1-\frac{v}{c}}} \qquad \frac{1+\frac{v}{c}}{1-\frac{v}{c}} = 2.243$$

$$v = \frac{2.243-1}{2.243+1}c = 0.383c \quad \text{or} \quad 38.3\% \text{ the speed of light, } 1.15\times10^8 \text{ m/s}$$ ◊

(b) Using Equation 31.7, $v = HR$, we find

$$R = \frac{v}{H} = \frac{(0.383)(3.00\times10^8 \text{ m/s})}{1.7\times10^{-2} \text{ m/s}\cdot\text{lightyear}} = 6.76\times10^9 \text{ lightyear}$$ ◊

---

33. The energy flux carried by neutrinos from the Sun is estimated to be on the order of 0.4 W/m^2 at Earth's surface. Estimate the fractional mass loss of the Sun over 10^9 years due to the radiation of neutrinos. (The mass of the Sun is 2×10^{30} kg. The Earth-Sun distance is 1.5×10^{11} m.)

Solution

Since the neutrino flux from the Sun reaching the Earth is 0.4 W/m^2, the total energy emitted per second by the Sun in neutrinos is

$$(0.4\ \mathrm{W/m^2})(4\pi r^2) = (0.4\ \mathrm{W/m^2})(4\pi)(1.5\times 10^{11}\ \mathrm{m})^2 = 1.13\times 10^{23}\ \mathrm{W}$$

In a period of 10^9 y, the Sun emits a total energy of

$$(1.13\times 10^{23}\ \mathrm{J/s})(10^9\ \mathrm{y})(3.156\times 10^7\ \mathrm{s/y}) = 3.56\times 10^{39}\ \mathrm{J}$$

in the form of neutrinos. This energy corresponds to an annihilated mass of

$$mc^2 = 3.56\times 10^{39}\ \mathrm{J}$$

$$m = \frac{E}{c^2} = \frac{3.56\times 10^{39}\ \mathrm{J}}{\left(3.00\times 10^8\ \mathrm{m/s}\right)^2} = 3.96\times 10^{22}\ \mathrm{kg}$$

Since the Sun has a mass of about 2×10^{30} kg, this corresponds to a loss of only about 1 part in 50 000 000 of the Sun's mass over 10^9 y in the form of neutrinos. ◊

37. Calculate the kinetic energies of the proton and pion resulting from the decay of a Λ at rest:

$$\Lambda^0 \rightarrow p + \pi^-$$

Solution

We first look up the energy of each particle:

$$m_\Lambda c^2 = 1115.6 \text{ MeV}$$

$$m_p c^2 = 938.3 \text{ MeV}$$

$$m_\pi c^2 = 139.6 \text{ MeV}$$

The difference between starting mass-energy and final mass-energy is the kinetic energy of the products:

$$(K_p + K_\pi) = (1115.6 - 938.3 - 139.6) \text{ MeV}$$

$$= 37.7 \text{ MeV}$$

In addition, since momentum is conserved, $|p_p| = |p_\pi| = p$

Applying conservation of relativistic energy:

$$\left(\sqrt{(938.3)^2 + p^2c^2} - 938.3\right) + \left(\sqrt{(139.6)^2 + p^2c^2} - 139.6\right) = 37.7 \text{ MeV}$$

Solving the algebra yields $p_\pi c = p_p c$ = 100.4 MeV.

Thus $$K_p = \sqrt{(m_p c^2)^2 + (100.4)^2} - m_p c^2 = 5.35 \text{ MeV} \quad \lozenge$$

and $$K_\pi = \sqrt{(139.6)^2 + (100.4)^2} - 139.6 = 32.3 \text{ MeV} \quad \lozenge$$

39. If a K^o meson at rest decays in 0.900×10^{-10} s, how far will a K^o meson travel if it is moving at $0.960c$ through a bubble chamber?

Solution The motion of the K^o particle is relativistic. Just like the spaceman who leaves for a distant star, and returns to find his family long gone, the Kaon appears to us to have a longer lifetime. That time-dilated lifetime is:

$$T = \gamma T_o = \frac{0.900 \times 10^{-10}\ \text{s}}{\sqrt{1 - v^2/c^2}} = \frac{0.900 \times 10^{-10}\ \text{s}}{\sqrt{1 - (0.960)^2}} = 3.21 \times 10^{-10}\ \text{s}$$

During this time, we see the kaon travel at $0.960c$. It travels for a distance of

$$d = vt = \left[(0.960)(3.00 \times 10^8\ \text{m/s})\right](3.21 \times 10^{-10}\ \text{s})$$

$$d = 0.0926\ \text{m} = 9.26\ \text{cm} \quad \Diamond$$